ASE Test Preparation Series

Specifications for Transit Bus

Compressed Natural Gas Engines (Test H1)

DELMAR
CENGAGE Learning

Australia • Brazil • Japan • Korea • Mexico • Singapore • Spain • United Kingdom • United States

DELMAR
CENGAGE Learning

Delmar's ASE Test Preparation Series: Specifications for Transit Bus Compressed Natural Gas (CNG) Engines Test (H1)

Joe Drouin

Vice President, Technology and Trades Professional Business Unit: Gregory L. Clayton

Director, Professional Transportation Industry Training Solutions: Kristen L. Davis

Product Manager: Lori Bonesteel

Director of Marketing: Beth A. Lutz

Marketing Manager: Jennifer Stall

Production Director: Carolyn Miller

Production Manager: Andrew Crouth

Senior Content Project Manager: Andrea Majot

Art Director: Benjamin Gleeksman

Production Technology Analyst: Thomas Stover

Cover Design: Michael Egan

Production Service: Pre-PressPMG

Library of Congress Control Number: 2009941420

ISBN-13: 978-1-4354-3939-9

ISBN-10: 1-4354-3939-2

Delmar Cengage Learning
5 Maxwell Drive
Clifton Park, NY 12065-2919
USA

Cengage Learning is a leading provider of customized learning solutions with office locations around the globe, including Singapore, the United Kingdom, Australia, Mexico, Brazil, and Japan. Locate your local office at: **international.cengage.com/region**

Cengage Learning products are represented in Canada by Nelson Education, Ltd.

For more information on transportation titles available from Delmar, Cengage Learning, please visit our website at www.trainingbay.com

For your lifelong learning solutions, visit **delmar.cengage.com**

Visit our corporate website at **www.cengage.com**

Printed in the United States of America
2 3 4 5 15 14 13 12

Contents

Preface v

Section 1 The History and Purpose of ASE

Section 2 Take and Pass Every ASE Test

How Are the Tests Administered? 3
Who Writes the Questions? 3
Objective Tests 4
Preparing for the Exam 5
During the Test 6
Your Test Results! 6

Section 3 Types of Questions on an ASE Exam

Multiple-Choice Questions 7
EXCEPT Questions 8
Technician A, Technician B Questions 8
Most-Likely Questions 9
LEAST-Likely Questions 9
Summary 10
Testing Time Length 10

Section 4 Overview of the Task List

Specifications for Transit Bus Compressed Natural Gas (CNG) Engines (Test H1) 11
Task List and Overview 12
A. General Engine Diagnosis (20 Questions) 12
B. Cylinder Head and Valve Train Diagnosis and Repair (11 Questions) 24
C. Engine Block Diagnosis, Repair, and Overhaul (18 Questions) 34
D. Lubrication and Cooling Systems Diagnosis and Repair (15 Questions) 47
E. Air Induction and Exhaust Systems Diagnosis and Repair (7 Questions) 55
F. Fuel System Inspection, Diagnosis, and Repair (12 Questions) 59
1. High-Pressure System (8 Questions) 60
2. Low-Pressure System (2 Questions) 65
3. Fuel Controls (2 Questions) 67

G. Starting and Charging System Diagnosis and Repair (8 Questions) 68
H. Ignition Systems Diagnosis and Repair (8 Questions) 73

Section 5 Sample Test for Practice

Sample Test 77

Section 6 Additional Test Questions for Practice

Additional Test Questions 93

Section 7 Appendices

Answers to the Test Questions for the Sample Test Section 5 147
Explanations to the Answers for the Sample Test Section 5 148
Answers to the Test Questions for the Additional Test Questions Section 6 168
Explanations to the Answers for the Additional Test Questions Section 6 171

Glossary 229

Preface

Delmar, a part of Cengage Learning, is very pleased that you have chosen our ASE Test Preparation Series to prepare yourself for the Transit Bus, Specifications for Transit Bus Compressed Natural Gas (CNG) Engines Test (H1) ASE Examination. This guide is designed to introduce you to the Task List for the Specifications for Transit Bus Compressed Natural Gas (CNG) Engines Test (H1) test that you are preparing to take, give you an understanding of what you are expected to be able to do in each task, and take you through sample test questions formatted in the same way the ASE tests are structured.

If you have a basic working knowledge of the discipline you are testing for, you will find the Delmar Learning's *ASE Test Preparation Guide* to be an excellent way to understand the "must know" items to pass the test. This book is not a textbook. Its objective is to prepare the technician who has the requisite experience and schooling to meet the challenge of ASE testing. It cannot replace the hands-on experience or the theoretical knowledge required by ASE to master vehicle repair technology. If you are unable to understand more than a few of the questions and their explanations in this book, it could be that you require either more shop-floor experience or further study.

This book begins with item-by-item overview of the ASE Task List with explanations of the minimum knowledge you must possess to answer questions related to the task. Following each explanation are two sets of sample questions followed by an answer key that includes explanations of the answers to each question. A few of the questions are not strictly ASE format but were included because they help teach a critical concept that will appear on the test. We suggest that you read the complete Task List Overview before taking the first sample test. After taking the first test, score yourself and read the explanation to any questions that you were not sure about, including the questions you answered correctly. Each test question has a reference back to the related task or tasks that it covers. This will help you to go back and read over any area of the task list that you are having trouble with. Once you are satisfied that you have all of your questions answered from the first sample test, take the additional test and check it. If you pass these tests, you will do well on the ASE test.

Our Commitment to Excellence

Delmar, Cengage Learning has sought out the best technicians in the country to help with the development of each part of this first edition of the *Transit Bus, Specifications for Transit Bus Compressed Natural Gas (CNG) Engines Test (H1) ASE Test Preparation Guide.*

About the Author

Joe Drouin has been an engine/electronic/electrical instructor since 2003 at a transit company with one of the largest CNG fleets in the United States. Prior to that, he worked as a technician for the same employer for more than two decades before being promoted to his current position. His work experience background includes engine rebuilding/diagnosis for CNG, diesel, methanol/ethanol, and gasoline engines; Allison/World Transmissions; HVAC; and body repair. He also teaches courses on transportation technology at Los Angeles Trade Technical College in Los Angeles, and in the past has provided training at various community colleges in the Los Angeles area as well as throughout the state of California for private contractors on testing and reducing smoke emissions of heavy-duty diesel engines. He holds a Bachelor of Vocational Education in Vocational Arts from the California State University, Los Angeles, and an Associate in Science in Automotive Technology from Citrus College of Glendora, California.

Thanks for choosing Delmar Learning's ASE Test Preparation Series. All of the writers, editors, and Delmar staff have worked very hard to make this Test Preparation Guide second to none. We know you are going to find this book accurate and easy to work with. It is our objective to constantly improve our products at Delmar by responding to feedback.

If you have any questions concerning the books in this series, please visit us on the web at http://www.trainingbay.com.

1 The History and Purpose of ASE

ASE began as the National Institute for Automotive Service Excellence (NIASE). It was founded as a nonprofit, independent entity in 1972 by a group of industry leaders with the single goal of providing a means for consumers to distinguish between incompetent and competent technicians. It accomplishes this goal by testing and certification of repair and service professionals. From this beginning it has evolved to be known simply as ASE (Automotive Service Excellence) and today offers more than 40 certification exams in automotive, medium/heavy-duty truck, collision, engine machinist, school bus, transit bus, parts specialist, automobile service consultant, and other industry-related areas. At this time there are more than 400,000 professionals with current ASE certifications. These professionals are employed by new car and truck dealerships, independent garages, fleets, service stations, franchised service facilities, and more. ASE continues its mission by also providing information that helps consumers identify repair facilities that employ certified professionals through its Blue Seal of Excellence Recognition Program. Shops where a minimum of 75 percent of the repair technicians are ASE certified and meet other criteria can apply for and receive the Blue Seal of Excellence Recognition from ASE.

ASE recognized that educational programs serving the service and repair industry also needed a way to be recognized as having the faculty, facilities, and equipment to provide a quality education to students wanting to become service professionals. Through the combined efforts of ASE, industry, and education leaders, the nonprofit National Automotive Technicians Education Foundation (NATEF) was created to evaluate and recognize training programs. Today more than 2,000 programs are ASE certified under the standards set by the service industry. ASE/NATEF also has a certification of industry (factory) training program known as CASE. CASE stands for Continuing Automotive Service Education and recognizes training provided by replacement parts manufacturers as well as vehicle manufactures.

ASE certification testing is administered by the American College Testing (ACT). Strict standards of security and supervision at the test centers ensure that the technician who holds the certification earned it. Additionally ASE certification also requires that those passing the test be able to demonstrate that they have two years of work experience in the field before they can be certified. Test questions are developed by industry experts who are actually working in the field being tested. The next section provides more detail on how the test is developed and administered. Paper and pencil tests are administered twice a year at over 700 locations in the United States. Computer-based testing is now available with the benefit of instant test results at certain established test centers. The certification is valid for five years and can be recertified by retesting. So that consumers can recognize certified technicians, ASE issues a jacket patch, certificate, and wallet card to certified technicians and makes signs available to facilities that employ ASE certified technicians.

You can contact ASE with the following information:

National Institute for Automotive Service Excellence
101 Blue Seal Drive S.E.
Suite 101
Leesburg, VA 20175
Telephone: 703-669-6600
Fax: 703-669-6123
http://www.ase.com

ASE CERTIFIED
WE SUPPORT VOLUNTARY TECHNICIAN CERTIFICATION THROUGH
National Institute for
AUTOMOTIVE SERVICE EXCELLENCE

2 Take and Pass Every ASE Test

Participating in an Automotive Service Excellence (ASE) voluntary certification program gives you a chance to show your customers that you have the know-how needed to work on today's vehicles. The ASE certification tests allow you to compare your skills and knowledge to the automotive service industry's standards for each specialty area.

If you are the "average" automotive technician taking this test, you are in your mid-30s and have not attended school for about 15 years. That means you probably have not taken a test in many years. Some of you, on the other hand, have attended college or taken post-secondary education courses and may be more familiar with taking tests and with test-taking strategies. There is, however, a difference between the educational tests you may be accustomed to and the ASE test you are preparing to take.

How Are the Tests Administered?

ASE tests are administered at over 750 test sites in local communities. Paper and pencil tests are the type most widely available to technicians. Each tester is given a booklet containing questions with charts and diagrams where required. You can mark in this test booklet but no information entered in the booklet is scored. Answers are recorded on a separate answer sheet. You will enter your answers, in Number 2 pencil only. ASE recommends you bring four sharpened Number 2 pencils that have erasers. Answer choices are recorded by marking (bubble in) a block by the question number on the answer sheet. These sheets are scanned electronically and the answers tabulated. For test security purposes, test booklets include randomly generated questions. Your answer key must be matched to the proper booklet, so it is important to correctly enter the booklet serial number on the answer sheet. All instructions are printed on the test materials and should be followed carefully.

ASE has introduced Computer Based Testing (CBT) at some locations. While the test content is the same for both testing methods the CBT tests have some unique requirements and advantages. It is strongly recommended that technicians considering the CBT tests go the ASE Web site (www.ase.com) and review the conditions and requirements for this type of test. There is a test demonstrator page that allows you to experience this type of test before you register. Some technicians find this style of testing provides an advantage while others find operating the computer a distraction. One significant benefit of CBT is the availability of instant results. You can receive your test results before you leave the test center. CBT testing offers an increased degree of flexibility in scheduling. Testing costs for CBT tests are slightly higher than paper and pencil tests and the number of testing sites is limited. The first-time test taker may be more comfortable with the paper and pencil tests but each technician has an option.

Who Writes the Questions?

The ASE test questions are written by service industry experts in the area being tested. Each area will have its own technical experts. Questions are entirely job-related. They are designed to test the skills you need to be a successful technician. Theoretical knowledge is important and necessary to answer the questions, but ability to apply that knowledge is the basis of ASE test questions.

Each question has its roots in an ASE "item-writing" workshop where service representatives from automobile manufacturers (domestic and import), aftermarket parts and equipment manufacturers, working technicians, and vocational educators meet in a workshop setting to share ideas and translate them into test questions. Each test question written by these experts must survive review by all members of the group.

The questions are written to deal with practical application of soft skills and system knowledge experienced by technicians in their day-to-day work.

All questions are pretested and quality-checked on a national sample of technicians.

Those questions that meet ASE standards of quality and accuracy are included in the scored sections of the tests; the "rejects" are sent back to the drawing board or discarded altogether.

Each certification test is made up of between 40 and 80 multiple-choice questions.

Note: Each test could contain additional questions that are included for statistical research purposes only. Your answers to these questions will not affect your score, but since you do not know which ones they are, you should answer all questions in the test. The five-year Recertification Test will cover the same content areas as those listed above. However, the number of questions in each content area of the Recertification Test will be reduced by about one-half.

Using multiple criteria, including cross sections by age, race, and other background information, ASE is able to guarantee that a question does not bias for or against any particular group. Any question that shows bias toward any particular group is discarded.

Objective Tests

A test is called an objective test if the same standards and conditions apply to everyone taking the test and there is only one correct answer to each question.

Objective tests primarily measure your ability to recall information. A well-designed objective test can also test your ability to understand, analyze, interpret, and apply your knowledge. Objective tests include true-false, multiple choice, fill in the blank, and matching questions. ASE's tests consist exclusively of four-part multiple-choice objective questions.

The following are some strategies that may be applied to your tests. Before beginning to take an objective test, quickly look over the test to determine the number of questions, but do not try to read through all of the questions. In an ASE test, there are usually between forty and eighty questions, depending on the subject. Read through each question before marking your answer. Answer the questions in the order they appear on the test. Leave the questions blank that you are not sure of and move on to the next question. You can return to those unanswered questions after you have finished the others. They may be easier to answer at a later time after your mind has had additional time to consider them on a subconscious level. In addition, sometimes you might find information in other questions that will help you recall the answers to some of them.

Do not be obsessed by the apparent pattern of responses. For example, do not be influenced by a pattern like D, C, B, A, D, C, B, A on an ASE test.

There is also a lot of folk wisdom about taking objective tests. For example, there are those who would advise you to avoid response options that use certain words such as *all*, *none*, *always*, *never*, *must*, and *only*, to name a few. This, they claim, is because nothing in life is exclusive. They would advise you to choose response options that use words that allow for some exception, such as *sometimes*, *frequently*, *rarely*, *often*, *usually*, *seldom*, and *normally*. They would also advise you to avoid the first and last option (A or D) because test writers, they feel, are more comfortable if they put the correct answer in the middle (B or C) of the choices. Another recommendation often offered is to select the option that is either shorter or longer than the other three choices because it is more likely to be correct. Some would advise you to never change an answer since your first intuition is usually correct.

Although there may be a grain of truth in this folk wisdom, ASE test writers try to avoid them and so should you. There are just as many A answers as there are B answers, just as many C answers as

D answers. As a matter of fact, ASE tries to balance the answers at about 25 percent per choice A, B, C, and D. There is no intention to use "tricky" words, such as outlined previously. Put no credence in the opposing words *sometimes* and *never*, for example.

Multiple-choice tests are sometimes challenging because there are often several choices that may seem possible, and it may be difficult to decide on the correct choice. The best strategy, in this case, is to first determine the correct answer before looking at the options. If you see the answer you decided on, you should still examine the options to make sure that none seem more correct than yours. If you do not know or are not sure of the answer, read each option very carefully and try to eliminate those options that you know are incorrect. That way, you can often arrive at the correct choice through a process of elimination.

If you have gone through the entire test and you still do not know the answer to some of the questions, then guess. Yes, guess. You then have at least a 25 percent chance of being correct. If you leave the question blank, you have no chance. Your score is based on the number of questions answered correctly.

Preparing for the Exam

The main reason we have included so many sample and practice questions in this guide is simply to help you learn what you know and what you don't know. We recommend that you work your way through each question in this book. Before doing this, carefully look through Section 3; it contains a description and explanation of the type of questions you'll find in an ASE exam.

Once you understand what the questions will look like, move to the sample test. Answer one of the sample questions (Section 5), then read the explanation (Section 7) to the answer for that question. If you don't feel you understand the reasoning for the correct answer, go back and read the overview (Section 4) for the task that is related to that question. If you still don't feel you have a solid understanding of the material, identify a good source of information on the topic, such as a textbook, and do some more studying.

After you have completed all of the sample test items and reviewed your answers, move to the additional questions (Section 6). This time answer the questions as if you were taking an actual test. Do not use any reference or allow any interruptions in order to get a feel for how you will do on an actual test. Once you have answered all of the questions, grade your results using the answer key in Section 7. For every question that you gave an incorrect answer to, study the explanations to the answers and/or the overview of the related task areas. Try to determine the root cause for your missing the question. The easiest thing to correct is learning the correct technical content. The hardest things to correct are behaviors that lead you to an incorrect conclusion. If you knew the information but still got it incorrect there is a behavior problem that will need to be corrected. An example would be reading too quickly and skipping over words that affect your reasoning. If you can identify what you did that caused you to answer the question incorrectly you can eliminate that cause and improve your score. Here are some basic guidelines to follow while preparing for the exam:

- Focus your studies on those areas you are weak in.
- Be honest with yourself while determining if you understand something.
- Study often but in short periods of time.
- Remove yourself from all distractions while studying.
- Keep in mind the goal of studying is not just to pass the exam, the real goal is to learn!
- Prepare physically by getting a good night's rest before the test and eat meals that provide energy but do not cause discomfort.
- Arrive early to the test site to avoid long waits as test candidates check in and to allow all of the time available for your tests.

During the Test

On paper and pencil tests you will be placing your answers on a sheet where you will be required to blacken (bubble) in your answer choice. Stray marks or incomplete erasures may be picked up as an answer by the electronic reader, so be sure only your answers end up on the sheet. One of the biggest problems an adult faces in test taking, it seems, is in placing an answer in the correct spot on a bubble sheet. Make certain that you mark your answer for, say, question 21, in the space on the bubble sheet designated for the answer for question 21. A correct response on the incorrect line will probably result in two questions being marked incorrectly, one with two answers (which could include a correct answer but will be scored incorrectly) and the other with no answer. Remember, the answer sheet on a written test is machine scored and can only "read" what you have bubbled in.

If you finish answering all of the questions on a test and have remaining time, go back and review the answers of those questions that you were not sure of. You can often catch careless errors by using the remaining time to review your answers. Carefully check your answer sheet for blank answer blocks or missing information.

At practically every test, some technicians will invariably finish ahead of time and turn their papers in long before the final call. Some technicians may be doing a recertification test and others may be taking fewer test than you. Do not let them distract or intimidate you.

It is not wise to use less than the total amount of time that you are allotted for a test. If there are any doubts, take the time for review. Any product can usually be made better with some additional effort. A test is no exception. It is not necessary to turn in your test paper until you are told to do so.

Your Test Results!

You can gain a better perspective about tests if you know and understand how they are scored. ASE's tests are scored by American College Testing (ACT), a nonpartial, unbiased organization having no vested interest in ASE or in the automotive industry.

Each question carries the same weight as any other question. For example, if there are fifty questions, each is worth 2 percent of the total score. The passing grade is 70 percent. That means you must correctly answer 35 of the 50 questions to pass the test.

The test results can tell you:

- where your knowledge equals or exceeds that needed for competent performance, or
- where you might need more preparation.

Your ASE test score report is divided into content areas and will show the number of questions in each content area and how many of your answers were correct. These numbers provide information about your performance in each area of the test. However, because there may be a different number of questions in each content area of the test, a high percentage of correct answers in an area with few questions may not offset a low percentage in an area with many questions.

It should be noted that one does not "fail" an ASE test. The technician who does not pass is simply told "More Preparation Needed." Though large differences in percentages may indicate problem areas, it is important to consider how many questions were asked in each area. Since each test evaluates all phases of the work involved in a service specialty, you should be prepared in each area. A low score in one area could keep you from passing an entire test.

There is no such thing as average. You cannot determine your overall test score by adding the percentages given for each task area and dividing by the number of areas. It doesn't work that way because there, generally, are not the same number of questions in each task area. A task area with twenty questions, for example, counts more toward your total score than a task area with ten questions.

Your test report should give you a good picture of your results and a better understanding of your strength and weaknesses for each task area.

If you fail to pass the test, you may take it again at any time it is scheduled to be administered. You are the only one who will receive your test score. Test scores will not be given over the phone by ASE, nor will they be released to anyone without your written permission.

3 Types of Questions on an ASE Exam

ASE certification tests are often thought of as being tricky. They may seem to be tricky if you do not completely understand what is being asked. The following examples will help you recognize certain types of ASE questions and avoid common errors.

Paper-and-pencil tests and Computer Based Test (CBT) questions are identical in content and difficulty. Most initial certification tests are made up of 40 to 80 multiple-choice questions. Multiple-choice questions are an efficient way to test knowledge. To answer them correctly, you must think about each choice as a possibility, and then choose the one that best answers the question. To do this, read each word of the question carefully. Do not assume you know what the question is about until you have finished reading it.

About 10 percent of the questions on an actual ASE exam will use an illustration. These drawings contain the information needed to correctly answer the question. The illustration must be studied carefully before attempting to answer the question. Often, technicians look at the possible answers and then try matching the answers to the drawing. Always do the opposite; match the drawing to the answers. When the illustration is showing an electrical schematic or another system in detail, look over the system and try to figure out how the system works before you look at the question and the possible answers.

Multiple-Choice Questions

The most common type of question used on ASE Tests is the multiple-choice question. This type of question contains three "distracters" (incorrect answers) and one "key" (correct answer). When the questions are written, effort is made to make the distracters plausible to draw an inexperienced technician to one of them. This type of question gives a clear indication of the technician's knowledge. If you encounter a question that you are unsure of, reverse-engineer it by eliminating the items that it cannot be. Consider the following example:

A technician is testing a circuit for voltage during a road test when the probe accidentally touches ground and burns the fuse, stalling out the vehicle. The technician does not have a fuse of the same rating available at hand but has higher and lower rated fuses. The next best solution would be to:

A. Use a higher rated fuse and install the correct fuse after returning back to the shop.
B. Use a jumper wire and install the correct fuse after returning back to the shop.
C. Use a lower rated fuse and install the correct fuse after returning back to the shop.
D. Have someone bring the correct fuse out to the location. (H1)

Answer A is incorrect. Installing a higher than manufacturer approved fuse for circuit protection could damage the electronic component it is designed to protect. A higher-rated fuse could potentially not protect the component from too much current draw and lead to expensive repairs.
Answer B is incorrect. Using a jumper wire to bypass the circuit protection would not provide any protection to the electronic component the fuse is designed to protect and should be avoided.
Answer C is correct. Installing a lower-rated fuse should allow the electronic component to function normally and offer protection to the component in case of an overcurrent condition. The fuse rating is usually designed with 10 to 15 percent excess capacity. The correct fuse should be installed as soon as possible upon arriving back to the shop area.
Answer D is incorrect. Having someone bring a fuse to the location would be time-consuming and expensive. Because a lower rated fuse is available it should be used and is the best possible solution.

EXCEPT Questions

Another type of question used on ASE tests has answers that are all correct except one. The correct answer for this type of question is the answer that is incorrect. The word "EXCEPT" will always be in capital letters. You must identify which of the choices is the incorrect answer. If you read too quickly through the question, you may overlook what the question is asking and answer the question with the first correct statement. This will make your answer incorrect. An example of this type of question and the analysis is as follows:

All of the following components are used for NGV ignition systems EXCEPT

A. Points
B. Coil
C. Spark Plugs
D. Ignition Control Module (ICM) (H6)

Answer A is correct. Points are not utilized for NGV ignition systems. The ignition systems are controlled through the use of sensors and electronic modules.
Answer B is incorrect. Coils are used in the ignition system of NGVs.
Answer C is incorrect. Spark plugs are used for the ignition system of NGVs.
Answer D is incorrect. An Ignition Control Module (ICM) is used to amplify the signal from the ECM and control the spark discharge of the coils.

Technician A, Technician B Questions

The type of question that is most popularly associated with an ASE test is the "Technician A says . . . Technician B says . . . Who is right?" type. In this type of question, you must identify the correct statement or statements. To answer this type of question correctly, you must carefully read each technician's statement and judge it on its own merit to determine if the statement is true.

Sometimes this type of question begins with a statement about some analysis or repair procedure. This is often referred to as the stem of the question and provides the setup or background information required to understand the conditions on which the question is based. This is followed by two statements about the cause of the concern, proper inspection, identification, or repair choices. You are asked whether the first statement, the second statement, both statements, or neither statement is correct. Analyzing this type of question is a little easier than the other types because there are only two ideas to consider, although there are still four choices for an answer.

Technician A, Technician B questions are really double true-or-false questions. The best way to analyze this type of question is to consider each technician's statement separately. Ask yourself, is A true or false? Is B true or false? Then select your answer from the four choices. An important point to remember is that an ASE Technician A, Technician B question will never have Technician A and B directly disagreeing with each other. That is why you must evaluate each statement independently. An example of this type of question follows:

A transit bus with a Compressed Natural Gas (CNG) engine cranks normally but will not run. Technician A says the ¼ fuel shut-off valve should be checked to ensure it is turned on. Technician B says the batteries are probably weak and need to be recharged or fresh batteries installed before further testing can continue. Who is correct?

A. A only
B. B only
C. Both A and B
D. Neither A nor B (F1.4)

Answer A is correct. Technician A is correct. The ¼ turn fuel shut-off valve should be checked to ensure it is turned on. Sometimes Technicians will turn the ¼ turn valve off for safety reasons when performing maintenance and forget to turn the valve back on.
Answer B is incorrect. Only Technician A is correct. Because the engine cranks over normally, the batteries probably are not the cause of the engine not running.
Answer C is incorrect. Only Technician A is correct.
Answer D is incorrect. Technician A is correct.

Most-Likely Questions

Most-Likely questions are somewhat difficult because only one choice is correct while the other three choices are nearly correct. An example of a Most-Likely question is as follows:

An operator complains the NGV transit bus is slow accelerating all the time but performs relatively well once the vehicle reaches 35 to 40 miles per hour. Which of these is the Most-Likely cause?

A. Poor quality natural gas
B. Low battery charging rate
C. High delivery fuel pressure
D. A partially stuck open or maladjusted turbocharger waste gate (H2)

Answer A is incorrect. Poor quality natural gas would perform poorly throughout the entire operating range of the engine.
Answer B is incorrect. A low battery charging rate would be evident when the vehicle is being started but would not affect the operation of the engine only during acceleration.
Answer C is incorrect. High fuel delivery pressure would not affect the acceleration of the vehicle.
Answer D is correct. A partially stuck open or misadjusted turbocharger waste gate is the most likely cause because this would prevent the turbocharger from spooling up or rotating faster when the engine is accelerated and delivering more air to the intake manifold and the boost sensor to allow the ECM to deliver more fuel to the engine.

LEAST-Likely Questions

Notice that in the Most-Likely questions there is no capitalization. This is not so with LEAST-Likely questions. For this type of question, look for the choice that would be the LEAST-Likely cause of the described situation. Read the entire question carefully before choosing your answer. An example is as follows:

An NGV transit bus is brought to the shop for the low fuel pressure indicator light coming on and logging a low fuel pressure code when the engine is under a high load. Which of the following is the LEAST-Likely cause?

A. A manual fuel tank shut-off valve is turned off at one of the cylinders.
B. The secondary fuel filter is dirty.
C. The vehicle has just been started on a cold morning.
D. The high-pressure regulator is malfunctioning. (F3.1)

Answer A is correct. A manual fuel shut-off valve would be the LEAST-Likely problem because CNG vehicles are equipped with numerous fuel cylinders and therefore would be able to maintain

fuel delivery. A shut-off valve at one of the cylinders being shut off would be indicated by a reduced operating range of the vehicle.
Answer B is incorrect. A dirty secondary fuel filter can be a cause of low fuel pressure.
Answer C is incorrect. Starting the vehicle on a cold morning and driving off before the engine coolant has time to warm up can cause a low fuel pressure condition due to the effect of cooling. During the change from a high pressure to a lower pressure gas and the lack of heat in the heat exchanger part of the high pressure regulator due to cold coolant circulating and not being able to assist in preventing the cooling effect can cause lower than normal fuel pressure in the low-pressure fuel delivery side of the fuel system as evidenced by icing on the fuel lines and colder than normal fuel temperatures to the engine.
Answer D is incorrect. A defective high pressure regulator can cause an intermittent low fuel pressure condition.

Summary

There are no four-part multiple-choice ASE questions having "none of the above" or "all of the above" choices. ASE does not use other types of questions, such as fill-in-the-blank, completion, true/false, word-matching, or essay. ASE does not require you to draw diagrams or sketches. If a formula or chart is required to answer a question, it is provided for you. There are no ASE questions that require you to use a pocket calculator.

Testing Time Length

An ASE written test session is four hours. You may attempt from one to a maximum of four tests in one session. It is recommended, however, that no more than a total of 225 questions should be attempted at any test session. This will allow for just over one minute for each question.

Visitors are not permitted at any time. If you wish to leave the testing room for any reason, you must first ask permission. If you finish your test early and wish to leave, you are permitted to do so only during specified dismissal periods.

You should monitor your progress and set an arbitrary limit to how much time you will need for each question. This should be based on the number of questions you are attempting. It is suggested that you wear a watch because some facilities may not have a clock visible to all areas of the room.

Computer Based Tests are allotted a testing time according to the number of questions ranging from 1/2 hour to 1 1/2 hours. Advanced level tests are allowed 2 hours. These tests are by appointment and you should be on time to ensure that you have all of the allocated time available. If you arrive late for a CBT test appointment, you will only have the amount of time remaining in your appointment.

4 Overview of the Task List

Specifications for Transit Bus Compressed Natural Gas (CNG) Engines (Test H1)

The following section includes the task areas and task lists for this test and a written overview of the topics covered in the test.

The task list describes the actual work you should be able to do as a technician and on which you will be tested by the ASE. This is your key to the test, and you should review this section carefully. We have based our sample test and additional questions upon these tasks. The overview section will also support your understanding of the task list. ASE advises that the questions on the test may not equal the number of tasks listed; the task lists outline the specific tasks that ASE expects you to know how to perform and which you will be tested on.

At the end of each question in the Sample Test for Practice and Additional Test Questions for Practice sections, a letter and a number are presented as a reference back to the related task for additional study, as seen in the following example.

A NGV engine vehicle is brought to the shop for an overheating complaint. It is equipped with a belt driven hydraulic pump to drive a hydraulic radiator fan. Technician A says that if the drive belts were loose it would be obvious because the belts would screech. Technician B says that slipping belts do not always screech, it depends on the severity of the slippage and the belt tension should be checked. Which Technician is correct?

A. A only
B. B only
C. Both A and B
D. Neither A nor B (D7)

Answer A is incorrect. Slipping belts are not always symptomatic with the familiar screeching associated with slipping belts. Sometimes belts such as a power steering and hydraulic fan drive pumps only screech when a large load is applied such as when the steering wheel is turned or the hydraulic fan is actuated.

Answer B is correct. Only Technician B is correct. Slipping belts are not always symptomatic with the familiar screeching associated with slipping belts. Engines equipped with a belt driven hydraulic driven fan, should always have the belt tension checked as a starting diagnostic point. The pulleys should be checked to make sure they are not worn to the point where the V of the belts bottom out in the pulley. The alignment of the pulleys should also be checked as this condition will cause pre-mature wear of the belts.

Answer C is incorrect. Only Technician B is correct.

Answer D is incorrect. Technician B is correct. (D7)

Task List and Overview

A. General Engine Diagnosis (20 Questions)

Task A1 Verify the complaint, check for and retrieve fault codes, and road test vehicle; review operator service request and past maintenance documents (if available); determine further diagnosis.

Natural Gas Vehicle (NGV) Technicians must be familiar with the following basic diagnostic procedures: The customer/operator complaint must be listened to carefully. Technicians should question the customer/operator to obtain more information regarding the problem whenever possible. To help identify the complaint, ask questions such as: Are the Check Engine Light (CEL) or the Stop Engine Light (SEL) or any other warning light or buzzer illuminating? Is the defect intermittent or constant? How long after operating the vehicle does the trouble occur?

Attempt to determine the root cause of the problem. Is the problem engine performance or transmission related, an electrical defect, or related to another system that could be giving the symptoms? Check the engine Electronic Control Module (ECM), the transmission Electronic Control Unit (ECU) for Diagnostic Trouble Codes (DTCs). Whenever possible, review the vehicle's repair records to determine past repair history. Check the coolant and engine/transmission oil levels before any tests are performed. Sometimes, a simple stall test of the vehicle if equipped with an automatic transmission will reveal the root cause of the complaint. Road test or dyno the vehicle using diagnostic tools such as a Diagnostic Data Reader (DDR) or a laptop with the appropriate diagnostic software and create a snapshot of operating parameters to be able to review for possible defects back at the shop, if necessary. Use the data retrieved from the road test to determine possible causes of the problem. Perform diagnostic tests to locate the exact cause of the problem. Always start with the easiest, quickest test first. The diagnostic software or the manufacturer's service manuals will contain troubleshooting trees that should be applied whenever diagnosing symptoms.

After the repair/s has been made, be sure that the customer's/operator's complaint is eliminated. Clear the Diagnostic Trouble Codes (DTC) and road test the vehicle again if necessary.

Task A2 Inspect engine assembly and engine compartment for natural gas, oil, coolant, exhaust, or other leaks; determine needed repairs.

A Technician must understand the basic fuel, lubricating, exhaust, and cooling systems and components. The location of all possible leaks in these systems must be identified. Sight, sound, and smell can be used to determine possible defects that could be detrimental to the safe operation of the vehicle.

Natural gas (methane) is lighter than air, odorless and colorless. Shops need to be retrofitted with special methane detectors at the roof area as well as heaters that are explosion proof. Leaks must be detected and repaired quickly for safety. Mercaptan, an odorant, is added by the gas companies to help identify the presence of methane. After prolonged exposure to the smell of mercaptan, a Technician can become desensitized to the odor. Therefore, an electronic methane leak detector should be utilized whenever a leak is detected by the onboard gas detection system or suspected. A soapy water solution or commercially available testing solutions such as Snoop can assist in pinpointing the leak once the general area has been located. Only an approved sealant, like Swak, should be utilized on pipe threads to assure that methane will not attack the sealant and cause it to fail. Whenever an O-ring is used to seal connections, an approved O-ring lubricant such as Dow Corning 55 should also be used to ensure the proper sealing of the mating surfaces. The surge tank should also be routinely checked for the presence of methane. If the high-pressure regulator fails, it can allow gas into the cooling system from the reheat part of the regulator. Unusual high pressures can be encountered when opening the surge tank along with a strong smell of mercaptan.

External coolant leaks can usually be identified by the presence of coolant on the ground. Because the engine and the cooling system of a transit bus is usually located in the rear, long lines bring coolant to the front of the vehicle for the driver's heater and defroster as well as passenger compartment heating. Additionally, the Heating, Ventilation and Air Conditioning (HVAC) system is usually located at the rear top or roof areas. The length of piping, the need of control valves and multiple connections can be sources of leakage that can affect the engine as well because of the common coolant usage. However, coolant leaks may be internal to the engine as well as external. Technicians should have a good working knowledge of how coolant is routed throughout the vehicle as well as through the engine and be familiar with potential internal engine leakage. Cylinder heads and gaskets, cylinder liners and seals, turbocharger housing seal failures, spark plug tube adaptors, and other parts can introduce coolant into the engine combustion chamber with no telltale signs other than a loss of coolant and cause catastrophic failures if undetected or left in disrepair. Sometimes internal coolant leaks show up as a concentration of condensation on the oil filler cap as well as coolant in the crankcase, giving the engine oil a milky appearance.

Many times when an engine is being diagnosed for the location of an oil leak, the engine has been leaking for an indefinite period and the engine oil has accumulated dirt. In this case, the engine must be cleaned to make the leak more visible. Another way to find an oil leak is to add a dye to the engine oil that reacts with an ultraviolet lamp. When an engine leaks oil, it can be a fire hazard as well as an environmental hazard. An NGV exhaust operates in the range of 1050 to upward of 1200° F in comparison to a Diesel engine, which can operate at 750 to 950° F when under load. Oil residue in the engine compartment could be from an overfilled engine crankcase or may indicate oil caught up in the air-pressure stream of the fan and thrown about. Oil on the exhaust manifold could indicate a leaking rocker housing cover gasket, and if the engine is equipped with a hydraulic cooling fan, the fan motor seal or connections could be a potential point of leakage. The oil leaks could be engine, power steering, fan controls, or even from the transmission. The color of the fluid can many times direct the Technician to the component that needs attention.

Task A3 Inspect engine compartment wiring harness, connectors, seals, and locks; check for proper routing and terminal/connector condition; determine needed repairs.

Modern engine electrical/electronic wiring is complex. The bus manufacturer and the engine manufacturer have to interface at many connection points. If connectors or wires fail, the circuit sends faulty inputs from the monitoring sensors that operate on milliamps of current. This, in turn, causes computerized circuits using these inputs to make inappropriate adjustments to the engine controls. Some of the common problems associated with circuits are corrosion and other foreign substances in connectors, loose or improperly mated connectors, and improper grounding. Some of these problems will not affect engine operation until several other problems exist, by which time they can complicate the troubleshooting process. If a visual inspection of a connector reveals corrosion, the connector should be cleaned or replaced. Wiring that is visibly lumpy or has nicked insulation should be repaired or replaced to prevent diagnostic faults and protect from corrosion.

The engine rear apparatus box and cover should be intact and properly secured with all the fasteners. Switches should have the protective caps in place and not be cracked. Connectors should be tightly installed to seal them from the environment and corrosion. It should be possible to manipulate wiring without the insulation breaking and exposing the copper conductor.

Plastic weather pack connector locks should be intact to prevent the connector from falling off the component. The joint seals should be installed and in good condition.

Loose or scale-coated grounds, as well as power cables, should be removed, cleaned, treated for corrosion, reinstalled, and coated with a corrosion inhibitor. All connections to a ground must be free of rust, undercoating, and/or paint to ensure a good electrical connection.

Some bus manufacturers place fuse panels and circuit breakers in the battery compartment that supply voltage to the engine ECM and associated engine controls. These fuse panels are exposed to battery electrolyte fumes and can be a source of corrosion. The panels should be checked for corrosion and repaired or replaced as deemed necessary. Many times, intermittent sensor low-voltage codes can be traced back to these panels.

Task A4 Listen for and diagnose engine noises; determine needed repairs.

A Technician who develops a discriminating ear for engine noises can simplify troubleshooting or find problems before they can cause a breakdown. The origination point of the noise needs to be determined to enable the Technician to work on the correct component. Transmissions, air compressor, A/C compressors, and power steering pumps can all be sources of noises. By using a tool similar to a stethoscope, listening through a metal bar or heater hose applied to each cylinder in sequence can expose a misfiring cylinder. Electronic sound amplification tools can also be employed. Sometimes the absence of change in the sound may provide confirmation of a faulty condition. If one of the cylinders appears to be misfiring, perform a cylinder cut-out test utilizing diagnostic software. Every time a good cylinder is turned off, the sound of the engine should change and the diagnostic software will usually indicate a change in the amount of pulse width in degrees of crankshaft revolution necessary to keep the engine at a predetermined speed. On the other hand, shutting down a dead cylinder will provide no noticeable change in sound or pulse width as read on the diagnostic software.

Often the frequency of an unusual noise is relevant to the defective component's operating speed. A loud knocking that increases at the same rate as the engine speed could probably be associated with the crankshaft or balance shaft assembly. Whereas, a sound that is half the engine speed might be created by a camshaft or something attached to it. Other components may produce high-frequency sounds like a squeal and may not be as discernible using this method. In these cases, turning your head from side to side to determine the direction of the loudest sound may point to the general location. Then, look for components in that area that operate at a high speed. If the sound diminishes or stops. such as when the air compressor unloader valves unload, it usually indicates the air compressor is the source and can be further diagnosed. Squeals may also be produced when two surfaces are rubbing each other, like belts slipping. Sometimes sounds appear random in nature. Usually the operator accelerating or turning on a system such as the HVAC, an automated operation, or a load change, like a power steering pump squealing when the steering wheel is turned or when the hydraulic cooling fan turns on, causes these randomly occurring noises. Spraying water on the belt is a helpful method to diagnose a belt squeal from a bearing squeal and will not damage the belt.

Task A5 Check engine exhaust for odor, smoke, or excessive temperatures; determine needed repairs.

The color and consistency of the exhaust smoke can reveal problems associated with the combustion of the engine. Natural Gas engines burn methane gas that has a very low carbon composition compared to its cousin the diesel engine. Black smoke is usually not associated with Natural Gas combustion. Clear, blue, or white smoke will be the most common colors of exhaust smoke. A clear pale haze is normal. Exhaust temperature is electronically monitored to ensure the engine is not burning excessively rich mixtures of Natural Gas. The normal temperature range is 1050 to 1200° F, depending on engine load and speed. High temperature warnings and eventual shutdown will usually occur at temperatures above 1350° F. Blue colored exhaust smoke indicates excessive oil in the combustion chamber. Blue smoke on acceleration is indicative of a defective turbocharger seal, worn piston rings, or excessive oil pressure/volume directed onto the cylinder wall area, and could indicate a bent piston cooling nozzle squirting on the cylinder wall and skirt of the piston instead of at the underside of the crown of the piston. Blue smoke at startup or deceleration usually indicates excessive wear of valve guides/seals. White smoke is an indication of steam being produced by the combustion process on cold startups due to natural condensation from the exhaust and especially on damp mornings. Continuous white smoke after warm-up is usually an indication of coolant leakage into the combustion chamber, a cracked cylinder head or blown head gasket, leaking spark plug tube seals, or in the case of a water-cooled turbocharger center section a failed turbocharger.

Task A6 Perform fuel supply system tests; check fuel for contamination and consumption; determine needed repairs.

NGV engines utilize methane as a fuel stored as Compressed Natural Gas (CNG), a high-pressure gas with storage pressure up to 3600 PSI after fueling. The CNG is regulated at the high-pressure regulator to a pressure of approximately 110 psi.

Today's engines are electronically controlled and can be monitored with tools such as a DDR, or a laptop with appropriate software. Sensors monitor fuel system pressures and temperature. The information is fed back to the ECM to ensure that in the event the pressure in the low-pressure system or the temperature of the fuel becomes too high or too low, the ECM can compensate fuel delivery to the engine. If needed, the ECM can utilize the information from the sensors to activate indicator lights to warn the operator of the condition and even shut down the engine to prevent damage from occurring.

Fuel pressure(s) and temperature can be monitored with the appropriate software to ensure that the filters and the high-pressure regulator are capable of supplying the recommended fuel pressure. The high-pressure regulator also utilizes circulated hot engine coolant to reheat the fuel passing through it to prevent low fuel temperatures that cause metering valves to malfunction and incomplete combustion in the combustion chamber. Fuel filters for NGVs are of the coalescent type and usually remove contaminants down to one micron. If the filters are suspected of being contaminated with oil due to a low fuel pressure condition, some manufacturers have a drain in the bottom of the housing that can be opened to allow the oil to be drained out. Some designs require the filters to be removed from their housings to be drained and checked for contamination. All safety procedures pertaining to depleting the fuel out of the system before attempting to remove the filters must be followed to prevent injury to yourself or others. Excessive amounts of oil or contamination indicates the filters have to be replaced. A further investigation as to the source of the contaminants needs to be performed.

Task A7 Perform air intake system restriction and leakage tests; determine needed repairs.

When an NGV engine suffers from a loss of power or rough operation, the Technician should go back to the basics. To obtain optimum performance from an engine, the proper quantity and quality of fuel must be supplied to the cylinder with the correct amount of air and correct spark timing.

Air starvation will generally result in incomplete combustion due to a lack of oxygen, resulting in a decrease of power, higher heat buildup in the cylinder and the exhaust system due to a lack of oxygen necessary for complete combustion. The electronic controls in conjunction with the Oxygen sensor (O_2) will command the fuel rate leaner; however, lower power output due to the amount of fuel in the cylinder is directly responsible for the amount of push on the piston during the power stroke and resulting in lower power output and poor performance complaints. The diagnostic tools usually can be useful in determining the lack of air supply as a low boost pressure defect. Some of the reasons for air starvation include a clogged air cleaner and restricted intake hoses and pipes on the suction side of the turbocharger. Downstream of the turbocharger, a clogged or partially clogged charge air cooler (CAC) should not be overlooked as a source of lack of air. Loose clamps and punctured hoses or pipes can allow air to escape. High-pressure leaks can usually be heard as a high-pitched whistling. Use of a soapy water solution or commercially available testing solutions such as Snoop can also assist the Technician in finding high-pressure leaks.

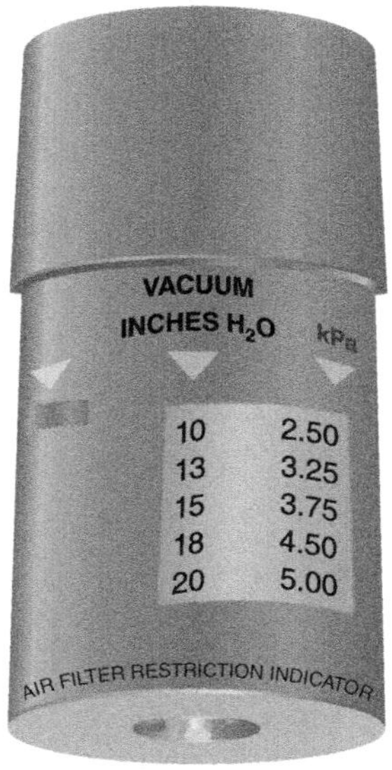

Air inlet restriction may be measured in a number of ways. Most NGVs are equipped with a resettable inlet restriction gauge that is mounted directly on the air intake piping or remotely in the engine compartment.

These gauges are not as accurate as test equipment but do provide an indication of when the element should be serviced. The following gauge is a more accurate and a more accurate device.

Inlet restriction is measured in inches of water vacuum. Most manufacturers recommend the air cleaner be replaced when a reading of 20" of water column is reached. To obtain an accurate measurement with test equipment, such as a water manometer or a magnehelic gauge, the engine should be under full load (stall test) and the gauge or manometer should be connected just downstream from the filter. When it is not possible to place the engine under a full load, a snap throttle test will give a close approximation.

Visually check the intake pipe for evidence of dirt entry and make immediate repairs if any is found. Dirt can quickly destroy and is the number one enemy of the engine. Last but not least, low-power complaints under load may also result from collapse of the rubber hose from the air cleaner to the turbocharger that starves the engine of air.

Task A8 Perform intake manifold pressure tests; determine needed repairs.

All modern NGV engines are turbocharged and therefore have pressurized intake piping between the turbocharger and the throttle valve at idle, and between the turbocharger all the way to the intake valves in the cylinder head once a load has been applied to the engine causing the turbocharger to spool up and pressurize the system. The intake manifold contains a boost pressure sensor or air intake pressure sensor that sends a signal back to the ECM to determine operating conditions. The sensor can measure both pressure and vacuum because it is calibrated in absolute pressure. This pressure sensor can be monitored using the diagnostic DDR or the software. Placing the vehicle under a stall test will allow the turbocharger to develop boost as it is driven faster by the pressure and heat of the exhaust as a result of burning more fuel. Consult the manufacturer's service manual for the specification of the minimum and maximum intake manifold pressures.

Low intake manifold pressure can be caused by a clogged air cleaner or collapsed intake piping on the vacuum side of the turbocharger. On the pressure side of the turbo, a clogged CAC (charge air cooler), a defective or partially stuck open turbo waste gate, or leaking hoses, pipes, or defective exhaust manifold gaskets can lead to lower than normal manifold pressure.

High intake boost pressure can be caused by stuck closed waste gate and defective electronic or air controls of the waste gate. A high boost pressure must not be allowed to persist as this can cause pre-ignition and ultimately engine damage.

Although the ECM can compensate for high manifold pressure by derating the engine and logging a fault code in conjunction with illuminating the CEL or SEL, depending on the severity, high intake manifold pressure can cause knock codes, detonation, and can potentially damage the engine if left uncorrected. Suspected air-intake system pressure leaks can be checked using a soapy water solution or commercially available testing solutions such as Snoop. Any dirt in the intake system should make the Technician suspicious of a possible air leak in the air cleaner or air intake piping before the turbocharger.

Task A9 Perform exhaust back pressure and temperature tests; determine needed repairs.

A slight back pressure in the exhaust system, usually no more than 2.5" mercury (Hg), is normal. However, excessive exhaust back pressure seriously affects engine operation. Because the engine will not be able to breathe correctly, exhaust will not be allowed to be scavenged, causing higher than normal emissions, high cylinder temperatures, high exhaust temperature, and shortened spark plug life. High back pressure also can lead to engine damage, a damaged catalyst, and eventually costly repairs.

High exhaust back pressure is usually caused by a collapsed exhaust pipe or a defective catalyst that has become contaminated by coolant, excessive oil consumption from worn rings or valve guide seals, or high exhaust temperatures due to maladjusted valves (exhaust valves too tight, causing them to open too soon and allowing unburned fuel into the exhaust stream that is still burning).

Exhaust back pressure, measured in inches of mercury, is checked with a manometer or magnehelic gauge rated in inches of mercury (Hg). Connect the manometer in the exhaust pipe after the turbocharger, usually 6 to 12" from the turbine outlet, by removing the temperature probe or O_2 sensor and using the appropriate adaptor to connect the manometer hose. Run the engine at full load to determine the back pressure. Compare to manufacturer specifications.

Task A10 Perform crankcase pressure test; determine needed repairs.

Crankcase pressure is created by a controlled amount of cylinder pressure passing by the piston rings and to a smaller degree past the valve guides and seals into the crankcase. A slight pressure in the crankcase is desirable to prevent the entrance of dust and to expel moisture as a byproduct of changing temperatures. Today's NGV has the crankcase breather routed though a coalescing filter assembly and plumbed though the air intake to prevent harmful emissions from escaping into the atmosphere. Excessive oil in the coalescing filter and high crankcase pressure are indicators of worn piston rings, a cracked piston or pistons, defective valve guides and seals, possible defective cylinder head gasket and even high exhaust back pressure. The breather tube should be checked for obstructions or a collapsed area to prevent high crankcase pressures. The coalescing filter must be replaced at normal servicing intervals or when excessively contaminated with engine oil to prevent the oil from being carried into the air intake system and contaminating the metering valves and sensors and the CAC.

To check the crankcase pressure, attach a manometer or magnehelic gauge rated in inches of water column to the engine oil dipstick tube. The Technician may have to drain some oil out of the oil pan to ensure the oil level is below the opening of the dipstick to allow an accurate reading. Check the readings obtained at various engine speeds with the manufacturer's specifications. Typically, the maximum pressures range at about 2" of water column under full load.

Common causes of excessive crankcase pressure are: excessively worn exhaust valve guides and/or seals; a blown cylinder head gasket; excessive exhaust back pressure; a cracked or burnt hole in a piston; a badly scored piston and/or cylinder wall; broken compression rings; and an obstructed crankcase breather hose.

Task A11 Diagnose no-cranking, cranks but fails to start, hard starting, and starts but does not continue to run problems; determine needed repairs.

Diagnosing a no-crank situation requires an in-depth knowledge of the vehicle being diagnosed. Today's NGV engines are controlled using not only electronic controls but also multiplexing through one of the following: Dinex's Input/Output (I/O) system, Allen Bradley's Programmable

Logic Controls (PLC), or Vansco's controls. The associated logic controls and diagrams must be available to the Technician for accurate and timely diagnostics. How to interpret the diagrams and quickly determine which system, switch, or sensor is malfunctioning or has not met the logic requirements to be able to crank the engine is the key to successful repairs. Accurate electrical schematics are a must as well. The ECM can cause a no-crank situation due to fault codes that are recognized as causing an unsafe condition and can potentially damage the engine. A defective charging and battery system can be another source of a no-start condition. The Technician must be prepared to perform the easiest and fastest task to repair the defect. A good quality high-impedance multimeter should be available to the Technician for diagnostic purposes. The Technician should have a good working knowledge of voltage checks, voltage drop checks, amperage demand checks, and resistance checks.

The basic checks of a no-start condition include the condition of the batteries to ensure enough voltage and amperage is available to keep these systems working properly as well as being able to deliver enough power to the starter solenoid and the starter. The ECM requires a minimum voltage of 9.6 volts during cranking to be able to function. A check of the SEL and CEL indicators in the dash must be made to ensure they cycle on and off for a duration of five seconds and remain off after the ignition switch has been turned on. If the indicators do not cycle this could be an indication the ECM has not received an ignition signal or a low-voltage situation exists. Ladder logic and electrical schematics must be consulted for determining what component(s) need to be checked.

Once the engine cranks, the basics, just like any other engine, come into play. The basic triangle of fuel, air, and ignition must take place. If one is missing, the combustion process cannot take place. Diagnostic tools can monitor fuel pressures being sent to the engine. If the fuel demand is not met, the Technician must determine if the engine is withholding the 60 rpm signal to energize the fuel solenoid. If the fuel demand is met, it must be in the correct ratio to the air mixture. If too much fuel is introduced into the intake manifold, a flooded condition occurs. Not enough fuel will have the opposite effect and will cause too lean a condition. Both of these conditions can cause the engine to crank but not start. Again, ladder logic and electrical schematics have to be consulted to properly diagnose the defect.

The Technician needs to break down a hard-starting condition into two categories. Hard starting due to starter dragging or hard starting due to a fuel or spark-related condition.

Starter Dragging

The battery potential can be low. Battery cables can be corroded or loose at their attachment points. The starter can be mechanically defective. The brushes and armature could be worn out and not allowing enough current to flow. Bushing wear can cause the armature to ground against the pole shoes and also cause dragging. Excessive starter usage due to a hard-starting engine or stalling could be the root cause of this situation as well as age. If the starter is unable to produce 60 rpms, the ECM will not allow the low-pressure solenoid to activate the fuel.

Fuel or Spark Related Condition

Relays are used to transfer the current needed to operate the fuel high-pressure solenoids as well as the low-pressure solenoid. Fuse protection for each tank solenoid is provided in case a solenoid fails shorted to ground in order to prevent all the solenoids from turning off and disabling the engine. A dedicated circuit to provide power to the ignition coils also has a fuse protection for the circuit. Any overamperage of the circuit could cause the fuse to open. The engine could crank but the lack of spark will prevent it from running.

If the engine starts but shuts off, the amount of time for the engine to shut down could be an indication of the problem. Many times, if 30 to 60 seconds is required for the engine to shut down, the engine shutdown system through the ECM could be the cause. This could be an indication of low coolant, high coolant temperature, or low oil pressure condition. Fault codes should be checked to determine the cause of the shutdown and the condition corrected.

The low-pressure regulator, the balance line, if equipped, or low-pressure shutoff valve could be defective, not allowing enough fuel to be delivered to the engine and causing the engine to shut off.

Task A12 Diagnose surging, rough operation, misfiring, low power, slow deceleration, slow acceleration, and shutdown problems; determine needed repairs.

When an NGV engine suffers from a loss of power or rough operation, a Technician should go back to the basics. To obtain the optimum performance from an engine, the proper quantity and quality of fuel must be supplied to the cylinders with the right amount of air and heat provided by a spark plug at the correct time.

Air starvation will generally result in incomplete combustion, a decrease in power, and heat buildup in the cylinders and exhaust system, due to insufficient scavenging. Air starvation can also lead to knock codes because insufficient scavenging and incomplete combustion can leave unburned fuel in the cylinders and pre-ignition can occur from the excess fuel in the chambers. Some of the reasons for air starvation are a dirty air cleaner, restricted intake hoses and pipes, a defective turbocharger, excessive exhaust back pressure, or high altitude operation. Leaks at the CAC or pipes and hoses can also be a cause or air starvation. Defective spark plugs can cause an engine to miss under a load but does not have to miss at idle or at light loads.

Misfiring cylinders may be located in several ways. One method is to compare exhaust temperatures of each cylinder. They should be uniform. When a cylinder is located that has notably cooler exhaust temperature, it is Most-Likely misfiring. A variety of equipment is available to measure exhaust temperature of individual cylinders, including temperature probes and noncontact infrared heat guns. Some of the modern NGV engines that are electronically controlled provide a method through the software to perform a cylinder balance test or cut-out of the cylinders individually. A weak cylinder will usually not show a change in the degrees of pulse width the other cylinders have to change to compensate for the weak cylinder because it was already compensated for. Follow the software's instructions and prompts to perform these tests.

Surging is an uneven power output from the engine that is usually the result of either fuel starvation or misfiring spark plug(s). A lean AFR (air fuel ratio) is an indication of fuel starvation and can be determined by using diagnostic software. In the case of the Cummins Insite software if the fuel rate has to be increased consistently, verses varying between lean and rich, this would indicate a lean condition. For the Detroit Diesel Series 50G, the AFR reading indicates the ratio; a lean condition usually is greater than 1.50 under load. Dirty fuel filters, defective fuel control valves, high/low-pressure regulators can all be causes of lean conditions. An oxygen sensor (O_2) can cause a lean condition. As a sensor ages, it can drift and give false readings. Some of the engine manufacturers' softwares compensate for this situation whereas others do not. If the software cannot compensate for the drift, the condition can be checked by installing a new sensor. If the engine performs better, then the oxygen sensor replacement repaired the defect even though no DTCs were set. The Cummins Insite software allows the engine to be set into open loop, removing the O_2 sensor from the system. If the engine power increases, the O_2 sensor should be suspected as a possible source of trouble and replaced with a good known sensor.

Task A13 Isolate and diagnose engine related vibration problems; check engine mounts; determine needed repairs.

Often the frequency of occurrence of an unusual vibration is a direct or multiple of the operational speed of the device creating the vibrations. A strong vibration that increases in occurrence rate with engine speed would probably the crankshaft or something directly attached to it. A vibration that has an occurrence rate half the engine speed might be created by a camshaft or something directly attached to it.

Other components, such as an out-of-balance air-conditioning compressor or alternator, may produce high-frequency vibrations that seem to have a constant occurrence rate. In these cases, use a stethoscope to find the strongest concentration of that vibration. Then look for components that operate at a high speed. The suspected component should be turned off if possible to see if the vibration or noise changes. If the drive belt can be removed, run the engine and check to see if the vibration remains or has stopped.

High-frequency vibrations can also be produced by bearings that are failing or two surfaces that are rubbing each other. These are usually destructive in nature, have a constant occurrence rate, progressively grow in intensity, and last for a short time before they become more apparent. Some automated operations, like the cycling of the air-conditioning compressor, may have a vibration occurrence rate that seems random. Two or more vibrations of different frequency at the same time may produce a rhythmic vibration.

Task A14 Check cooling system for temperature protection level, contamination, coolant type and level, temperature, pressure, supplemental coolant additive (SCA) concentration, filtration, and fan operation; determine needed repairs.

Regular maintenance requires the coolant be kept with a minimum freeze protection to inhibit the system from corrosion, acid formation, erosion, and to increase the boiling point. The recommended concentration is usually a 50/50 concentration of coolant and distilled water, which can provide protection to about −40° F. A refractometer is a reliable tool to check the level of protection, although other methods are also available. The correct scale of the refractometer, propylene-glycol (PG) or ethylene-glycol (EG) must be used. Too high a concentration level of glycol can cause heat retention and overheating. Too low a concentration level can cause freeze-up and engine damage in freezing climates, and cause boilover in high ambient temperatures because, when mixed in proper concentration, glycol helps to increase the boiling point of coolant.

A water filter can also be used to remove any grit or dirt left over when the engine block was cast from the cooling system, and the filter should be replaced at scheduled PM. The dirt, if not removed, can accumulate against cylinder liners and the seals and start wearing away at the liner and engine block, eventually causing a failure. Water pump seals can be scratched or damaged as well by dirt, causing leakage.

A pressurized cooling system is a requirement of NGV engines. Pressure in the cooling system assists gravity to ensure a constant supply of coolant at the water pump. The water pump is a centrifugal design and cannot pump in a vacuum. Pressure will raise the boiling point approximately 3° F per 1 psi at sea level. The radiator cap will maintain the cooling system pressure at a specific value, usually 9 psi. However, newer engine models with exhaust gas recirculation (EGR) systems are using caps that regulate the pressure as high as 15 psi. During heavy loads and prolonged periods of idling, the cooling system pressure may exceed the rating of the radiator cap, overcoming the vent valve and allowing the excess pressure to vent through a vent tube. When back at the rated pressure the valve's spring tension will close the vent line. The pressure cap will also prevent a vacuum from developing in the cooling system. A spring-loaded valve will open and allow the cooling system to stabilize back to ambient pressure and not allow a vacuum to develop and potentially damage the radiator and the HVAC heater cores.

The surge tank is usually equipped with a sight glass below the opening of the filler cap. Sometimes two sight glasses are used to accommodate a low and full level to allow for expansion, especially in vehicles with larger systems. A constant overflow of coolant from the vent tube may indicate a defective pressure cap, air pockets from improper bleeding of air from the cooling system, uneven heating, or hot spots in the engine. Repetitious and rhythmic excessive surges in the surge tank level usually indicate a leak from the combustion chamber such as a blown head gasket or leaking seal at the spark plug tube adaptor.

If the engine overheats spontaneously, the following should be checked as a unit in this manner:

- Check tightness of hose clamps, especially before the water pump.
- Check for soft or sometimes mushy-feeling hoses.
- Check for collapsed and kinked hoses.

The cooling system should be inspected for leaks. A common method of leak testing is to use regulated shop air set to 9 psi. Air pressure is introduced into the surge tank by tapping in with a fitting. During this time, all areas of the bus should be examined to ensure there are no leaks. Special attention should be paid at the water pump seal, coolant hoses, heater pipes and hoses, the driver's heater area, the high-pressure regulator and coolant connections, marine pump, and the HVAC area.

The radiator should also be checked for external clogging due to dirt, paper, and other foreign debris. Internal clogging can also occur. Scale buildup due to improper coolant maintenance can plug the tubes of the core gradually. Symptoms will begin to appear under heavy pulls on hot days but eventually just light engine loads will cause engine overheating. An infrared pyrometer is a useful tool in checking for internal clogging. The engine is warmed up until the thermostats open. A cardboard in front of the radiator will assist in speeding up the process. The pyrometer is aimed at the radiator core from side to side beginning at the top and working downward. Slight variances of temperature can be encountered from side to side but getting cooler as the pyrometer is worked downward. If large fluctuations in temperature are found, the radiator internal passages may be plugged and in need of either cleaning or core replacement. Remove and test the thermostats for proper opening and closing temperatures, and replace as necessary following the manufacturer's recommended practices.

EG and PG coolants require close monitoring and the addition of supplemental coolant additives (SCAs). Monitoring is performed during a preventive maintenance (PM) service of the engine. SCA test strips are dipped into the coolant and compared to the chart supplied.

Many manufacturers now equip their cooling systems with extended life coolants (ELC). ELCs are low maintenance; coolant life can be as long as six years, with only one addition of SCA during that period. ELC is usually dyed red but can be light green or gold as well, depending on the manufacturer, for identification. Make up coolant should be a premix of ELC and should not be mixed with EG- and PG-type coolants.

Task A15 Check lubrication system for contamination, oil level, condition, temperature, pressure, filtration, and oil consumption; determine needed repairs.

The engine oil level should be checked daily, either after fueling the vehicle or at the beginning of the day. If the level is low, it should be corrected to the correct level. If the level is excessively low, then a defect ticket should be created for further evaluation and repairs if needed. If the level is high, again a trouble ticket should be created for evaluation and for repairs if needed.

Examination of the lubrication oil can reveal problems associated with the engine. Moisture droplets on the oil dipstick and the oil filler cap or milky-colored oil is an indication of coolant contamination. This can occur for any of the following causes: a cracked engine block or cylinder head, a blown head gasket, a leaky oil cooler, or a leaky cylinder liner sleeve O-ring.

Excessive oil consumption can be an indicator of a defective turbocharger oil seal, of improperly seated or defective rings, a damaged piston, or defective valve stems and seals. These problems will usually be accompanied by blue smoke in the exhaust.

A low oil pressure complaint should be investigated thoroughly. Verify with a known good gauge connected into the main oil galley to determine the actual pressure of the lubricating system. Low oil pressure may indicate a low oil level, improper or too long of a dipstick causing low oil level, a worn oil pump, a restricted pickup screen, cracked pickup tube, a partially stuck-open or improperly adjusted relief valve, clogged oil filter, clogged oil cooler, or worn bearings. The condition of the oil should also be examined, as thinned out oil can cause low oil pressure. Regardless of the initial problem, if a low oil pressure problem is not corrected, problems that are more serious and costly to repair will develop.

Task A16 Check, record, and clear electronic diagnostic codes; monitor electronic data; determine needed repairs.

Current NGV engines use computerized engine controls systems. According to the SAE Standards, diagnostic codes are known as diagnostic trouble codes, or DTCs. The DTC extraction method varies with the engine and vehicle manufacturer. The Technician needs to understand how to check, record, and clear DTCs. The typical process is to use a scan tool, diagnostic reader, or laptop computer to check for, record, and clear DTCs.

Modern engines are controlled and monitored by electronic modules. These modules are capable of storing fault codes. While each manufacturer may use its own codes, each must also design its software so that SAE-sanctioned codes are displayed. This means that one manufacturer's diagnostic hardware and software is capable of at least reading any others. The term message identifier, or

MID is used to identify the chassis electronic system; the engine would be one of these systems, the transmission and antilock brakes others. Within the electronic system or MID would be branches, components, and subcircuits that are each divided into codes known as parameter identifiers (PIDs) and subsystem identifiers (SIDs). Within each PID or SID, faults are identified by common codes known as fault mode indicators, or FMIs. This makes it easier for a Technician trained on one type of NGV brand of equipment to troubleshoot another brand. For instance, FMI 4 tells the Technician the component or circuit is voltage below normal or shorted low. This would be so on any PID or SID on any NGV engine.

Electronic service tools (ESTs) connect to the vehicle's electronic systems by means of a data connection, either a six-pin Deutsch connector (SAE 1587/1708 protocol) or more recently a nine-pin Deutsch connector known as an SAE J1939 protocol connector. These connectors are usually referred to as the diagnostic connectors. Most electronic systems will also blink codes out using dash-mounted diagnostic lights.

Diagnostic codes should be checked and recorded as necessary. The Cummins Insite software allows a Technician to create an image of the ECM parameters. It can also be utilized as a diagnostic tool. The software has to be initially set to create an image at start-up. Other software will allow data logging and play back later for diagnostic purposes. The Technician should be thoroughly familiar with the procedures to start the software, open the screens to code retrieval, data logging, converting files to comma separated values (.csv) files, and accessing troubleshooting of the codes through the software.

Task A17 Perform visual inspection for physical damage and missing, modified, or tampered components; determine needed repairs.

Experience teaches the Technician how to inspect components visually with the kind of detail that enables the detection of tampering and modifications. Technicians who regularly work on a North American Bus Industries (NABI) transit bus or a New Flyer (NF) bus who at a glance can identify tampering on a Detroit Diesel Series 50G engine, may find it difficult to locate a similar tampering abuse on a Cummins L-Gas Plus, ISL-G, or B5.9G engine. There is no substitute for knowing the equipment you are working with.

Technicians should be especially aware of tampering with any devices or components that have a bearing on engine emissions, as this is illegal. Penalties for emission system infractions can be severe. Improper or defective emission devices should be replaced or repaired when discovered. NGV vehicle emission devices include crankcase coalescing filters, exhaust catalyst, spark plugs approved for a certain application, turbocharger waste gate adjustments, and fuel control valves to name a few.

All manufacturers are required to affix emission certification decals to the engine with the engine model and serial numbers, engine timing information, and the certifying agencies. Valve adjustment specifications are usually found on these decals as well. These decals will outline the mandated equipment for controlling emissions and the certification date. The decals should be in place and legible. The engine manufacturer should be contacted for a replacement when the decals are defective or illegible. In the case of the Cummins engine, the critical parts list (CPL) number will also be included on the decal. The CPL is an important tool when replacement parts need to be purchased.

Task A18 Research applicable vehicle and service information, service precautions, and technical service bulletins; determine needed actions.

Technicians today use a variety of sources for service information to keep their skill current. Technological advancements are frequent, so service procedures also need frequent revisions. The volume of information published each year is more than can be committed to memory. Today, the same information is available on CDs or DVDs, whereas once a shelf of books was needed. A single CD is capable of containing over 250,000 pages of text. New service information systems have been introduced to help Technicians. Below is a partial list of sources for service information.

Maintenance manuals: These contain routine service intervals, procedures, and lube requirements for vehicle components and systems. Look here for fluid recommendations, capacities, and charts. Torque values and adjustments will be found here. Detailed information about most components is available in maintenance manuals but these do not usually contain engine overhaul manuals or transmission diagnostics and rebuilding procedures.

General and specialty repair manuals: The manuals and databases are produced by engine and transmission manufacturers. The manuals outline the procedures for troubleshooting each manufacturer's engines and transmissions as well as detail overhaul information for their components.

Service bulletins: Service bulletins contain the latest information providing service tips, precautions, product improvements, recalls, field service modifications, and related service information previously published in service manuals, and take precedence over service manual information. Service bulletin manuals are usually available through the dealers, who receive them from the manufacturers. Other times, bulletins are sent directly to the transit properties as needed. These bulletins will contain the latest information available as changes are made.

Parts manuals: Parts manuals not only provide information regarding which parts are correct for a particular application but also contain valuable illustrations that show the correct assembly and/or parts. Often exploded views show things that are not illustrated in the service manuals. These are useful guides for reassembling.

Internet-based information: A variety of information is available over the Internet. Service information databases compiled by engine, transmission, and drive systems manufacturers and suppliers, and replacement parts for air systems, electrical devices and many more are available online as a means of supporting the Technicians. Some manufacturers offer the transit properties Technician training courses, service bulletins, and software upgrades, which are downloaded into electronic management systems or technical information kiosks (TIKs) via the internet. Most component suppliers to the OEM have Web sites where information is also available.

Task A19 Inspect accessory belts, drives, tensioners, and pulleys for tension and condition; determine needed repairs.

Many of the engine accessories are driven by belts. The two most common types of belts in use on the NGV CNG engines are the serpentine and V-belts. The belts are to be inspected at the normal PM cycles for cracks, splits, missing pieces, glazing, and wear. The larger the load placed on belts such as those for the alternator and A/C compressor the faster wear occurs. The drive pulleys and tensioners have to be checked for wear, alignment, and damage to the pulleys, and the bearings for wear as well. A tensioner should be checked to ensure it is able to compensate for wear or stretch of the belts and that the tensioner bearings are in good condition. A screeching beltis an indication of wear of the belt, pulley, tensioner, or a combination of these conditions and should be remedied as quickly as possible to prevent further damage to the components.

Defective drive pulley or tensioner bearings can cause knock codes in the engine and should not be overlooked when diagnosing knock codes. Whenever possible, the drive belt of the suspected system can be removed and road tested for operation. If the symptom goes away, the defect can be further isolated to the component that is causing the problem.

It is important that whenever replacing belts, tensioners, and pulleys the alignment be checked. Improper alignment will cause premature wear and result in the condition reoccurring quickly.

Task A20 Perform engine compression test; determine needed repairs.

Most of today's modern NGV engines operate at a compression ratio of 10:1. In the event of engine malfunctions such as low power, rough running, high crankcase pressure, or other suspected cylinder defects, a cylinder compression test might be necessary to perform engine diagnostics. Compression pressure will help to assist the Technician in determining the soundness of the engine. The manufacturer's specifications should be checked, but as a rule of thumb minimal pressure is usually 200 to 240 psi with a maximum of four puffs of the piston. Valve adjustment should be checked prior to performing a compression test to ensure the valve train is operating properly and an accurate compression pressure and interpretation of the engine condition is obtained. The air inlet pipe should be disconnected at the throttle valve to eliminate the possibility of air inlet restriction. The throttle valve should not be forced open. Because the valve is electronically operated, forcing it open could damage it.

A compression tester must be equipped with an accurate gauge rated to at least 300 psi. The engine must be brought up to operating temperature, a minimum of 180° F. The engine should be depleted of all natural gas by turning the ¼ turn fuel shutoff valve to the off position while the engine is running. When the fuel is depleted between the shutoff valve and the engine, the engine will turn off and the Technician should check again to make sure the high and low fuel pressure gauges reads 0 psi at this time. The coils, the extenders, the boots, and the spark plugs should be removed from the engine and set aside in a manner so they can be reinstalled on the same cylinder from which they were removed. Install the compression gauge adaptor in the cylinder to be tested. The number of cylinder puffs should be observed for diagnostic reasons. Usually three to four cylinder puffs are required for the cylinder to develop maximum pressure. If the engine requires more than three or four puffs to reach maximum pressure this could be an indication that the valves could be weak or starting to leak, or the rings and cylinder walls are allowing pressure to bypass.

If the pressure is low, the Technician should listen for abnormal sounds coming from either the intake or exhaust manifold to determine possible areas that should be investigated further. It is important for the Technician to remember that low compression is also an indication of worn compression rings. Reinstall the spark plugs and associated accessories back on the cylinders they were removed from as per the manufacturer's specifications.

B. Cylinder Head and Valve Train Diagnosis and Repair (11 Questions)

Task B1 Remove, inspect, disassemble, and clean cylinder head assembly(s).

Inspection of the cylinder head begins by checking the torque of the cylinder head bolts before removing the head. This can indicate the potential for a warped or cracked head. Inspect the head gasket and head surface for signs of deposits or coolant stains between cylinders and passages before steam cleaning the head. Use a steam cleaner or high-pressure washer to remove oil, coolant, and carbon residue.

Then sand the surface or use ScothBrite buffing pads as necessary to remove any remaining carbon and gasket material. Use of sharp objects like knives or scrapers to remove residue may impair inspection for cracks and damage. Magnetic or dye crack detection may be required to highlight cracks that are suspected but not apparent.

Some manufacturers recommend pressure testing with compressed air after blocking off the coolant passages and dunking the cylinder head in a hot water tank. The hot water allows the head to expand and make the cracks more pronounced, allowing air to escape out of the crack. Pressure testing the head can also be used to locate a crack in the oil passages. Visually reinspect the head for cracks in the casting especially between the valves and at the spark plug adaptor area, scale buildup in the coolant passages, and pitting of machined surfaces. Excess accumulation of scale in water passages can usually be removed by placing the head in a hot soak tank for about two hours. If a cylinder head is to be hot tanked, all the coolant and oil plugs have to be removed to allow cleaning of the passages. In the case of an overhead cam engine, the cam bearings should also be removed before hot tanking the engine.

Task B2 Inspect threaded holes, studs, and bolts for serviceability; service or replace as needed.

When removing or installing a component, carefully inspect all bolts, studs, holes, and nuts for damage to threads. Stripped, cross-threaded, nicked, rolled, and rusty threads can produce errors in the torque values. Use compressed air to remove foreign matter from holes. Obstructions lodged in holes, such as metal and liquids, can cause a bolt to bottom out and reach torque value without obtaining the required clamping force. Examine the shank of bolts and studs for signs of twisting or over-torque fracturing. Replace defective bolts and studs. Ensure that nuts are not distorted or cracked. Replace self-locking and defective nuts. Clean and dress threaded holes. If threads are damaged beyond use, drill and tap the hole and install approved helicoil. Some manufacturers do not allow the repairs of threads and require the component to be replaced. Use an approved stud extractor to remove broken studs and bolts.

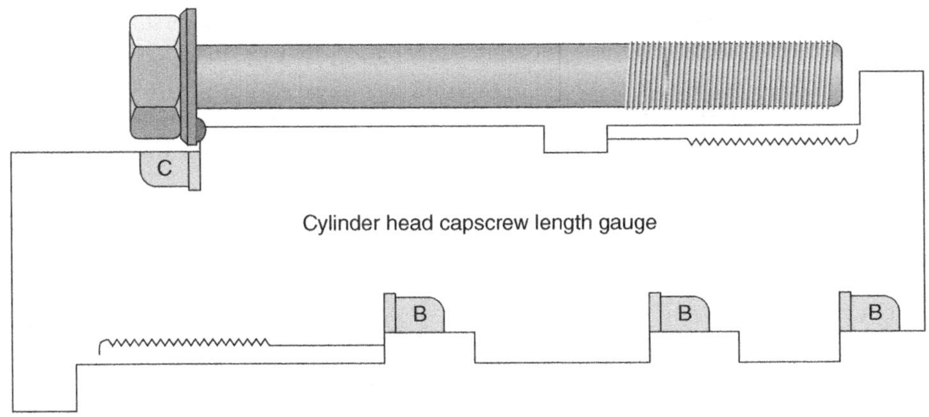

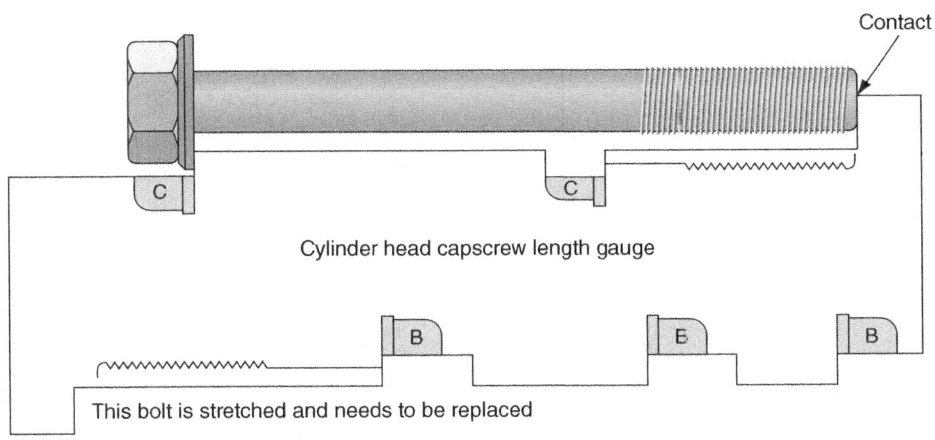

Many modern engine manufacturers use bolts that are installed with torque-to-yield for the cylinder head. These bolts are usually discarded and new ones need to be installed. Some manufacturers require the bolts to be gauged with a stretch gauge to determine if the bolts are reuseable. In either case, the manufacturer's recommendations need to be followed to prevent serious engine damage when the engine is placed back in service.

Task B3 Measure cylinder head-to-deck thickness, and check mating surfaces for warpage and surface finish; inspect for cracks and damage; check condition of passages; inspect core and gallery plugs; service as needed.

The cylinder head should be checked for warpage. Using a straightedge and feeler gauge, measure the face (fire deck) for flatness. Check for transverse (across the width of a cylinder head) warpage at each end and between each cylinder. Check for longitudinal (across the length of the cylinder head) warpage at points above and below each cylinder, between valve centerlines, and the outer edge of the fire deck. Compare measurements to manufacturer's specifications to determine the need for resurfacing. Resurfacing may necessitate removing inserts, such as valve guides, and/or deburring of coolant ports. Check cylinder head thickness, called deck-to-deck thickness, to determine the maximum allowable machining limit.

All core and threaded plugs need to be checked to make sure they were not leaking if they are not to be removed. If the plugs are removed, remove any debris that may have accumulated during normal operation and replace with new plugs. Install core plugs to the correct depth to prevent damage or leakage and apply appropriate sealers to the edges. Threaded plugs need to have thread sealant applied and torqued to the correct torque. Overtorqued plugs can lead to cracked components.

Depicted below are the differences in pipe threads vs. straight threads. Pipe threads are cut at an angle, not to be mistaken with thread pitch, from the first thread to the last, whereas the straight threads are cut straight. The threaded holes they are installed into are also different. Straight threads can require an O-ring if the mating surface is equipped with an undercut area to accept an O-ring or a gasket to seal the mating surfaces. Pipe threads are sealed with either Teflon tape or an anaerobic thread sealant. The tape or sealant is used to lubricate the threads as well as to prevent galling, especially in the case of stainless steel fittings.

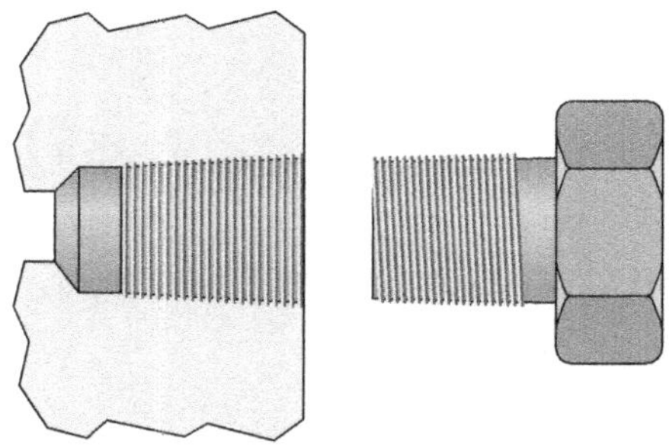

Tapered pipe threads

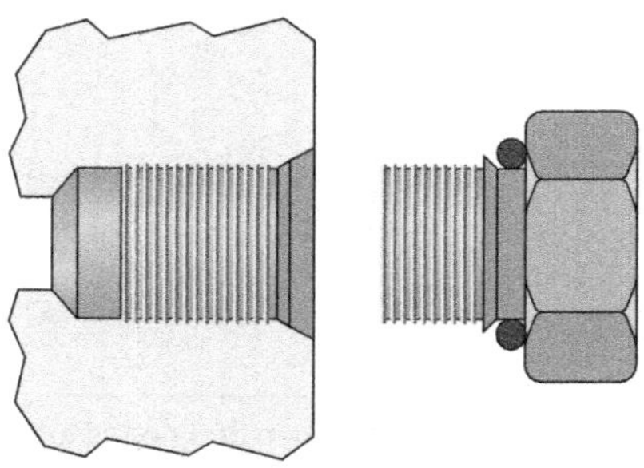

Straight threads with an o-ring seal

Task B4 Inspect valves, guides, seats, springs, retainers, rotators, locks and seals; determine serviceability and needed repairs.

Check valve spring straightness with a T-square. This will help prevent stems from binding in the guides as the valves are operated. Uneven side pressure causes the valve stems and guides to wear faster. Also check the unloaded (or free length) and loaded length (or tension) of the spring using a spring-tension gauge. Replace valve springs that do not meet specifications to prevent damage to the piston or crankshaft.

Valve springs using the same valve bridge must be replaced together to prevent unbalanced valve operation. This can damage the bridge from uneven stress, can cause valve guides and stems to wear faster, and can result in insufficient scavenging of the cylinder. Valve spring retainers keep the valve spring on the valve, and valve rotators are used to rotate the valve each time it is actuated. Some engines also have seals. Careful inspection of these components can prevent damage and extend the life of the engine. Retainers, also called keepers, that become worn or damaged may allow the spring to release the valve. If this happens, the valve drops into the cylinder and jams the piston, a condition called dropping a valve. Free release rotators use engine vibration and exhaust gases to rotate, because the spring tension is removed when it is depressed. Mechanical rotators apply a torque to rotate the valve each time the valve is opened. Both will help eliminate carbon deposits from forming on the valve face and seats. Seals between the valve guide and valve stem prevent oil from entering the cylinder on the intake stroke. Defective seals can cause oil consumption, leading to excess smoking, carbon buildup, formation of hot spots, and precombustion or knocking.

After a cylinder head has been checked or resurfaced and is considered usable, the valve guides should be checked for wear as follows:

- Check the guide inside diameter with a snap, ball, or dial gauge in three different locations throughout the length of the guide. If guides are the replaceable types and are worn excessively, they should be replaced. In most cylinder heads, the valve guides can be removed by using a driver. Select a driver that fits the valve guide. After selection of a driver, the guide should be removed as follows:
- Support the cylinder head on a block or stand.
- If the guide needs to be pressed out, place the cylinder head in the press and align the driver with the press ram.
- Note the position of the guide in relation to the cylinder head for reference when installing a new guide.
- Drive or press the guide out of the cylinder head, making sure that the driver is driven or pressed straight. If it is not pressed straight, damage to the valve guide or cylinder head may result.
- After the guide has been removed, check the guide bore for scoring. A badly scored guide bore may have to be reamed out to accommodate the next larger size guide.
- After the guide bore is checked, select the correct guide (intake or exhaust) and insert it in the guide bore.
- Insert the guide driver and drive or press the guide into the cylinder head until the correct guide position is reached. If manufacturer's specifications are not available, drive the guide into the same position as the old guide.
- Some engine manufacturers recommend that a new guide be reamed after installation to ensure that the guide inside diameter did not change during the installation process. Never ream a surface-hardened guide.

A common practice during major maintenance is to replace all valves if they have been in the head for several thousand operating hours. Valves must be changed if the head shows signs of carbon buildup from oil leaking around the valves. In this case, also consider changing valve guides. Valves with scratched, scored, pitted, or worn stems or stems with worn keeper grooves should be replaced. If the valve head is cupped, nicked, or marred, it should be replaced. If a valve passes the visual inspection, it must be cleaned using a wire buffing wheel or glass bead blaster, if available. Use a valve refacing tool or valve grinder to obtain the correct face angle.

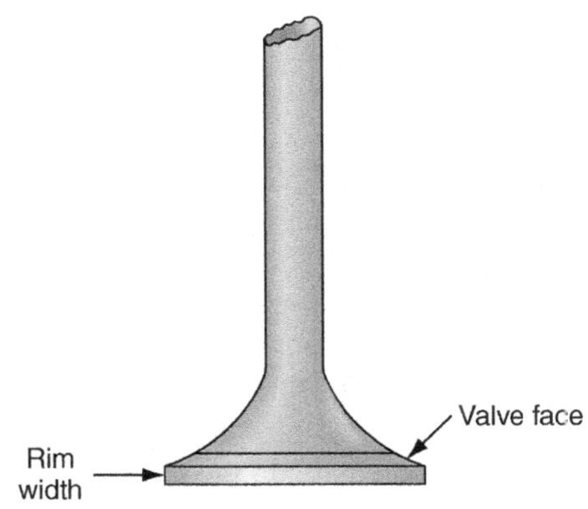

Inspect the valve to ensure that the rim width or valve margin is within specifications. Insufficient valve margin may result in premature burning and subsequent failure. Finally, valve stem ends are ground flat and with a small taper; usually about 1/32 to 1/16" is applied.

Valve seats are used to absorb the forces of valves opening and closing, and the effects of expansion and contraction. Valve seats are also used to transfer the heat from the valves to the cylinder head and the cooling system. It is important to properly inspect, clean, and recondition or replace valve seats to maintain a good valve to seat contact. Loose valve seats are located by lightly tapping each seat with the peen end of a ball-peen hammer. A loose seat will produce a sound that differs from the ring sound created by tapping the head adjacent to the seat. The seat may even move a little. Leaks may occur if the seat has visible cracks, uneven burns or discoloration, or an opening that is oblong or excessive in diameter. Replace any seats that cannot be reconditioned. To recondition a valve seat, a mandrel pilot is installed in the valve guide to keep a specially dressed grinding wheel centered in the seat. New seats must be dressed for the valve that will use them. Repeat the inspection process after reconditioning the valve seats. Also, check the concentricity (roundness) of the seat opening. If the seat is not round, it will not seal during compression and combustion strokes. Verify that the valve head height is correct. If the head height is out of specification, seats may need to be replaced, as this may alter the compression ratios or injector spray pattern. It is important not to remove any more material than is necessary to refinish the valve seat.

After any of the guides have been replaced or reconditioned, the valve seat should be checked and reconditioned. If the seat is solid, check it for cracks and excessive width. If the seat passes all checks it should be reconditioned as outlined, using a specially designed valve seat grinder. Most valve seat grinders are similar to the one in the figure below.

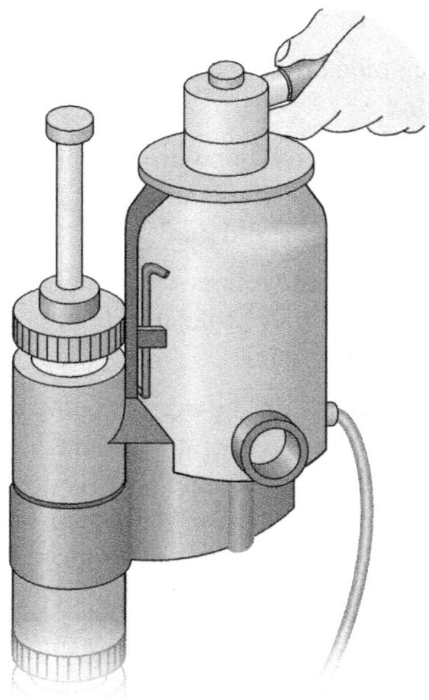

Select the proper mandrel pilot by measuring the valve stem with a micrometer or caliper or by referring to the manufacturer's specifications. After experience has been gained in this area, selection of the pilot is easily done by a visual check. Pilots are usually one of two types: expandable or tapered. Expandable pilots are not considered as accurate as tapered pilots and should be used only when a tapered pilot is not available. Some Technicians prefer expandable pilots because they compensate for guide wear better than tapered pilots do; by expanding the guide the pilot tightens and adjusts the guide size for wear. A tapered pilot does not have an expanding screw and relies on the taper of the pilot to tighten it into the guide.

Before installing valves, check the valve-to-seat contact. One method of doing this is to coat the face of the valve with Prussian blue and lightly snap the valve into its seat. Remove the valve and check its face. If the valve face does not have an even seating profile, the seat is not concentric and must be reground. Another way to check contact is to place pencil marks around the face of the valve about 1/8" apart and then snap the valve against its seat. Remove the valve and check the face. If any of the marks are unbroken, the valve seat must be reground. When the valves have been seated, check the height of the valve heads. This checks the distance the valve protrudes above or below the machined surface of the head. Place a straightedge across the cylinder head and use a feeler gauge to measure the distance between the valve head and straightedge. If the valve is designed to protrude above the head, place the straightedge on the valve. Use a feeler gauge between the valve head and the cylinder head machined surface. A depth micrometer may also be used for this procedure as mentioned previously.

Task B5 Inspect, reinstall or replace seals and spark plug adaptors/tubes; inspect spark plug threads; service as required.

If a leak is detected in a spark plug tube's adaptor seals or crush washer at the base, they must be replaced to prevent further damage to the engine. Telltale signs of coolant on the spark plug tip and/or electrodes can be evident sometimes, whereas at other times coolant can be detected coming from the exhaust port of the affected cylinder when the exhaust manifold is removed. When replacing the adaptor(s), follow the manufacturer's recommendations.

The spark plug adaptor wells must be inspected for contamination from foreign debris. The threads must be inspected to ensure they are in good condition with no burrs and no sign of contamination from engine oil and no dirt present. If the adaptors are suspect, they should be replaced.

If extender tubes are used, use O-ring approved lubricant such as Dow Corning 55 to lubricate them and install the extenders in the adaptors, making sure the bolt holes will align with the covers.

Install the spark plugs and torque to appropriate torque. The boots or the extenders and the coils should now be installed. Ensure that no contamination is introduced in the spark plug area. In the case of the Cummins Engines, an approved dielectric compound is already applied to the boots and should not be added to or removed. In the case of the Detroit Diesel, the boots and coils should be applied dry. Improper use of dieelectric compounds could severely affect the spark plug life by reducing it. When the compound heats up, it becomes a conductor, allowing the spark to flash over the ceramic and go to ground externally, and not allowing the spark plug to fire. Modern NGV engine ECMs are capable of detecting this condition and will set a DTC as a misfire.

Task B6 Inspect, reinstall or replace valve bridges (crossheads) and guides; adjust bridges (crossheads).

Some CNG engines are designed with only one intake valve and one exhaust valve per cylinder and do not require valve bridges. Other engines use four valves per cylinder and are designed to use one rocker arm per two intake and two exhaust valves and utilize the overhead cam design; these engine designs also do not require valve bridges.

Valve crossheads are used on some NGV engines that have four valves per cylinder. Two valves are the intakes and two valves are the exhaust. The crosshead is a bridge device that allows a single rocker arm to open two valves at the same time. The crossheads (bridge) must be checked for wear as follows:

- Visually check crossheads for cracks and with a magnetic crack detector if available.
- Check the crosshead inside diameter for out-of-roundness and excessive diameter.
- Visually check for wear at the point of contact between the rocker lever and the crosshead.
- Check the adjusting screw threads for damaged or worn threads.
- Check the crosshead guide pin diameter with a micrometer to ensure that it meets the manufacturer's specifications.
- Check the crosshead guide pin to ensure that it is at right angles to the head-milled surface.
- Visually check the point of contact between the valve stem and the crosshead for signs of wear.
- Replace any defective crosshead or guide pins that are borderline or close to needing replacement.

Some crossheads are of the adjustable type. Others are the self-compensating type. The self-compensating bridges need no adjustment as they are floating between the two valves. The adjustable types are adjusted by utilizing 0.0015" feeler gauge strips between the top of each valve stem and the bridge. Slight pressure needs to be applied on the top of the bridge while pulling the feeler gauge strips out and observing the amount of drag on each strip. If adjustment is needed, loosen the jam nut and turn the screw in or out as needed to make the correct adjustment. Tighten the jam nut and recheck the tension on the feeler gauge strips to ensure the tension is equal. Caution must be used to prevent from damaging the crosshead guide pin when torquing the jam nut.

Task B7 Clean components; reassemble, check, and install cylinder head assembly.

The assembly of the cylinder head or heads onto the engine block will involve many different procedures that are distinct to a given engine. The following procedures are general in nature and are offered to supplement the manufacturer's service manual.

Before attempting to install the cylinder head(s), make sure the cylinder head and block surfaces are free from all rust, dirt, old gasket materials, and grease or oil. The bolt hole threads should be in good condition, free of burrs and dirt to allow the bolts to turn freely and achieve accurate torque measurements. Before placing the head gasket on the block, blow out all bolt holes in the block with compressed air to make sure they are free of oil and dirt. Eye protection must always be worn when using compressed air.

Most modern engines will have dowels or locating pins in the block to aid in holding the cylinder head and gasket in place during cylinder head assembly. If there are any dowels in the cylinder block or head, make sure they are installed correctly and protruding to the specified depth. The dowels should not be damaged and square with the engine block.

Select the correct head gasket and place it on the cylinder block, checking it closely for an "up" or "front" mark. In most cases, head gaskets will be installed dry with no sealer. Take note of the recommendations in the service manual.

After placing the head gasket on the block, place coolant and oil O-rings (if used) in the correct positions. Head gaskets of NGV engines are usually of a one-piece solid composition type with metal and gasket materials bonded together. The installation of some head gaskets requires the use of dowels (threaded rods) that are screwed in the head bolt holes to hold the head gasket in place during head installation. If dowels are recommended, they can be made from bolts by sawing off the heads and machining a taper on the end.

After the gasket and all O-rings are in place, check the cylinders to make sure no foreign objects have been left in them, such as O-rings and bolts or nuts.

Place the head(s) on the cylinder block carefully to avoid damage to the head gasket. On engines with more than one cylinder head, it may be necessary to line the heads together by placing a straightedge across the intake or exhaust manifold surfaces.

Clean and inspect all head bolts or cap screws for erosion or pitting. Lightly coat bolt threads, washers, and the head bolts with engine oil or International Compound no. 2 and place the bolts or cap screws into the bolt holes in the head and the block. Using a speed wrench and the appropriate-size socket, start at the center of the cylinder head and turn the bolts down snug.

When all the bolts are snug, torque each bolt in sequence with a calibrated torque wrench and the appropriate size socket, according to the manufacturer's requirements until the final torque is achieved. After the cylinder is torqued down to the final specification, wait 30 minutes and recheck the final torque. This will allow the cylinder head to properly seat and ensure uniform torque.

In the case of an overhead cam engine, install the bearings, the camshaft, and then the caps and torque them in place. Install and torque the cam retaining bolt and check the cam end play.

The rocker arms and shafts can now be installed and torqued to specifications in both the in-block and overhead cam engines. Adjust the valve clearances to manufacturer specifications.

Clean the valve rocker cover(s). Install a new gasket or gaskets and install the cover on the engine. Die cast covers have a grooves in them. The groove must be clean and the silicone gasket must be installed in the groove correctly to seal against oil leakage. On NGV engines, the valve covers should be installed dry to ensure sealing of the gasket and prevent oil leakage.

Task B8 Inspect, measure, reinstall, or replace pushrods, rocker arms, rocker arm shafts, and supports for wear, bending, cracks, looseness, and blocked oil passages; determine needed repairs.

Engine designs are of two basic designs: overhead cam and in-block-cam.

Overhead Cam:

After the cylinder head has been tightened to the correct specification, the cam bearings, the camshaft and the caps may be installed and torqued to specs. The cam gear can be installed over the timing guide pin and cam nose. The cam and gear retaining bolt needs to be torqued to specification. At this time, the adjustable idler gear to the cam gear lash must be checked and adjusted if needed before the rocker arm assemblies are installed on the engine to allow the cam to be moved fore and aft to accurately determine end play. The manufacturer's directions and specifications must be followed to prevent serious engine damage from occurring.

Check the rocker arm bushings for wear. The valve adjustment screw pivot buttons that contact the valve stems should not be grooved. The rocker arm shafts and rocker arms have to be installed paying close attention to the rocker arm shafts for proper oil whole alignment. Improper alignment will cause premature wear of the rocker shafts and rocker arm bushings. Install the rocker arms and shafts and torque to specs. The roller followers should be checked for correct alignment on the camshaft lobe. Pre-lubing of the cam bearings and rocker arms and shafts is a must to ensure no damage will occur on engine start-up.

In-block-cam:

After the cylinder head(s) have been tightened to the correct specification, the pushrods or tubes and rocker arm assemblies may be installed. Check each push tube for straightness. Straightening of push tubes is discouraged. Bent ones should be replaced. Push tubes must be checked for breaks or cracks where the ball socket on either end has been fitted into the tube. Place rods or tubes in the engine, making sure that they fit into the cam followers or tappets. The rocker arm assembly or rocker arm is one component that is occasionally overlooked during engine overhaul. Check each boss for wear. Oil consumption occurs because the increased clearance allows excessive oil to splash or leak on the valve stem. This oil will run down the valve and end up in the combustion chamber, especially if the valve seals are worn even slightly.

Task B9 Inspect, install, and adjust cam followers.

Tappets, lifters, and cam followers are components that are positioned to ride, or be actuated, by a cam profile. The function of cam followers is to reduce friction and evenly distribute the force imparted from the cam's profile to the train it is responsible for actuating. NGV engines using cylinder block–mounted camshafts usually are designed with tappet-type lifters. NGV engines with overhead camshafts are usually fitted with roller-type cam followers. Contact surfaces of the tappet lifter should be checked for wear and pitting and grooves and flat spots at the contact area of the camshaft as well as the outside diameter where it rides inside the block. Roller surfaces of the roller-type followers should be checked for pitting, grooves, or flat areas. The mounting pins and bushings should be checked for excessive side-to-side (axial) and up-and-down (radial) movement. The manufacturer's specifications should be closely followed to determine if the followers are reusable or need replacement.

Task B10 Adjust valve clearances.

When valves are properly adjusted, there should be clearance between the pallet of the rocker arm and the top of the valve stem. NGV engine combustion temperatures are higher than a diesel engine. It is not unusual to see exhaust temperatures in excess of 1100° F. Valve lash is required because as the moving parts heat up they expand. If clearance was not factored somewhere in the valve train, the valves could remain open by the time an engine reached operating temperature. Actual valve lash values depend on factors such as the length of the push tubes or overhead cam design and the materials used in valve manufacture. Exhaust valves are subject to more heat. As a consequence,

OEMs require a valve lash setting for exhaust valves that is usually greater than the intake valve lash setting. Under normal operating conditions, an NGV engine's valves will tighten from normal wear, whereas the diesel engine's valves will usually loosen between adjustments.

Maladjusted Valves:

A loose valve adjustment will retard valve opening and advance valve closing, decreasing the cylinder breathing time. Actuating cam geometry is designed to provide some "forgiveness" to the train at valve opening and valve closure to reduce the shock loading the valve train. When valves are set loose, the valve train is loaded at a point on the cam ramp beyond the intended point. The same occurs at valve closure when the valve is seated. High valve opening and closing velocities subject the valve and its seat to hammering, which can result in cracking, failure at the head to stem fillet, and scuffing to the cam and its follower.

A tight valve adjustment will advance valve opening and closing. Because the exhaust valves will open sooner, the air fuel mixture could still be burning when scavenging occurs as the valves open due too short of a burn time. This will increase the temperature of the valves and seats, leading to premature valve burning. Higher exhaust temperatures will also follow, damaging the turbocharger as well. Eventually the valves will wear in their seats more quickly, affecting adjustment and accelerating wear, increasing the chance of premature failure.

Valve Adjustment Procedures:

The following steps outline the valve adjustment procedures on the typical four-stroke-cycle six-cylinder engine. Valves should always be adjusted using the OEMs specifications and procedures. The engine should be 140° F or less.

Caution: Shortcutting the engine OEM-recommended valve setting procedures can result in engine damage.

The firing order of the six-cylinder engine models with an in-block cam is 1-5-3-6-2-4 with cylinder pairings of 1-6, 2-5 and 3-4. Cylinder throw pairings are sometimes called companion cylinders. In other words, when the #1 piston is at TDC completing its compression stroke, the #6 piston (its companion) is also at TDC having just completed its exhaust stroke. To determine #1 cylinder TDC on the compression stroke, bar the engine in the direction of rotation, or clockwise as viewed from the front of the engine, until the #6 cylinder exhaust valves move downward or open. Continue rotation until the #6 cylinder exhaust valves are closing and the intake valves just begin to open. The #6 cylinder is now in the valve overlap position. Valve overlap is the time or position that both exhaust and intake valves are partially open at the end of the exhaust stroke and the beginning of the intake stroke. This allows turbocharged air to be pushed into the cylinder, pushing any remaining exhaust out. Because this is a very short period of time when the engine is running, no fuel or very little fuel can escape the cylinder. In the #6 cylinder valve overlap the #6 piston is at TDC, ending the exhaust stroke and beginning the intake stroke. The #1 cylinder is at TDC of the compression stroke, beginning the power stroke, and thus both sets of valves are closed.

In this position, the #1 cylinder intake and exhaust valves can be adjusted, as can the #2 cylinder intake valves, the #3 cylinder exhaust valves, and the #4 cylinder intake and #5 cylinder exhaust valves can also be adjusted. Mark the vibration damper to the engine front cover and rotate the engine 360° clockwise. The #6 cylinder intake and exhaust valves, the #5 cylinder intake valves, the #4 cylinder exhaust valves, the #3 cylinder intake valves, and the #2 cylinder exhaust valves are now in position to be adjusted. This procedures allows the engine to be rotated only once for the valve adjustments.

The following steps outline the valve adjustment procedures on the typical four-stroke cycle four cylinder engine. Valves should always be adjusted using the OEMs specifications and procedures. The engine should be 140° F or less.

Caution: Shortcutting the engine OEM-recommended valve setting procedures can result in engine damage.

On a four-cylinder OHC engine, the firing order is: 1-3-4-2 with cylinder pairings of 1-4 and 2-3. Cylinder throw pairings are sometimes called companion cylinders. In other words, when #1 piston is at TDC completing its compression stroke, #4 piston (its companion) is also at TDC having just completed its exhaust stroke. To determine #1 cylinder TDC on the compression stroke, bar the engine in the direction of rotation or clockwise, as viewed from the front of the engine, until the #4 cylinder exhaust valves move downward or open. Continue to rotation until the #4 cylinder exhaust valve are closing and the intake valves just begin to open. The #4 cylinder is now in the valve overlap position. Valve overlap is the time or position that both exhaust and intake valves are partially open at the end of the exhaust stroke and the beginning of the intake stroke. This allows turbocharged air to be pushed into the cylinder, pushing any remaining exhaust out. Because this is a very short period of time when the engine is running, no fuel or very little fuel can escape the cylinder. In the #4 cylinder valve overlap the #4 piston is at TDC ending the exhaust stroke and beginning the intake stroke. The #1 cylinder is at TDC of the compression stroke beginning the power stroke, thus both sets of valves are closed.

In this position, the #1 cylinder intake and exhaust valves can be adjusted, the #2 cylinder intake valves and the #3 cylinder exhaust valves can also be adjusted. Mark the vibration damper to the engine front cover and rotate engine 360° clockwise. The #4 cylinder intake and exhaust valves, the #3 cylinder intake valves, and the #2 cylinder exhaust valves are now in position to be adjusted. This procedures allows the engine to be rotated only once for the valve adjustments.

Adjustment:

1. Locate the valve lash dimensions on the emission certification label. The lash specification for the exhaust valve(s) will usually (but not always) be greater than that for the intake valves.
2. After positioning the engine in the correct position as outlined previously, set the valves to their correct setting by inserting the feeler gauge in between the rocker arm and the valve stem or bridge. If the feeler gauge is too tight or loose, proceed to adjust it as follows:
   A. Loosen the jam nut on the adjustment screw.
   B. Adjust the screw in or out as needed until a slight resistance is felt on the feeler gauge when being moved in or out on the stem or bridge. Tighten the screw another flat of the nut, or 60°.
   C. Tighten jam nut and torque to manufacturers specifications.
   D. Check the feeler gauge to ensure that it is not too loose or tight when being slid in or out between the adjuster and bridge or stem.
   E. Proceed to perform the next valve adjustment using the steps and sequences referenced above until all the valves are adjusted.

Task B11 Inspect, measure, and reinstall or replace camshaft and bearings; measure and adjust endplay and backlash.

A careful inspection of the camshaft will ensure that proper valve operation and timing can be obtained. Because the camshaft is constantly supplied with oil, cleaning should be as simple as rinsing with dip tank solution and blowing out with compressed air. Oil passages may require cleaning with a wire brush. Discard any camshaft that does not pass a visual inspection. If the camshaft passes the visual part of the inspection, take measurements and compare with manufacturer's specifications. Inspect the cam lobes for signs of pitting, scoring, or flat spots. Use the manufacturer's procedures for taking lobe measurements. The heel-to-toe measurement is equal to the diameter of the circular portion of the lobe and maximum amount of lift created by the rise at the top of the lobe. Typically, the cam follower will ride in the center of the lobe travel, causing a groove to form. This groove should not be greater than 0.003" deep. If either the top of the lobe or the circular portion is worn (grooved), the amount of lift will be reduced. If an intake or exhaust cam lobe fails to produce enough lift, the valve it controls may not open enough to supply adequate air for combustion or scavenging to occur. Visually inspect bearing journals for bluing, scoring, or wear. Measure the diameter of journals with an outside micrometer and inspect for out-of-round condition by checking the journal in several places. Inspect the shaft and gear keyway for cracks or distortion. Ensure the keyway in the gear and shaft matches the woodruff key. Any movement between these three can randomly change the timing of the valves. A shift of as little of 0.004" in the keyway could equate to almost one degree in timing shift.

Some manufacturers use an offset key when installing the cam gear to the camshaft to obtain desired timing, and this should not be confused with a key that is partially sheared. When an offset timing key is used, make sure that the original key or an identical replacement is installed with the offset facing in the correct direction.

Once the camshaft has been installed in the engine for an in-block cam or the cylinder head for an overhead cam, the end play must be measured and recorded. A dial indicator and base is the usual method of measurement. Place the indicator on a solid surface and apply the pointer on the end of the cam or gear. Depress the pointer halfway of the travel to preload the indicator. Pry the cam in one direction or the other and zero the indicator. Pry the cam in the opposite direction and observe the movement of the indicator. Compare to the manufacturer's specifications.

C. Engine Block Diagnosis, Repair, and Overhaul (18 Questions)

Task C1 Remove, inspect, service, and reinstall pans, covers, breathers, gaskets, seals, and wear rings.

Engine oil pans are usually located in the airflow under the frame rails and are vulnerable to damage from objects and rocks on the road, from manhole covers on city streets, speed bumps, and small animals on the highway. Most NGV engines have oil pans that can be removed from the engine while it is in chassis. It is advisable to remove the oil filter(s) and drain the oil sump prior to removing it. Oil pans that seal to the engine block using gaskets can be difficult to remove especially where adhesives have been used to ensure a seal. A 4-lb. rubber mallet may facilitate removal. Avoid driving screwdrivers between the oil pan and the engine block mating flange because the result will be damage to the oil pan flange and the cylinder block mating flange. Oil pans that use rubber isolator seals tend not to present removal problems.

After removal, the oil pan should be cleaned of gasket residues and washed with a pressure washer. Inspect the oil pan mating flanges, check for cracks, and test the drain plug threads. Cast aluminum oil pans that fasten both to the cylinder block and to the flywheel housing are prone to stress cracking in the rear. With these oil pans, always meticulously observe torque sequences and values. Cast aluminum oil pans can be successfully repair-welded without distortion using the Tungsten Inert Gas (TIG) process, providing the cracks are not too large and the oil-saturated area of the crack is ground clean. The ends of the cracks should be drilled through to prevent to cracks from redeveloping. It is more difficult to execute a lasting repair weld on a stamped-steel oil pan due to the distortion that results. Steel oil pans may rust through, in which case they should be replaced rather than welded. Whenever an oil pan has indications of porosity resulting from corrosion, it should be replaced. In recent model years, some manufacturers have developed composite oil pans that are less prone to corrosion, but caution should be observed when installing the drain plug to prevent the threads from being damaged by cross-threading or overtorquing. The Technician should always be aware that failure of the oil sump or its drain plug usually results in the loss of the engine due to severe damage cause by the lack of lubrication.

Oil pan gaskets should be installed dry to prevent the possibility of oil leaks. The engine and oil pan gasket surfaces should be dry and free of old gasket material or dirt. The pan should be torqued following the manufacturer's recommendations. Neoprene isolators should be checked for damage from previous overtorquing damage and replaced if necessary.

Task C2 Disassemble, clean, and inspect engine block for cracks; check mating surfaces for damage or warpage and surface finish; check deck height; check condition of passages, core, and gallery plugs; inspect threaded holes, studs, dowel pins, and bolts for serviceability; service, reinstall, or replace as needed.

Most NGV engine designs are based on the same cylinder block as their cousin, the diesel engine. Therefore, most NGV engine blocks have the same testing procedures that are applied to the diesel blocks.

If the engine block is removed from the chassis, some manufacturers recommend the coolant ports be plugged and be pressure tested while submerged in a hot water tank, if possible, for two hours. Other manufacturers recommend electromagnetic flux testing to check for cracks. Any cracks that are detected mandate the replacement of the block.

The block counterbore area must be completely cleaned of all rust, scale, and grease. Most scale and/or rust can be removed using a piece of crocus cloth or a 100 to 120 grit emery cloth. Inspect the counterbore closely for cracks and ensure that no rough or eroded areas exist to ruin a sleeve or cut its O-rings or crevice seal. Damage of this nature usually requires the engine block be discarded.

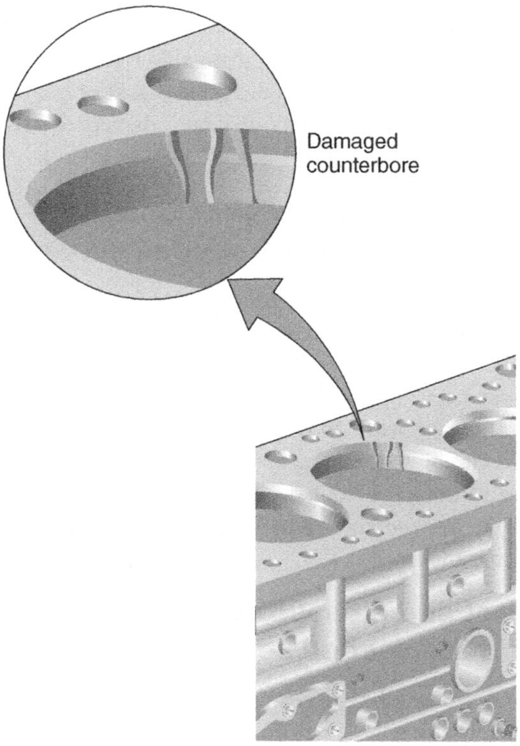

If the counterbore is in good shape, use a dial indicator mounted on a fixture to measure the counterbore depth. A depth micrometer can be used in place of the dial indicator if firm contact is maintained with the block surface. Counterbore depth and uniformity are crucial to attain proper cylinder liner protrusion before installing a cylinder head.

The cylinder liner bores in the engine block must be checked for wear and out of round as specified by the engine manufacturers. The usual areas to be checked are at the top of the block, just below the counterbore area. The O-ring seal areas at the base of the liners are also important areas to check. The out-of-round measurements are usually made perpendicular and longitudinal to the engine block.

Check for deck warpage using a straightedge and thickness gauge. The deck warpage must be checked longitudinally and side to side between cylinders and at the ends of the block.

Check the main bearing bore and alignment. Check the engine service history to ensure that the engine was not previously line bored. A master bar is used to check alignment for a specific engine series. Check that the correct bar has been selected for the engine tested and then clamped to position by torquing down the main caps minus the main bearings. The master bar should rotate in the cylinder block main bearing line bore without binding. If it binds, the cylinder block should be aligned-bored. Check the main bearing bore diameters to make sure they are in specification and are not out of round. Again, if the block is out of specification, align-boring will be necessary. Some manufacturers recommend removing a minimal amount of material from the bearing caps and align-boring the bores to factory original size. This could change the distance of the center line of the crankshaft and the main bearings to the top of the block deck and result in a slight change to the compression ratio. Other manufacturers provide oversized outside-diameter bearings to accommodate the machining process of the main bearing bores to ensure the center line of the main bearings and crankshaft to the top of the deck is not changed and maintaining the original compression ratio of the engine. If the compression ratio is increased, pre-ignition of the fuel-air mixture could be more difficult to control. This process also assures the gear train gear centerlines and bearing clearances are maintained.

If the block is an in-block cam, check the cam bushing bore dimensions. If the bushings were removed and need replacing, due to being out of specification or having been removed due to hot tanking the block, install the cam bushings with the correct cam bushing installation equipment. Use drivers with great care as the bushings easily may be damaged. Make certain the oil holes are lined up prior to driving the bushing home. Cam bushings should only be installed once. If for any reason any bushing must be removed, a new bushing must be installed.

Gallery and expansion plugs are either threaded or interference fitted and sealed with thread sealants, silicone, and hydraulic dope. Close attention must be paid to the gallery bosses and plug bores to ensure they are not cracked or damaged. Any damage or cracks could lead to leaks when the engine is placed back in service after the overhaul.

Task C3 Inspect cylinder sleeve counterbore and lower bore; check bore distortion; determine needed service.

Check the cylinder liner counterbore for the correct depth at several points around the bore and its circumference. Counterbore depth typically should not vary by more than 0.001" to 0.002" in depth between cylinders. Counterbore depth can be subtracted from the liner flange dimension to calculate sleeve protrusion. If the cylinder block deck surface is machined for any reason, the liner counterbores must be machined the same amount to maintain the correct liner protrusion once installed in the block. As a common practice, machine shops usually stamp the amount of material removed from a previous deck resurfacing operation next to the oil pan rail at the front of the block. OEM recommended tools and specifications must be followed when cutting counterbores and/or shimming the liners.

Task C4 Inspect and measure cylinder walls or liners for wear and damage; determine needed service.

Before reinstalling used wet sleeves, they should be cleaned by immersion or glass bead blasting and closely inspected. Before immersing pistons in a cleaning solution, it must be assured the solution will not damage the aluminum or overlaid areas of the piston. If bead blasting the piston, the piston pin bore must be protected from damage by the glass beads. Sleeves that fail the inspection should be discarded to ensure maximum overhaul life.

The inspection should include checking for taper using an inside micrometer, snap gauge, or dial bore gauge. An out-of-round condition can be detected by measuring the diameter at one point inside the cylinder and rotating the inside micrometer around the cylinder 90°. This process is performed at the top of the piston fire ring travel, and several other areas to the bottom of the piston travel and checking with the manufacturer's specifications. These measurements will also determine the taper of the cylinder. Excessive out-of-round or taper requires liner replacement.

Some manufacturers recommend the cylinder be honed or glaze busting. Others recommend that the cylinders be reinstalled in the engine without deglazing as the crosshatch pattern cannot be duplicated in the field.

Glaze busting: When checked to be within serviceability specs, the liner should be deglazed if the OEM approves this method. Deglazing involves the least amount of material removal. A power-driven flex hone or rigid hone with a 200 to 250 grit stone is used. The best type of glaze buster is the flex hone, typically a conical (Christmas tree) or cylindrically shaped shaft with flexible branches with carbon/abrasive balls. The objective of glaze busting is to machine away the cylinder ridge above the ring belt travel and reestablish the crosshatch. The drill should be set at 120 to 180 rpm and used in rhythmic reciprocating thrusts: short sequences, stopping frequently to inspect the finish produces the best results. Observe a 60° crossover angle, 15 to 20 micro inch crosshatch.

Honing: Use a hone in which the cutting radius of the stones can be set in a fixed position to remove irregularities in the bore rather than following the irregularities as with a spring-loaded hone. Clean the stones frequently with a wire brush to prevent stone loading. Follow the hone manufacturer's instructions regarding the use of oil or kerosene on the stones.

Insert the hone in the bore and adjust the stones snugly to the narrowest section. When correctly adjusted, the hone will not shake in the bore but will drag freely up and down the bore when the hone is not running.

Start the hone and "feel out" the bore for high spots that will cause an increased drag on the stones. Move the hone up and down the bore with short overlapping strokes about one inch long. Concentrate on the high spots in the first cut. As these are removed, the drag on the hone will become lighter and smoother. Feed the hone gently to avoid an excessive increase in the bore diameter. Some stones cut rapidly even under low tension. To check if the honing process is complete, remove the hone and inspect and measure the bore. Determine which spots must be honed most. Moving the hone from the top to the bottom of the bore will not correct an out-of-round condition. To remain in one spot too long will cause the bore to become irregular. The feel of the hone, as it is worked in the cylinder, will help to judge where and how much to hone the liner. A heavy cut in a distorted bore produces a steady drag on the hone and makes it difficult to feel the high spots. Therefore, us a light cut with frequent stone adjustments. Overhead boring tools can be set to produce the required stroke rate for the specified crosshatch, but if using a handheld tool remember that a few short strokes with a moderate radial load tend to produce a better crosshatch pattern than many strokes with a light radial load. Observe a 60° crossover angle pattern 15 to 20 micro inch crosshatch. It should be clearly visible by eye.

Wash the cylinder liner thoroughly with hot soap and water after the honing operation is completed. Wipe the inside of the liner with a clean white towel. If the towel shows dark signs reclean the liner and check again. Apply light oil to the liner and let stand for 20 or more minutes to allow the oil to absorb. Wipe clean. Stand the liners on end, not on their sides, to prevent the liners from becoming egg shaped.

Task C5 Replace/reinstall cylinder liners and seals; check and adjust liner heights.

Liner to block height, or protrusion, must be checked before the liner seals are installed on the liners. Install the liners in the same location from which they were removed, with no seals. If new liners are installed, they can be installed in any location. Install the liner clamping device and torque to specifications to hold liner in place. Next, measure the liner protrusion with a sled gauge in three or four places as per specifications. Perform the same operation for the remainder of the liners. The liner protrusion must meet the manufacturer's specifications; usually with no more than 0.002" variance between cylinders typical. If liner height varies from cylinder to cylinder, sometimes liners can be interchanged to try to adjust the height of the flange. If liner height cannot be adjusted, then the cylinder block counterbore must be reinspected and corrected as needed.

Once the liner heights have been measured and determined to be within specifications, some engine types such as the Cummins L-10G require the lower pads to liner clearance be checked as well. A clearance of 0.010" is required to prevent the liners from bottoming out. The liners and their respective cylinders must be marked to be installed in the same position as measured. Remove the liners. Install seals as necessary. Lubricate the seals as per manufacturer's specifications. Reinstall liners in the cylinder block, ensuring they seat all the way down. Recheck the protrusion before proceeding with the remainder of the engine reassembly.

Task C6 Inspect in-block camshaft bearings for wear and damage; replace as needed.

The in-block camshaft is supported by pressure-lubricated friction bearings at the main journals. The camshaft is subjected to loading whenever a cam lope is actuating the valve train that rides its profile. Normally, cam bushings are routinely replaced during engine overhaul. If they are to be reused, they must be measured with a dial bore gauge or telescoping gauge and micrometer to ensure that they are within OEM reuse parameters. Interference-fit, cylinder block located, camshaft bushings are removed sequentially starting usually from the front to the back of cylinder block using a correctly sized bushing driver (mandrel) and slide hammer. Interference-fit cam bushings are installed using the same tools used to remove them in the reverse order.

Care should be taken to properly align the oil holes. When installing bushings to a cylinder block that is still in the chassis, where access and visibility are restricted, painting the oil whole location on the bushing rim with whiteout (type of correction fluid) may help align the bushing before driving it. Ensure that the correct bushing is driven into each bore. Bushings and their support bores vary dimensionally, and new bushings rarely survive being driven into a bore and then removed.

Bearing clearances should be measured after installation of the bushings. Bearing clearance is the camshaft bushing dimension measured with a dial bore gauge or telescoping gauge and micrometer subtracted from the camshaft journal dimension measured with a micrometer.

Task C7 Inspect, measure, reinstall or replace in-block camshaft; measure and adjust end play; inspect, reinstall or replace, and adjust cam followers (if applicable).

A careful inspection of the camshaft ensures that the specified valve timing can be obtained. Because the camshaft is constantly supplied with oil, cleaning should be as simple as rinsing with dip tank solution or pressure washing and blowing out with compressed air. After washing the cam and blowing it dry, a light coat of engine oil should be applied to the cam and then wiped clean to prevent it from corroding. Oil passages may require cleaning with a stiff bristle round brush. Discard any camshaft that does not pass a visual inspection. If the camshaft passes visual inspection, take measurements and compare with the manufacturer's specifications. The cam lobes must be inspected for signs of pitting, scoring, or flat spots.

Use the manufacturer's procedures for taking lobe measurements. The heel-to-toe measurement is equal to the diameter of the circular portion of the lobe and the maximum amount of lift created by the rise at the top of the lobe. Typically, the cam follower will ride in the center of the lobe travel, causing a groove to form. This wear groove should not be greater than 0.003 in deep. If either the top of the lobe or the circular portion is worn (grooved), lift will be reduced. If an intake or exhaust cam lobe fails to produce enough lift, the valve(s) it controls may not open enough to supply adequate air for combustion to occur.

Visually inspect bearing journals for bluing, scoring, or wear. Measure the diameter of the journals with an outside micrometer, and inspect for out-of-round condition by checking around the journal in several places.

Inspect the shaft and gear keyway for cracks or distortion. Ensure the keyway in the gear and shaft matches the woodruff key. Any movement between the three can randomly change the timing of the valves. A shift of as little as 0.004 in. in the keyway could equate to almost one degree in timing shift.

In-block camshafts require the tappets to be installed prior to installing the camshaft. The tappets should be lubricated with engine oil, installed, and held up if the engine cannot be inverted upside down to allow the camshaft to be installed without contacting the tappets.

When installing a camshaft in an engine the camshaft end play must be measured and corrected to manufacturer's specifications if needed. Camshaft end play is defined either by free or captured thrust washers/plates. Thrust loads are not normally excessive unless the camshaft is driven by a helical toothed gear. In this case, it is more likely to be wear at the thrust faces of the cam and thrust washers. As a load is applied from acceleration, the helical gear will cause the gears to walk on each other, placing thrust in one direction and reversing it when the load is removed when deaccelerating. End play is best measured with a dial indicator. The camshaft should be gently levered longitudinally rearward then forward and the travel measured and checked to specifications. If necessary, the end play must be corrected and then measured again.

Task C8 Clean and inspect crankshaft and journals for surface cracks and damage; check condition of oil passages; check passage plugs; measure journal diameters; check mounting surfaces; determine needed service.

An inspection of the crankshaft (crank) begins with the connection rod and main bearing shells. Bearing shells will usually show signs of damage before the crank. If the bearings indicate damage, inspect the crank journal associated with that shell before it is cleaned. Clean the crank and visually inspect for obvious damage. This includes cracked or worn front hub key slots, the main bearing journal for scoring or bluing, dowel pins and holes for cracks and wearing, oil supply holes for cracks, and the shaft for seal grooving. The front gear and reluctor gear need to be inspected for damage due to contact with other objects, nicks, and worn teeth. In addition, a careful inspection of the crank is performed by taking measurements. An out-of-round journal can be detected by using an outside micrometer to measure in at least two places around the journal's diameter. Generally, this measurement should not exceed 0.002". To identify a tapered journal, measure the diameter nearest each side and the middle of journal. Examples of these measurements are no more than 0.0015" between any two, no more than 0.0005" between tapers in the journal and bearing. Check thrust surfaces visually for scoring, nicks, and gouges. Visually inspect for cracks or breaks. If any cracks are suspected the magnetic flux detection can be used to find smaller, less obvious one.

Crankshaft Failures:

Abnormal bending stresses occur when

- Main bearing bores are misaligned. Any kind of cylinder block irregularity will cause abnormal stress over the affected crankshaft area.
- Main bearings fail or become irregularly worn.
- Main caps are broken or loose.
- Standard specification main bearing shells are installed where an oversize is required.
- The flywheel housing is eccentrically (relative to the crankshaft) position on the cylinder block. A possible outcome of failing to check and adjust, using a dial indicator, a flywheel housing on reinstallation.
- The crankshaft is not properly supported either out of the engine prior to installation or while replacing main bearings in-chassis. The latter practice is more likely to damage the block line bore but may deform the crankshaft as well.

Bending failures tend to initiate at the main journal fillet and extend through to throw journal fillet 90° to the crankshaft axis.

Torsional Failures:

Excessive torsional stress can result in fracture occurring from an initiation point in a journal oil hole extending through the fillet at a 45° angle or circumferential severing through a fillet. In an inline six-cylinder engine, the #5 and #6 journal oil holes tend to be vulnerable to torsional failure when the crankshaft is subjected to high torsional loads.

Causes of crankshaft torsional failures:

- Loose, damaged, or defective harmonic damper or flywheel assembly.
- Unbalanced engine-driven components: fan pulleys and couplings, fan assembly, idler components, compressors.
- Engine overspeeding. Even fractional engine overspeeding will subject the crankshaft to torsional stresses that it was not engineered for that may initiate a failure. When performing failure analysis, remember that the event that caused the failure and the actual failure may be separated by a considerable time span.
- Unbalanced cylinder loading. A dead cylinder can result in a torsional failure.
- Defective engine mounts. This can produce a "shock load" effect on the power train.

Task C9 Inspect, reinstall or replace main bearings; check cap fit and bearing clearances; check and correct crankshaft endplay.

Inspection of the main bearing bore for alignment and out-of-roundness can prevent possible damage to the crankshaft or premature bearing failure. If the bearing bore is out of alignment, uneven pressure is applied to each of the bearings, causing the crank to flex as it is rotated. This flexing and its effects can also be created by an operator that abuses the engine or when the engine becomes lugged down. In this case, the engine will heat unevenly, resulting in uneven expansion of the journal. Verify the alignment of the main bearing bores by using a master bar. If a master bar is not available, a straightedge can be used by placing it across the top of the main bearing bores and checking for a gap at any one bore. This flexing can cause the crank to fracture and will result in bearings with unusual wear patterns. Use an inside micrometer to check the bearing bore for correct diameter and an out-of-round condition. If the bearing bore is too large, bearing shells will not seat properly against the bore. This may cause inadequate heat transfer from the shells to the bore, turning or movement of shells in the bore, and misalignment of the crank when a load is applied. An out-of-round bore will cause uneven torque on the crank, which could lead to fatigue, fracturing, and failure or breakage of the crank. The bearing shells will not be lubricated evenly, will wear more on one side, while the distorted shell will create fractures in the bearing surface.

Correct installation of bearing shells is essential to any successful overhaul. After reconditioning the block and selecting proper-sized bearings, carefully install the top half of each bearing into the cylinder block. The shells should be installed dry, metal to metal. Ensure the locating lugs (identical notch in each shell) are fitted into the matching slot in the cylinder block bore. This type of error can cause movement of bearings, leading to metal in the oil and finally a spun bearing, resulting in oil starvation and seizing of the crank. Sight the oil passage through the bearing shell to ensure that it is aligned to the oil passage in the bore. Failure to align the oil passage will result in oil starvation, bearing fatigue, scoring, metal in the oil, and finally, could lead to failure of other bearings as many times oil travels to other journals through the main bearing journals. Install the crank carefully in the block, making sure the thrust washers are in place.

Use Plastigage to check all main bearing clearances. Install the other half of the main bearings in their respective main caps and torque them in place to the manufacturer's specifications. Remove the caps and check the Plastigage to determine the bearing to crank journal clearances.

If the clearances are not within specifications, check for improperly sized bearings, dirt or metal under the shells, or misalignment of the shells. If clearances are correct, remove the Plastigage, lubricate the bearings, reinstall bearing caps, and torque to specifications.

Rotate the crank by hand to check for binding after torquing each cap individually. If binding occurs, check and repair the reason for binding before proceeding to another journal. Check the end play with a dial indicator to ensure it is within specification. If end play is over or under specifications, usually anywhere from 0.004" to as much as 0.022" (check specifications), adjustment to the thrust washers can be made. It is important for the Technician to remember when changing the end play to add or subtract equal amounts from both thrust washers to keep the crankshaft centered in the main journals.

Spun Bearing(s)/Bearing Seizure:

A lubrication-related failure is caused by insufficient or complete absence of oil in one or all the crank journals. The bearing is subjected to high friction loads and surface-welds itself to the affected crank journal. This may result in a spun bearing in which the bearing friction welds itself to the affected journal in the bore or alternatively continues to scuff the journal to destruction. When a crankshaft fractures as a result of bearing seizure, the surface of the journal is destroyed by excessive heat and it fails because it unable to sustain the torsional loading.

Causes of spun bearings and bearing seizures include:

- Misaligned bearing shell oil hole.
- Improper bearing-to-journal clearance. Excessive clearance will result in excessive bearing oil throw-off, which starves journals farthest from the supply of oil. Insufficient bearing clearance can be caused by overtorquing, use of oversized bearing shells where a standard specification is required, and cylinder block line bore irregularities.
- Sludge in the lubricating oil causing restrictions in oil passages and can lead to inadequate lubrication.
- Contaminated engine oil. Coolant in lube oil will destroy its lubricity.
- The incorrect type oil for the engine application. NGV engines use 15W-40 GEO low ash oil with required additives for lubrication and corrosion protection. Different engine brands require different additives. The specific engine manufacturer's specifications must be followed. If a regular diesel-rated engine oil is added to the crankcase, the bearing life can be severely affected and shortened.

Etched main bearings are caused by the chemical action of contaminated engine lubricant. Chemical contamination of engine oil by coolant or sulfur compounds can result in high acidity levels, which can corrode all metal. However, the condition is usually first noticed in engine main bearings. It may result from extending oil change intervals beyond those recommended for the type of service. Etching appears initially as uneven, erosion pock marks or channels.

Task C10 Inspect, reinstall, and time the drive gear train (check timing sensors, gear wear and backlash of crankshaft, camshaft, balance shaft, auxiliary drive, and idler gears); service shafts, bushings, and bearings.

Inspect the timing gear train, which includes all gears that drive the camshaft as well as the balance shaft assembly, in the case of a four-cylinder engine, for damage and wear. The teeth on the gears must be in good shape to maintain the timing of the intake and exhaust valves. Excess tooth wear can affect engine performance and long-term operation as well as cause damage to the pistons, cylinders, and crank. Carefully inspect for a slight roll or lip on each gear tooth in addition to looking for chipping, pitting, and burring. If a defect is present on any gear, replace it and examine the mating gears for similar damage. Some manufacturers require the gears be changed out as a unit. Manufacturer's requirements should be followed. Measure the backlash using a dial indicator and compare to specifications.

Try to determine the cause of damage. If removal is necessary, generally a press or puller will be required for some of the gears. Ensure the gear is properly supported on the center of the hub next to the shaft to prevent cracking or breaking and protect the end of the shaft from which the gear is being removed. Refer to the manufacturer's service manuals for removal of intermediate and crankshaft gears. Before installing gears on the engine block, it is important to protect the seal area in case a gear comes in contact with it. A rubber hose fitted over the crankshaft will protect it. Installation of these gears is very critical as it will affect the timing of the engine. The marks must be aligned as required. The camshaft is generally keyed the same between the shaft and gear, although some manufacturers use offset keys to either retard or advance the timing. The specifications will call for a specific offset key utilizing an arrow to indicate the position the key must be installed in.

All bushings and bearings need to be inspected for wear, binding, and overheating due to lack of lubrication. Some of the bushings are not field serviceable and therefore have to be replaced as a unit.

Timing wheels (reluctor gear) need to be inspected to ensure the teeth have not been damaged from handling and the clearances for the timing and synchronous sensors are correct.

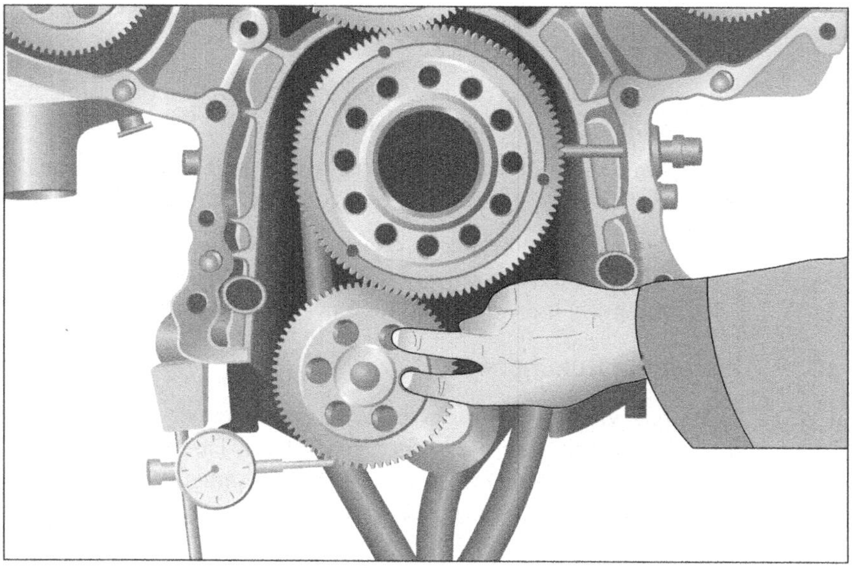

Task C11 Clean, inspect, measure, reinstall or replace pistons, pins, and retainers.

After removing each piston, carefully inspect it and its matching cylinder for possible signs of damage. If there is damage, try to determine the probable cause(s) before cleaning them. Check for scoring in the skirt of the piston and compare it to the matching side(s) of the cylinder. This could be caused by the engine overheating, overfueling, improper piston clearance, a defective or bent piston cooling nozzle, or insufficient lubrication. A crack in the skirt or lands may be accompanied by a bluing or gouges in the skirt and/or cylinder matching the same damage. This can be the result of excessive use of starting fluid, or introduction of starter fluid at operating temperatures, excessive piston clearance, or foreign objects in the cylinder, such as a dropped valve. A nick in the bottom of the skirt and connecting rod, or the cylinder and connecting rod, may be caused by improperly installed connecting rods. Uneven wear in the skirt matching a bluing or scuffing on the cylinder could be caused by abnormal operation at low temperatures, dirty or use of the wrong type of lubricating oil, a defective piston cooling nozzle, too little piston clearance, or dirty intake air. The effects of stuck or broken rings may appear on the cylinder as scrapes and gouges, scratches and carbon buildup, or glazing. The leading causes of stuck rings are overheating, insufficient lubrication, and overfueling. Worn piston pin bores can also be caused by normal wear for high-mileage engines, insufficient lubrication, use of the wrong type of engine oil, or dirty lubricating oil. Under normal conditions, during an overhaul the piston and rings are replaced along with the cylinder liner as a unit.

Piston fit problems:

- Excessive piston-to-bore clearance—results in piston knocking against cylinder wall—especially noticeable when engine is cold.
- Too little piston clearance—causes piston scoring and scuffing (localized welding)—the film of lube oil on the cylinder wall is scraped off.

Piston and Rings Inspection and Failure Analysis

When clamping a piston/connecting rod assembly in a vise, use brass jaws or a generous wrapping of rags around the connecting rod. The slightest nick or abrasion may cause a stress point, from which a failure could develop. When attempting to diagnose the cause of a piston failure, the following may be used as a very general guideline:

- Skirt scoring can be caused by overheating, overfueling, insufficient lubrication, adamaged piston cooling nozzle, or use of an incorrect type of engine oil for an NGV engine.
- A cracked skirt can be caused excessive use of starting fluid, excessive piston-to-bore clearance, or foreign objects in the cylinder.
- Uneven skirt wear can be caused by dirt in lube oil, abrasives in the air charge, or too little piston-to-bore clearance.
- Broken ring lands can be caused by excessive use of starting fluid, excessive piston to bore clearance, or foreign objects in the cylinder.
- Worn piston pin bosses can be caused by high mileage (normal wear), contaminated lube oil, or use of the wrong type of lube oil for an NGV engine.
- A burned or eroded piston crown can be caused by overfueling, detonation caused by high air intake temperature, defective spark plugs, deposits in the combustion chamber, advanced timing from overly tight valves or improper gear timing at overhaul, or high boost pressures.

Piston Pin Retainers

Pistons with full-floating wrist pins have a retainer at each end of the pin. Carefully examine the retaining ring groove area for signs of wear. This area may be eroded due to the excessive temperatures of improperly advanced engine timing, When installing the retaining rings, it is recommended that the open end of the ring be facing down, never at 90° to the connecting rod.

Task C12 Measure piston-to-cylinder wall clearances.

Use a micrometer to measure the piston diameter at right angles to the piston pin bore and 1 inch below the bottom edge of the lowest ring groove. Compare these measurements to the bore diameter of the parent cylinder bore. The difference between these two values becomes the piston-to-cylinder wall clearance. The average running clearance for larger bore engines is 0.006", but not less than 0.002" per inch or bore diameter. Insufficient clearance will cause premature piston or cylinder/liner failure due to scuffing and lack of lubrication. Too much clearance will cause a noisy piston and will rap when the engine is started or accelerated. The Technician must measure cam ground pistons at right angles (90°) to the piston pin as the piston crown is not round when it is cold.

Task C13 Check ring-to-groove clearances and end gaps; install piston rings.

Ensure the rings have the proper end gap before installing them on the piston. This measurement is taken with the rings in the cylinder in the area of their normal travel. Slide each ring into the cylinder and level them using a ringless piston. Measure the gap using a feeler gauge. Ring gap specifications vary but a rule of thumb is usually 0.003" to 0.004" per 1 in. of piston diameter.

Remove the rings from the cylinder. Follow the manufacturer's specifications for installing the specific rings on the piston. Ring side clearance is the installed clearance between the ring and the groove it is fitted to. The dimension is measured using feeler gauges in the case of rectangular rings and a groove gauge in the case of tapered or keystone rings. Detroit Diesel does not require the ring grooves be gauged in the case of the Series 50G engines due to the lower compression ratio compared to the diesel engine. To prevent overstressing the rings and damaging them the rings should never be spread more than is required to install them in their grooves. When installing the piston in the cylinder, ensure the ring gaps are spaced 90 to 120° apart, depending on the total number rings. Some manufacturers have specific gapping procedures to follow. Illustrated below is one OEM's recommendation.

Task C14 Identify piston, connecting rod bearing, and main bearing wear patterns that indicate connecting rod and crankshaft alignment or bearing bore problems; check bearing bore and bushing condition; determine needed repairs.

The reasons for piston and engine bearing failures are usually evident by the appearance of the components. When all of the main bearings have been removed from the engine, line up all the upper and lower bearings in order. The wear patterns should be uniform when compared with each other. Excessive wear in one area is an indication of localized stress of those bearings by something like a misalignment of the bearing bores in the block, or an out-of-balance condition, or out-of-round journal. Anything that subjects the engine to uneven loads, such as a faulty harmonic balancer, an out-of-balance or out-of-alignment flywheel or torque converter, will result in uneven wear. The engine is designed to withstand heavy loads, provided they are steady, and not excessive. Line up the upper and lower rod bearing shells, and make a similar comparison. When misalignment of the main bearing bores is suspected, begin by checking each bearing bore using an inside micrometer for an out-of-round condition. If any are found to be out of round, the block needs to line bored. Next, check for alignment of all the main bearing bores using a master alignment bar or a straightedge. Misalignment of any one bore will require the block to be line bored. Look for signs the connecting rods may be bent or twisted. The wear pattern on the piston skirt should be light, with no evidence of scoring, and should be centered on the skirt in a vertical line. Bent or twisted rods will result in a wear pattern on the skirt that is diagonal, not vertical.

Engine manufacturers produce failure analysis guides for Technicians. These guides use high-definition color photography so Technicians can determine the serviceability of internal engine components.

Due to the amount of cylinder combustion forces acting on a piston, the slightest misalignment of the connecting rod will create uneven loading on the wrist pin and compromise the lubricant hydrodynamics at the journal or big end of the rod. When this happens, the big end bearing will show the evidence of the misalignment.

Some failure analysis guides can be obtained from the major OEMs so the appearance of typical piston and bearing failures can be studied.

Task C15 Assemble pistons and connecting rods and install in block; check piston height; replace rod bearings and check clearances; check condition, position, and clearance of piston cooling jets (nozzles).

Installation of semi-floating piston pins into an aluminum piston can be made easier by preheating the piston in 200° F water. Some pistons require the piston pin to be pressed on. Some pistons are of one piece designs. Other pistons are two or more pieces and must be correctly assembled. Slide the piston pin through the bosses in the piston and connecting rod and install retaining rings. Other types of pistons have the connecting rod bolted to the piston pin. If the connecting rod is bolted to the piston pin and it has a drilled passage, ensure the oil passages line up and the bolts are torqued to specifications.

Clamp the piston assembly in a soft-jawed vise by the connecting rod to support the assembly while installing the rings. Ensure the rod number is correct for the cylinder bore in which it is being placed and facing in the correct direction in relation to the engine block. Properly stagger the rings. Lubricate the piston, rings, and ring compressor with a light coat of engine oil. Place the ring compressor on the piston and ring assembly. Place the rod journals at bottom dead center (BDC) to allow access to the caps. Protect the rod bearing and journal during assembly in the liner by covering rod bolts with plastic caps or rubber hoses. Carefully guide the rod onto the journal until the bearing shell seats. Remove the journal protectors off the bolts. Verify the bearing clearances are within specification using Plastigage and torquing the appropriate connecting rod cap in place. Remove the cap and determine the clearance. Clean the Plastigage, apply engine oil or Lubriplate grease on the bearing and re-torque the rod cap. Check to make sure the rod has a slight side to side movement on the journal and is not binding. Finish installing the remainder of the piston and rod assemblies.

Task C16 Inspect and measure crankshaft vibration damper; determine needed repairs.

A vibration damper reduces the amplitude of vibration and adds to the flywheel's mass in establishing rotary inertia. Its primary function is to reduce crankshaft torsional vibration. The damper consists of a damper drive or housing and inertia ring. The housing is coupled to the crankshaft and, using springs, rubber, or viscous medium, drives the inertia ring. The objective is to drive the inertia ring at average crankshaft speed. Viscous-type harmonic balancers have become almost universal in transit bus NGV engine applications. The annular housing is hollow and bolted to the crankshaft. Within the hollow housing, the inertia ring is suspended in and driven by silicone gel. The shearing of the viscous fluid film between the drive ring and the inertia ring effects the damping action. Although most OEMs recommend the replacement of the harmonic balancer at each major overhaul, this is seldom done because of the expense. Practice has shown these components frequently exceed OEM-projected expectations. The consequences of not replacing the damper when scheduled are economic as this may result in a failed crankshaft. The shearing action of the silicone gel produces friction, which is released as heat. This leads to eventual breakdown of the silicone gel, a result of prolonged service life or high mileage. Drive housing damage is another common reason for viscous damper failure and this is probably caused by careless service facility practices in most cases.

Rubber-type vibration dampers are not often found on today's NGV engines. This type consists of a drive hub bolted to the crankshaft. A rubber ring is bonded to the drive hub and the inertia ring. The rubber ring acts as the drive and damping medium. The inherent elasticity of rubber enables it to function as a damping medium, but the internal friction generates heat, which eventually hardens the rubber and renders it less effective and vulnerable to shear failures.

To ensure the balance of the crankshaft and the drivetrain is maintained during assembly of the engine, mark alignment of all drivetrain components before removal. Inspect the vibration damper, flywheel/flexplate, and the torque converter for damage that may cause an unbalanced condition. If damage is detected, there may also have been uneven wear on the main bearings and/or stress on the crankshaft. Inspect the flywheel for cracks, missing or damaged ring gear teeth, distortion, or oblong mounting holes.

Visually inspect the damper housing. Evidence of any dents or signs of warpage are reasons to reject the component. Use the following process:

- Using a dial indicator, rotate the engine manually and check for damper housing radial and axial runout against the OEM specification. This is a low-tolerance specification.
- While on the crankshaft, check for wobble using a dial indicator, and replace if it is out of manufacturer's specifications. Viscous dampers should be checked for nicks, cracks, or bulges. Cracks or bulges may indicate the internal fluid has ignited and caused the case to expand.
- Check for indications of fluid leakage, initially with the damper in place before removing from the engine. Trace evidence of leakage justifies replacement of the damper.
- Should the damper not be condemned using the above test, remove it from the engine. By hand, shake the damper. Any clunking or rattle is reason to replace it.
- Check the crankshaft to damper mounting flange to ensure there is no dirt, grease, or raised metal surfaces that could cause the damper to wobble.

Task C17 Inspect, install, and align flywheel housing.

Anytime the flywheel housing is removed from the cylinder block, the Technician must check the housing inner flange face concentricity with the crankshaft using a dial indicator. The maximum tolerance for the crankshaft to flywheel housing eccentricity is low. The consequences of installing a flywheel housing that exceeds the allowable specification are severe. When flywheel eccentricity is present, the drive axis is broken, resulting in possible engine mount, torque converter, transmission, and engine failures. The flywheel housing-to-crankshaft concentricity is preserved by interference fit cylindrical or diamond dowels. These dowels may be relied on to realign properly the flywheel housing on the cylinder block at each reinstallation, but the specification is so critical that it should be checked. The following procedure outlines a method of checking flywheel housing concentricity and a couple of typical strategies for correcting an out-of-specification condition.

1. Ensure that the engine is properly supported. Indicating the flywheel housing may be performed with the engine in or out of the chassis. Locate the flywheel inner flange face to crankshaft concentricity total indicated reading (TIR) tolerance in the OEM service manual. Either mechanically or electrically, ensure the engine fuel system will not function so the engine will not start.
2. Mount the flywheel housing to the engine cylinder block using the dowels to align the assembly and snug the flywheel fasteners at about half of the OEM-specified torque value.
3. Use fabricator's chalk to mark the flywheel flange face in four evenly spaced locations such as at 12, 3, 6, and 9 o'clock positions.

4. Next, thoroughly clean the inside face of the flywheel housing with emery cloth and affix a magnetic-base dial indicator to any position on the crankshaft, setting the plunger to contact the flywheel housing inside face. Using an engine barring tool, rotate the engine in its normal direction of rotation until the indicator probe is positioned at any one of the chalk indexes. Now set the indicator to zero.

5. Bar the engine through a full rotation, recording the indicator reading at each of the chalk indexes. The indicator should once again read zero when the revolution is completed. If the readings are A = 0, B = minus 0.003", C = minus 0.005", and D = plus 0.004", the TIR would be the highest negative value (minus 0.005") added to the highest positive value (0.004"), giving a reading of 0.009". If the OEM TIR maximum specification is 0.013", as in the case of the Detroit Diesel Series 50G, the reading taken would be within range. If the flywheel housing-to-crankshaft concentricity is outside of the specifications, the housing should be removed. Recheck the mounting surfaces for dirt, burrs, old gasket material, anything that could affect the positioning of the housing on the engine.

Task C18 Inspect flywheel or flexplate (including ring gear) and mounting surfaces for cracks, wear, and runout; determine needed repairs.

Checking Flywheel Runout

Mount a dial indicator with a magnetic base on the flywheel housing if the housing is made of steel or a clamp is made of aluminum. Check the flywheel runout at the clutch contact face. The typical allowable runout is 0.001" per inch of clutch radius. Measure the radius from the center of the flywheel to the outer edge of the clutch contact face. For example, a 14-inch clutch would have a 7-inch radius and allow a 0,007" total indicator reading of the dial indicator. If the measured values of this specification fall, remove the flywheel and check for dirt or burrs between the crankshaft butt end and the flywheel contact surface. Repeat the assembly process and remeasure until a cause can be determined.

Flywheels are commonly removed from engines for reasons such as clutch damage, leaking rear main seals, leaking cam plugs, etc. Care should be taken when inspecting and reinstalling the flywheel and the flywheel housing. Flywheels should be inspected for face warpage, heat checks, scoring, intermediate drive lug alignment and integrity, axial and radial runout using a dial indicator, straightedges, and thickness gauges. Damaged flywheel faces may be machined using a flywheel resurfacing lathe to OEM tolerances. Typical maximum machining tolerances range from 0.020" to 0.090" total. When resurfacing pot-type flywheel faces, the pot face must have the same amount of material ground away as the flywheel face. Only machining the clutch face will result in an inoperable clutch. Inspect the flywheel for cracks, missing or damaged ring gear teeth, distortion, of oblong mounting holes, or a pilot bearing with a loose or improper fit. Inspect the vibration damper, flywheel/flexplate, and clutch/torque converter for damage that may cause an unbalanced condition. If damage is detected, there may also have been uneven wear on the main bearings and/or stress on the crankshaft.

Shrunk-fit to the outer periphery of the flywheel is the ring gear, which is the means of transmitting cranking torque to the engine by the starter motor during startup. A worn or defective ring gear is removed from the flywheel by first removing the flywheel from the engine and then using an oxyacetylene torch to cut partially through the ring gear. Working from the outside, apply heat from the torch to expand the ring gear so that the removal can be completed using a hammer and chisel.

To install a new ring gear, place the flywheel on a flat, level surface and check that the ring gear seating surface is free from dirt, nick, and burrs. Ensure that the new ring gear is the correct one and if its teeth are chamfered on one side, that they will face the cranking motor pinion after installation. Next, the ring gear must be expanded using heat so that it can be shrunk to the flywheel once it cools down. Most OEMs specify a specific heat value because ring gears are heat treated and overheating will damage tempering and substantially reduce the hardness. A typical temperature specification would be around 400° F. Because of its size, the only practical method of heating ring gear for installation is using a rosebud type (high gas flow), oxyacetylene heating tip. To ensure that the ring gear is heated evenly to the specified temperature and especially to ensure that it is not overheated, the use of a temperature-indicating crayon such as Tempilstick is recommended. When the ring gear has been heated evenly to the correct temperature, it will usually drop into position and almost instantly contract to the flywheel.

D. Lubrication and Cooling Systems Diagnosis and Repair (15 Questions)

Task D1 Verify engine oil pressure and temperature; check operation of pressure/temperature sensors/switches, pressure gauge, and sending unit.

Oil pressure gauges are either electrically or mechanically operated. Low oil pressure lights are electrically operated indicators. These devices must be considered anytime wrong or faulty oil pressure readings are obtained. Usually mechanical gauges are reliable and seldom fail.

Mechanical gauges use the engine oil pressure routed through a tube to operate the meter movements. If the tube becomes punctured, kinked, or clogged, the fluid will not be able to operate the meter movements. The complete assembly must be replaced if the gauge or the tube malfunctions.

Common problems associated with electrical gauges are open or shorted wiring or circuitry, clogged passages to the sensor, or a defective sensor. The gauge can become damaged if too much voltage is applied, but this is usually a symptom of an electrical wiring problem. Use a digital multimeter (DMM) to check the resistance of the sensor and continuity of the wiring. If the sensor is at fault, replace it and check the oil passage for obstructions. If the wiring is at fault, repair or replace it and determine the cause.

Indicator lights use a method similar to the electrical gauge except they incorporate a pressure switch and a light bulb. Typically, the oil pressure switch provides a ground for the light bulb when low/no oil pressure is present. The switch then opens when oil pressure builds past a low level. When the switch opens, the lights turns off.

Modern NGV vehicles utilize multiplexing on the J1939 Controller Area Network (CAN) data bus to alert the driver of an unsafe oil pressure condition. The engine's oil pressure sensor sends data to the ECM, which is analyzed. The ECM broadcasts the oil pressure signal over the CAN, which in turn is picked up by the multiplexing system. The multiplexing system can in turn drive a display panel that is capable of displaying the oil pressure information in real time. If the ECM fails to receive a signal or it analyzes the oil pressure signal as inappropriate for the operating condition of the engine, it will broadcast a low oil pressure signal, which in turn will be received by the multiplexing system. The multiplexing system will illuminate a light on the driver's display panel to warn the driver of the unsafe engine condition and to start moving to the shoulder of the road. The ECM can also be programmed to provide a timed period to start shutting down the engine in the event of a loss of oil pressure. The engine power can also be ramped down (removed gradually) through the engine override switch to give the driver time to pull off the road safely and help prevent further damage to the engine.

Task D2 Inspect, measure, repair or replace oil pump, drives, pipes, and screens.

When removing and inspecting engine oil pumps, refer to the manufacturer's manual for specific instructions and required tools. Most NGV engine oil pumps are mounted internally to the engine. Some can be accessed from the crankcase by dropping the oil pan with conventional tools while access in other models requires the removal of the front engine gear train cover. Remove the bolts that mount the oil pump to the engine and then carefully remove the oil pump. The Detroit Diesel Series 50G oil pump, an external gear–type positive displacement pump, is mounted to the balance shaft assembly. The entire assembly must be removed for servicing. The Cummins CNG engines utilize the Gerotor internal gear type positive displacement oil pump that can be serviced after removal from the engine.

Remove the cover from the pump body and inspect it for wear. Remove and inspect the gears or rotors for pitting and wear. Check the diameter of the gears and mainshaft using a micrometer and compare with the manufacturer's specifications. Check all mating surfaces and the pump body for damage and wear. After reinstalling the gears in the body, check the gear-to-cover clearance with Plastigage. If it fails to meet specifications, replace the worn parts. If the pump has an integral or external bypass valve, it must also be checked for wear and cleaned. Because oil pumps are well lubricated by the fluid they pump, there will usually be little wear unless they are damaged by foreign particles passing through them. An engine that has been well maintained should have an oil pump showing very little wear.

Inspect the pick-up tube(s) for cracks and damaged mounting flanges. Replace the gaskets or O-rings with new ones. The clearances between the drive and the oil pump driven gears must be checked and adjusted if needed.

Task D3 Inspect, repair or replace oil pressure regulator valve(s), by-pass valve(s), and filters.

Oil filters trap particles that would cause damage to critical components, such as engine bearings. Regular replacement of filter elements and inspection of the housing is necessary. Engines use a variety of materials from cast iron to die cast aluminum for the housing and steel or plastic for machined parts. For this reason, the Technician is often limited to visual inspection and replacement of the filter assembly parts. Cracks anywhere in the housing and nicks on machined surfaces can cause engine failures. Some filter head assemblies incorporate a bypass valve in their housing. These valve assemblies must be removed and inspected for corrosion, wear, and other signs of damage during overhaul or any time the filter assembly contains metal particles.

Canister filters can be cut open and the element material examined for content. Metal particles indicate a potential bearing or gear failure. A milky gray sludge indicates water in the oil caused by a possible head or cylinder block leak, which may already have caused bearing damage. Ensure all gasket surfaces are straight and free of nicks, which could cause an improper seal of the assembly. Apply a small amount of clean engine oil to the seals to lubricate them when installing filter assemblies. Priming or prelubing of the filters, whenever possible, is recommended to ensure adequate oil is present during startup to lubricate internal engine components as well as the turbocharger.

Task D4 Inspect, clean, test, reinstall or replace oil cooler, by-pass valve, oil thermostat, lines and hoses.

Complete servicing of the oil cooler assembly should be done as part of a major engine overhaul, or whenever there is oil in the water, water in the oil pan, or a milky gray sludge in the filter. In either of these cases, specific manufacturer's instructions will apply. Look for places where the oil cooler core baffles may have vibrated against the housing. If you find signs of this, replace the cooler core. Look for gasket particles and debris that is obstructing fluid flow and clean thoroughly. The cooler bypass valve should have a slightly worn appearance and must have no signs of scoring. Check the lines and hoses for signs of leaking. Use a small light to inspect the inside of hoses, since they deteriorate from the inside first. Replace any hoses with internal cracks.

Task D5 Inspect turbocharger lubrication and cooling system; repair or replace components as needed.

To check the turbocharger lubrication, begin by checking for signs of leaks. Inspect hoses and tubes for damage, such as frayed or cut hoses and tubes for kinks or nicks. Remove hoses and tubes, especially the oil drain to check for obstructions. Install the lines after replacing any damaged ones. Install a tee the pressure feed line at the turbocharger and tighten all the connections. Start the engine and monitor the oil pressure gauges over a range of operating speeds. Stop the engine and ensure the pressure drops. All gauge readings must closely follow the instrument panel oil pressure gauge.

Most CNG engine turbochargers incorporate coolant flow through the center section for cooling purposes. Check to make sure there are no coolant leaks present. The lines should be inspected for kinks or worn areas due to rubbing. Any noted defect must be corrected.

Task D6 Change oil and filter; select oil type; verify oil level and condition.

Clean lubricating oil is the life blood of any engine. CNG engine oil color after being used for a period of time will be distinctly different in color from a diesel engine. The CNG engine's oil color will become dark brown whereas a diesel engine's oil will become black. The difference is the fuel. Diesel fuel has a high carbon chain in the fuel. Natural gas or methane is much lower in carbon. The engine oil maintenance cycles are predicted and recommended by the engine manufacturers as a result of analysis and testing. Most manufacturers of transit bus CNG engines recommend the engine oil be replaced at about every 6,000 miles. If the engines have to do a lot idling or slow speed driving, the oil change interval may have to be shortened because engine hours also are recommended by the manufacturers.

CNG engines require a very specific engine oil classification, Gas Engine Oil (GEO). Within this classification, some manufacturers also have specific additives that are necessary to prevent premature wear of the engine bearings. Zinc, phosphorous, and low ash are some of the special additives. The weight of the engine oil is 15W-40 multi-viscosity.

Engine oil filtration is accomplished by utilizing the same oil filters that a diesel engine uses. The oil should be drained out of the filter by either drilling a small hole or puncturing to allow most of the oil to drain before removal. Pre-filling of the filter, if possible, is always recommended to minimize the possibility of setting a DTC in the engine for low oil pressure. Additionally, pre-filling ensures that lube oil is directed to critical engine parts to minimize wear during start-up. The filter gasket or seal should be lubed with clean engine oil, installed hand tight and turned the additional required amount that is usually printed on the side of the oil filter housing.

The engine oil should be at or close to the operating temperature to allow all contaminants to drain with the oil. Place an appropriate drain container under the sump to capture the oil as it is being drained. Hand and eye protection should always be used due to the toxic materials that are emitted from crankcase along with the lube oil. The drain plug should be observed for any metal particles or debris that is drawn out as it is removed. The plug's gasket or seal should be checked to see if it needs to be replaced. Once the oil is drained, the drain plug should be re-installed back into the sump. Some manufacturers have installed composite sumps that are susceptible to damage from cross threading or over-torquing of the drain plug. A torque wrench is the most accurate method of tightening down a drain plug. If improperly installed, a drain plug falling off a sump can be catastrophic to an engine due to the lost of all lubrication necessary for the engine.

The required amount of oil should be added to the crankcase to bring it to the operating range on the dipstick taking in consideration the filters. The engine should be started while observing the oil pressure gauge to ensure the oil pressure is within the required range. While the engine is idling, the filter(s) should be checked for leaks. The drain plug should also be checked. The engine should be turned off after idling for four to five minutes and allowed to stand for another four to five minutes. This will allow the oil to drain back into the sump. Check and adjust the oil level as necessary.

The engine oil level should be checked and corrected at least once a day. Excessively dirty oil should be noted.

Task D7 Inspect, reinstall or replace drive belts, pulleys and tensioners; adjust drive belts and check alignment.

Today's modern NGV engines use basically two types of belts to drive accessories: V-drive belts and serpentine belts.

To install V-belts

When two or more identical V-drive belts are used on the same pulley, all of the belts must be replaced at the same time to ensure conformity in their lengths.

- Make sure the distance between the pulley centers is as short as possible when you install the belts. Do not roll the belts over the pulley. Do not use tools or pry the belts onto the pulley.
- The pulley must not be out of alignment more than 1/16" for each 12 inches of distance between the pulley centers.
- The belts must not touch the bottom of the pulley grooves. The belts must not protrude more than 3/32" above the outside diameter of the pulley.

- When identical belts are installed on a pulley, the protrusion of the belts must not vary more than 1/16".
- Make sure the belts do not touch or hit against any part of the engine or frame.

To adjust V-drive belts:

- Use the belt tension gauge to check the tension of the belts.
- The manufacturer's specifications should be followed. If no specifications can be found a good rule of thumb is to tighten the belts to obtain a reading of 110 to 160 lbs. as indicated on the gauge.
- After the engine has been running for at least one hour, stop the engine and check the belt tension. If the tension is less that the value given in the previous step, adjust the belt to the correct value.

Serpentine Belts

Many late-model NGV vehicles use a serpentine belt drive with V-ribbed or Multi-V-style belts. They are called serpentine because the belt follows a snake like path around the pulleys. The belt is thinner and more flexible so that it can bend around smaller pulleys and can also be bent backward, allowing the use of both sides of the belt to transfer power. One belt usually drives many or all of the belt-driven accessories, saving space and transferring power more efficiently. They also last longer than conventional V-belts and are easier to replace. A spring-loaded tensioner and idler pulleys are usually used to apply constant belt tension. Other methods of tensioning the belt include adjustable jackscrews and a tensioner with an off-center bolt.

Examine the underside of the belt; if there are 10 or more cracks per inch, replace belt. Do not confuse the cracks with diagonally cut grooves across the ribs. The diagonal grooves are to give the belt more flexibility and lengthen the life of the belt. The idler and tensioner pulleys should be checked for noisy, dry or worn bearings, and for wear to the pulley surface. If V-ribbed belts are noisy or walk off the pulleys, look for misalignment of the pulleys. The condition of the tensioner should be checked and not overlooked to ensure the tension spring or rubber block are not cracked or broken. If the tensioner is noisy, the knock sensor(s) on the engine can interpret this as a knock signal, give erroneous codes and affect engine performance. If the tensioner pulley can be observed jumping or deflecting, the cause should be found and corrected.

Task D8 Verify coolant temperature and pressure; check operation of temperature/pressure and level sensors/switches, temperature gauge, and sending unit.

Coolant temperature is used by electronically controlled NGV engines to monitor and control the engine operation, as well as keep the driver informed if an over-temperature condition should exist. Late model engines incorporate a coolant pressure sensor (CPS) in the temperature sensor assembly. Standard temperature gauges and indicator lights fall into one of two types: mechanical or electrical. Mechanical gauges use liquid-filled sensor that functions much like a thermometer. This type of gauge is self-contained and must be replaced as an assembly. Common reasons for failure of a mechanical gauge are corrosion on the sensor bulb, leaking sensor, pinched tube, and binding of meter movement of needle. Electrical temperature gauges use a sensor known as a thermister, which changes its resistance as temperature changes. This resistance is used to vary a return signal voltage value. Failures of an electrical gauge can be caused by a defective sensor, an open or short in the wiring, and open or defective ground points.

The CPS feeds information to the ECM to monitor the integrity of the cooling system. In the event the water pump is unable to direct coolant under pressure to the engine and EGR cooling system, a code will log, illuminating a warning light and de-rating the engine. If severe enough, it will eventually shut the engine down.

Indicator lights typically use a sensor that acts like a switch to turn on a light circuit. Probable causes for failure are the same as an electrical gauge except the light bulb itself can fail. With the introduction of electronics into the management of engines, sensors in many cases are no longer connected to display gauges. As such, a problem may be with a faulty module or circuit card. Other problems may be introduced by the software, programs or scrambled input circuit signals.

Modern NGV vehicles utilize multiplexing on the J1939 Controller Area Network (CAN) data bus to alert the driver of an unsafe coolant temperature condition. The engine's coolant temp sensor sends data to the ECM that is analyzed. The ECM in turn broadcasts the over-temp signal over the CAN, which in turn is picked up by the multiplexing system. The multiplexing system can in turn drive a display panel that is capable of displaying the coolant temperature information in real time. If the ECM receives a signal or it analyzes the coolant temp signal as inappropriate for the operating condition of the engine, it will broadcast a hot engine temperature signal which in turn will be received by the multiplexing system. The multiplexing system will illuminate a light on the driver's display panel to warn the driver of the unsafe engine condition and to start moving to the shoulder of the road. The ECM can also be programmed to provide a timed period to start shutting down of the engine in the event of a hot engine condition. The engine power can also be ramped down (removed gradually) through the engine override switch to give the driver time to pull off the road safely and help prevent further damage to the engine.

A low coolant probe is installed in the surge tank to warn the driver of a low coolant condition. This probe provides an electrical signal to either directly to driver's warning panel or to the ECM which in turn will illuminate the warning light and buzzer.

A gauge should be connected to the surge tank to ensure the pressure cap or relief is performing as needed. The cap should release the pressure at the preset pressure stamped on the cap. It should be checked to ensure it will hold pressure as well. A defective cap or relief valve must be replaced as soon as possible or serious engine damage could occur.

Task D9 Inspect and replace cooling system thermostat(s), bypasses, housing(s), and seals.

To function effectively a thermostat must:

- Start to open at a specified temperature.
- Be fully open at a set number of degrees (°) above the start to open temperature.
- Define a flow area through the thermostat in the fully open position.
- Permit zero coolant flow or a defined small quantity of flow when in the fully closed position.

NGV engines usually operate at higher coolant temperatures compared to the diesel engine and have higher rated temperature thermostats.

The cooling system thermostat is normally located either in the coolant manifold or in a housing attached to the coolant manifold. Its primary function is to permit a rapid warm up of the engine. When the engine has attained its normal operating temperature, the thermostat will open and permit coolant circulation. As the thermostat defines the flow area for circulating the coolant, there may be more than one. A heat-sensing element actuates a piston that is attached to the seal cylinder. When the engine is cold, coolant is routed to the coolant pump to be re-circulated through the engine. When the engine heats to operating temperature, the seal cylinder gates off the passage to the coolant pump and routes the coolant to the radiator. The heat-sensing element consists of a hydrocarbon or wax pellet into which the actuating shaft is forced outward in the pellet, opening the thermostat. Thermostats can be a full blocking or a partial blocking type.

Top Bypass Thermostat

The top bypass thermostat simultaneously controls the flow of coolant to the radiator and the bypass circuit. During engine warm up, all of the engine coolant is directed to flow through the bypass circuit. As the temperature rises to operating temperature, the thermostat begins to open and coolant flow is routed to the radiator, increasing incrementally with temperature rise.

Poppet or Choke Thermostats

Poppet-type thermostats control the flow of coolant to the radiator only and the bypass circuit is open continuously. Flow to the radiator is discharged through the top of the thermostat valve.

Side Bypass or Partial Blocking Thermostat

The side bypass thermostat functions similarly to the poppet type. It has a circular sleeve below the valve that moves with the valve as it opens. This serves to partially block the bypass circuit and direct most of the flow to the radiator.

Vented and Unvented Thermostats

Vented thermostats have a small orifice in the valve itself or a notch in the seat. Usually this must be positioned in an upright position on installation. The function of the vent orifice is to help deareate the coolant by routing air bubbles out of the bypass circuit. Positive dearation–type systems usually require nonvented thermostats.

Bypass Circuit

The term *bypass circuit* describes the routing of the coolant before the thermostat opens; that is through the engine cylinder block and head. The flow of bypass coolant permits rapid engine warm up to the required operating temperature and increasing fuel efficiency and reducing emissions.

Running without a thermostat is not recommended. It also contravenes EPA requirements regarding tampering with emission control components. Removing the thermostat invariably results in the engine running too cool. This can cause vaporized water in the crankcase to condense and cause corrosive acids and sludge in the crankcase. Engines that should use top bypass or partial bypass-type thermostats may overheat when the thermostat is removed as most of the coolant will be routed through the bypass circuit with little being routed in the radiator. Remember, there is a difference between the start-to-open and the fully opened temperature values.

Task D10 Flush and refill cooling system; bleed air from system; recover coolant.

Engine coolant is a mixture of water, antifreeze, and supplemental cooling additives (SCAs). To cool the engine, coolant must have an unobstructed flow throughout the cooling system. Often when the cooling system is being refilled, air is trapped inside the block. If this condition is not corrected and the engine temperature exceeds 212° F, water will turn to steam. This steam pressure forces more water out of the block, thereby causing a gas lock in the cooling system.

To prevent this from happening, manufacturers usually install bleeder valves and provide instructions on their use. Generally, bleeder valves are located near the top of the cooling system and the engine block. After a cooling system has been in operation for several years, its efficiency may be reduced due to the formation of scale and sludge. If this is the case, the radiator may need to be removed and flushed. While this can be performed in the vehicle, it generally is not effective.

Drain the coolant system into a container by opening the lower drain valves and upper bleed valves. Coolant should be tested before reusing. When the coolant must be replaced, it is a good practice to premix the coolant in a container before filling the system. Manufacturers usually recommend a 50/50 mix of coolant and distilled water. Once the radiator or some other component is replaced, refill the system, ensuring the bleeder valves are open. When all the air is out of the system, close the bleeder valves, start the engine, and check the coolant level. Test the antifreeze protection level at operating temperature. Antifreeze will prevent corrosion and raise the temperature at which the coolant will boil under pressure. This lengthens the life of the engine and cooling system components.

Task D11 Inspect, repair or replace coolant conditioner/filter, check valves, lines, shutoff valves, and fittings.

Coolant filters are designed to remove dirt and leftover sand during the casting process of the engine. The coolant filters should be replaced at regular intervals as recommended by the manufacturer's preventive maintenance schedule. Some filters contain SCAs and the Technician should be cautious to not over-service this type of coolant filter. Excessive amounts of SCAs will lead additive dropout and to clogging of the radiator assembly. When servicing the coolant filter, the shutoff valves should be checked to ensure they function correctly and move freely. The coolant lines should be inspected for damage from rubbing with other components and leaks. Sometimes the lines will weep slightly, indicating the beginning of a leak. All fittings need to be inspected regularly for leaks or signs that they are beginning to crack.

Task D12 Inspect, repair, or replace water pump, housing, hoses, idler pulley and drives.

Some water pumps are belt driven and some pumps are gear driven. Water pumps are classified as centrifugal-type pumps and need to be submerged in order to be able to pump effectively. For this reason a functioning pressure cap or valve and hoses that are in good shape, not collapsed or leaking, are required to assure the pump will function well.

Belt-Driven Water Pumps

Belt-driven water pumps utilize a tensioner and idler pulleys to maintain belt tension. Some belts are V-belts, while others are serpentine. As belts age they stretch and the tensioner pulley (on most engines) automatically adjusts the tension. If the belt is loose, check for binding in the idler pulley. A loose belt will cause the engine to overheat without any sound or squealing. Some manufacturers utilize a belt dedicated to drive the water pump.

A weep hole near the bearing area will allow the Technician to determine if the impeller seal is leaking. The engine must never be run without coolant in it. The water pump seal could be damaged because it requires coolant to remain lubricated and cool. The coolant must be drained from the radiator and engine before removing the pump. Some belt-driven pumps are mounted on stand-offs requiring inlet and outlet hose connections, while some are mounted directly into the engine and require no hose connections. With this approach, there is a gasket or an O-ring used between the block and the water pump. If the belt is damaged or cracked, it must be replaced.

Gear-Driven Water Pumps

Gear-driven water pumps are driven by the gear train and must be installed to the engine manufacturer's specifications. Gear backlash must be checked whenever a new pump is installed or when excessive gear lash is suspected due to noise coming from the gear case. The inlet and outlet hoses must be inspected for cracks and leaks at the regular preventive maintenance cycles. The seal area has a weep hole to indicate a seal failure. Because the pump is installed in the gear case and the gears are lubricated by the engine oil, it is also possible to observe engine oil leaking from the shaft seal. If any coolant or engine oil is observed leaking, the pump should be replaced. The pump should not be run dry because the impeller seal requires coolant to lubricate and cool it.

Task D13 Inspect radiator, pressure cap, and tank(s); determine needed service.

The radiator has a pressure control cap with a normally closed valve. A cap with a number 9 stamped on its top is designed to permit a pressure of approximately 9 psi in the system before the valve opens, while a cap with a number 15 stamped on its top requires 15 psi in the system before the valve opens. Typical NGV engine maximum system pressures seldom exceed 16 psi. This maximum pressure raises the boiling point of the cooling liquid and permits somewhat higher engine operating temperatures without loss of any coolant. Additionally when the system cools down, to prevent the collapse of hoses, the radiator core and the HVAC heater core, a second valve in the cap operated by vacuum opens. This prevents the cooling system pressure from becoming lower than 0.25 psi than the atmospheric pressure. Use extreme care when removing the pressure control cap as the cooling system could be pressurized. Use the vent cock, if installed to release the pressure from the surge tank and prevent injury from the hot coolant. Any sudden release of pressure from a heated cooling system can result in loss of coolant and possible injury (scalding) from the hot liquid. To ensure against possible damage to the cooling system from either excessive pressure or vacuum, check both valves periodically for proper opening and closing pressures. A regulated shop air pressure tester can be used to check for opening pressure. A handheld vacuum pump with a gauge calibrated in inches of Hg can be used to check the vacuum valve. If the pressure valve or the vacuum valve does not open at the specified pressures, replace the pressure control cap.

Task D14 Inspect, repair, or replace fan hub, fan, and fan clutch; inspect mechanical, hydraulic, and electronic fan controls, fan thermostat, and fan shroud.

Fan assemblies must be inspected. Clean the fan and related parts with soap and water and dry them with compressed air. Shielded bearings should be protected from high-pressure water as much as possible to prevent dirt from being washed into them. Check the fan blades for cracks. Replace the fan if the blades are bent, since straightening may weaken the blades particularly in the hub area. Fiberglass blades should be inspected for signs of being struck by foreign objects. Light colored spots on the blades and along the hub area are indicators that delaminating of the blade is occurring and the fan blade should be replaced.

Remove any rust or rough spots in the grooves of the fan pulley and crankshaft pulley. Examine and measure the fan hub shaft front and crankshaft pulley. If the grooves are damaged or severely worn, replace the pulley. Check the belts for signs of cracking, glazing or wear. The belts should be replaced as per the manufacturer's specifications.

Today's NGV engine fan is computer controlled. The fan may be turned on by the engine ECM when the intake air and/or the coolant exceed the preset temperature parameters, and the hydraulic brake retarding system activates. The fan system falls in two basic categories: belt/mechanically driven and hydraulically driven. In examining fan controls, verify the operation of computer input sensors such as the engine coolant temperature (ECT) sensor or the engine oil temperature (EOT) sensor, and the intake air temperature (IAT) sensor. Use the OEM recommended electronic service tools (EST) to verify the operation of ECM controlled components. The software allows the Technician to override the fan to ensure proper operation. Hydraulic fans should be closely inspected for oil leaks at the pump, motor and controller fittings. Any oil leak should be taken seriously. Hydraulic leaks have caused engine fires in the past, especially when the oil strikes the turbocharger or the exhaust system.

Task D15 Pressure test cooling system and radiator cap; determine needed repairs.

Cooling system leaks are a common problem. Even with the newest types of clamps and hoses, leaks can still occur without notice. The coolant level should be checked daily. Whenever it is necessary to replenish coolant frequently, a pressure test should be performed to locate the source of the loss. Pressure testing will easily locate most external leaks. Leaks are usually located by a good visual inspection, since antifreeze leaves a film wherever there is a leak. The location where the vehicle is parked is a good place to begin the inspection process. Coolant will drip on the ground under the area the inspection should begin.

Cold leaks occur due to contraction of sealing components, especially hose clamps. These leaks often cease to leak at operating temperatures. Silicone hoses are frequently used due to their longer life. Use the special clamps they require. The Technician should be careful to torque them properly since they are sensitive to overtightening.

A cooling system pressure-testing kit includes a hand-actuated pump and gauge, and the necessary adapters to fit different radiators, surge tanks, and pressure caps. Attach the pump using the necessary adapters to fit different radiators, surge tanks, and pressure caps. Pump up the pressure to the pressure range of the cooling system. Observe the system's ability to hold pressure while looking for visible signs of leakage. If a leak is detected during the test, it should be repaired before proceeding with the testing. The pressure cap and relief valve should also be tested to ensure their integrity. Radiator caps are rated by the pressure to overcome the cap's spring pressure and unseat the seal. When the cooling system cools after the engine is shut down, the pressure will drop. When it falls to a point where atmospheric pressure is about 0.25 psi higher than the coolant pressure, a vacuum valve in the cap unseats and atmospheric pressure is allowed back in to equalize the pressures. When the vacuum valve fails, the hoses, radiator core or heater cores can be collapsed and damaged. Internal leaks are harder to find and are usually located because coolant is found in places where it does not belong, such as in the oil or exhaust system.

E. Air Induction and Exhaust Systems Diagnosis and Repair (7 Questions)

Task E1 Inspect, service or replace air induction piping and clamps, air cleaner, and element; check for air restriction or contamination.

Most air filters are designed to route air to circulate around the filter. This cyclonic action causes airborne particles, like sand and dust, to be slung against the sides of the canister where they fall and collect in the bottom chamber. This reduces the clogging of the filter element and extends its life. The filter element is impregnated with a resin to trap most of the remaining particles. This type of air filter is the most common and has about a 99.6 to 99.9 percent efficiency rating. A vacuator valve is usually mounted in the bottom of the housing assembly to allow the contaminants to drop out when the weight overcomes gravity. The vacuator valve must be replaced in the event it is missing or has hardened to the point where it no longer is flexible. When it is not flexible, it could remain in the open position, allowing hot dirty air from the engine compartment to be introduced into the engine air intake and reducing the efficiency of the engine and possibly the life of the filter element itself.

Excessive restriction of the air inlet will affect the flow of air to the cylinders and result in poor combustion and lack of power. A normal maximum spec is 20 inches of water at full load. Consequently, inlet restriction must be kept within specifications. An obstruction in the air inlet system or a dirty air cleaner element will result in a high inlet restriction. Check inlet restriction with a water manometer connected to a removable fitting in the air inlet ducting. Check the normal air inlet vacuum (restriction) and compare the results with the manufacturer's service manual. Most manuals require the engine to be operated at full load while measuring the restriction. This method allows the turbo to be at full boost; therefore, the maximum amount of air is flowing through the air filter. Others require the restriction to be measured at high idle (no-load). The maximum restriction values will vary for each procedure.

When replacing an air filter element, the Technician must ensure the correct element is used for the vehicle and housing assembly. If the element seals are not working properly, dirt will be allowed to bypass the element and engine damage will occur. When the element is removed, the Technician must check the inlet piping directly next to the element area for signs of dirt. Any defect must be corrected.

Task E2 Inspect, test, and replace turbocharger, wastegate, actuator, and wastegate controls.

Turbochargers are used on all modern NGV engines to create intake manifold boost. That is, they pressurize the air delivered to the cylinders. Exhaust gas-driven turbine blades in the exhaust side of the turbocharger drive compressor blades in the intake side of the compressor. Because the rotor and compressor are connected by a common shaft and are built to tolerances, the assembly can achieve peak speeds from 60,000 to 150,000 rpm or more. Variable geometry and wastegate controlled turbochargers rotate more slowly and develop boost at lower rpms, reducing turbo-lag time. The resultant increase in air density increases engine power output.

Compressing air in a turbocharger heats it up. To reduce the effects of superheating, a heat exchanger, known as an air-to-air-charge (CAC) cooler, is located between the turbocharger and the cylinders. This CAC increases efficiency because cool air is denser and contains more oxygen for combustion.

The rotor on the exhaust side of the turbo is called the turbine because it is driven by exhaust gases. The impeller of the intake side is called the compressor because it compresses the air charge to the engine.

As exhaust gases drive the turbine, the compressor is rotated. Rotation of the compressor causes air to be drawn from the outside through the filter and into the inlet side of the compressor. Pressurization causes the air to heat and additional heat is transferred from the turbine through the shaft to the compressor and to the compressed air. This hot pressurized air is routed to the CAC where it is cooled and directed to the throttle valve and the intake manifold. After combustion, hot expanding gases are routed through the exhaust manifold to the turbine inlet. The hot gases turn the turbine and exit through the outlet to the exhaust and muffler/catalyst assembly.

NGV engine turbochargers use an electrically or mechanically controlled valve to route a portion of the exhaust gases around the turbine. This valve is referred to as a wastegate or overboost protection valve, and prevents the turbocharger from overboosting the cylinders at high loads. Overboosting can cause high cylinder temperatures and detonation because the fuel ignition cannot be controlled. Modern electronic systems incorporate a boost sensor to monitor the turbocharger and an electronic module to control the wastegate. Today's NGV engines also incorporate a turbocharger speed sensor to control the turbo speed and ensure the wastegate is functioning correctly as well.

When inspecting a turbocharger, begin with a visual inspection. Check the turbocharger outlet for the presence of oil. Oil in the outlet would indicate the turbocharger seals might be leaking. The seals may be worn or their leakage may be cause by restricted air cleaner, a turbocharger oil drain that is restricted or improperly routed, or by high crankcase pressure. Visually inspect the turbine and compressor wheels for damaged or eroded blades. Check the turbocharger shaft for axial and radial end play and compare to specification. Some movement is acceptable and the turbocharger should not be condemned without consulting the manufacturer's specification. When damaged blades or end play that is out-of-spec is found, check for damage to the housings. If the housings are also damaged, the turbocharger should be replaced as an assembly. If the housings are still serviceable, only the center housing rotating assembly (core) may be replaced. It is a pre-balanced assembly, so re-coring is merely a matter of transferring the housings, being careful to observe the correct alignment. When re-coring, or replacing a turbocharger, pour clean engine oil into the oil supply line fitting to pre-lube the turbocharger before starting the engine. Some of the causes of turbocharger failures are hot shutdowns that produce warped shafts and bearing failures due to lack of lubrication and cooling, turbocharger overspeeding from overfueling or a malfunctioning wastegate, lubrication-related failures, and ingestion of dirt or foreign objects.

Most NGV engines use a coolant cooled center section to further help protect the turbocharger bearings, seals and shaft from damage due to the high exhaust temperatures associated with a natural fueled engine. The lines should be inspected regularly and replaced when signs of wear appear or coolant starts to weep from the lines.

A mechanical wastegate should be checked for proper adjustment by using a dial indicator and an accurately regulated shop air source. Because the wastegate canister is spring loaded to keep the gate closed, air pressure is utilized to overcome the spring tension and cause the gate to open. The dial indicator is mounted against the external arm of the wastegate and zeroed out. When the specified air pressure is applied to the can of the wastegate, the dial indicator needs to move the specified amount of travel. The arm can be adjusted is the travel is out of spec.

An electronically controlled wastegate can be actuated through the diagnostic software to observe for correct operation of the gate or the variable geometry ring. The Technician should suspect a malfunction wastegate or variable geometry ring when high boost pressure codes are logged in the ECM for the engine.

High boost pressures are problematic for a natural gas fueled engine. During the intake cycle of the piston, fuel and air enter the combustion chamber unlike the diesel engine which is air only. Under normal conditions, due to a lower compression ratio (10:1) and lower compression pressure (200 to 240 psi) the fuel requires a spark to ignite the fuel. With high boost pressures, the amount or air and fuel entering the combustion chamber can be dramatically increased. During the compression stroke, the excess amounts of fuel and air will be squeezed more tightly, raising the temperature in the cylinder. When the temperature reaches the lower ignition limit of the methane, ignition will spontaneously occur even though the piston is still moving upward. The igniting fuel will attempt to push the piston downward. This condition is undesirable and can lead to severe engine damage if left uncorrected. Piston damage, bent connecting rods, a damaged crankshaft, and bearings or cracked cylinder heads are common symptoms.

Task E3 Inspect and replace intake manifold and gaskets; test temperature and pressure sensors; check connections.

Intake piping, tubing, and hoses are durable yet flexible to allow for movement of the engine. To reduce weight, solid piping is usually made of aluminum. Connections between the solid section of piping and major components are made using rubber sections clamped at both ends. When these connections become loose, unfiltered air is allowed to enter the intake system. Airborne abrasive or corrosive particulates can then cause damage to critical components. On the other hand, as the rubber components age or become exposed to extreme heat, they will collapse and cause restrictions in the airflow. To provide a visual warning to the operator of potential restriction in the air intake system, most manufacturers include an air intake inlet restriction indicator. This can be as simple as a vacuum tube attached to the intake piping that run to a gauge in the engine compartment, or an electronic sensor that is monitored by an electronic module. If the inlet restriction increases, check for a restriction such as a clogged air filter or a collapsed pipe. A drop in vacuum indicates a possible leak, clogged vacuum indicator tube or a defective sensor.

Two sensors are located on or near the intake manifold. The solid state boost pressure (BP) sensor monitors positive intake manifold pressure and is used by the ECM to control fuel metering. The intake air temperature (IAT) sensor monitors intake manifold temperature. The ECM uses this signal to control fuel metering for engine protection and to operate the engine fan to force air across the CAC in the event the intake air temperature gets too hot. Leaking manifold mounting gaskets will leak boost air and fuel and will lead to engine performance problems as well as leaking methane in the engine compartment that the gas detection can return a trace or significant lights on the fire gas detection panel.

The connectors should be inspected to ensure the locks are intact and the pins are free of corrosion that could impede the sensor's delivery of accurate signals.

Task E4 Inspect, test, clean, repair or replace aftercooler or charge-air cooler, clamps, and piping system.

Turbocharging intake air has the effect of causing friction between air molecules, raising the temperature of the intake air and possibly causing pre-ignition or detonation. A CAC has been incorporated into the air intake systems of NGVs as a necessity. Due to the lower ignition point of methane versus diesel and the mixing of the gaseous fuel with air in the intake system versus the cylinder, the CAC performs the important task of keeping the intake air below critical temperatures. Because the CAC is mounted in the rear of a transit bus, the ram air effect of air passing through it when the vehicle is moving is negligible. The ECM monitors the intake air by using a sensor in the intake manifold and will require the engine fan to run whenever intake air temperature reaches a predetermined level.

The CAC resembles a radiator except the fin tubes appear to be larger and the inlet and outlet connections are larger. This allows a large volume of air to pass through. As the hot air pressurized air passes through the small fin tubes, heat is transferred from the pressurized air to the tubes, then to the fins and finally to the outside air passing between the fins.

A CAC can usually be inspected visually while still in the vehicle. Wash fins with low-pressure water and straighten the fins as necessary. Inspect for missing or loose fins, dented or pinched tubes, corrosion, or holes in the tubes. Leaks can be detected by running the engine, spraying the aftercooler tubes with soap and water and watching for bubbles.

Task E5 Inspect, repair or replace exhaust manifold, gaskets, piping, mufflers, insulation/heat shield and mounting hardware; inspect, replace, or repair exhaust after-treatment devices and controls.

Most exhaust manifolds are manufactured of cast iron sections, flanged and bolted to the exhaust ports of the cylinder heads. Many are tuned. A tuned exhaust manifold is designed to deliver exhaust gas to the turbocharger so that as a pulse of exhaust from one cylinder is unloaded into the manifold, the vortex of the pulse before it helps pull it toward the turbine housing of the turbocharger. A tuned or pulse turbo is also designed to optimize the delivery of hot end gas onto the turbine vanes to produce better turbocharging efficiencies. Detroit Diesel Series 50G engine exhaust manifolds are wrapped with heat resistant material to control the amount of heat radiated from them into the engine compartment. The exhaust pipes are also covered in most NGV applications to help control the heat radiated from the pipes as well.

Most NGV exhaust piping components are manufactured from stainless steel or galvanized steel. When leaks occur, they are usually at the clamp joints are usually detectable by whitish looking telltale signs at the point of the leak as well as by being louder and noisier.

Mufflers are used in early NGV engine applications for both resonation and sound absorption principles to change the frequency of exhaust noise. The muffler is considered and emission device and as such, when one is replaced, it must meet the specifications of the original muffler.

Catalytic converters are now used exclusively on NGV engine applications. Today's catalytic converters are reduction type and help to reduce emissions. When catalytic converters are used, it is essential to adhere to the original piping geometry. Doing anything that changes the gas dynamics through the exhaust piping upstream from a catalytic converter can cause the device to overheat or not function at all. The integrity of an exhaust is extremely important to assure the vehicle continues to meet the levels of emissions as certified by the manufacturer at the time it was purchased. Exhaust leaks are a critical defect that will affect emissions. Catalyst mounting brackets have to be inspected as part of the maintenance program. If there is a broken bracket, the catalyst can move out of position and contact body panels. This is a fire danger and should be remedied as soon as possible because the catalyst can be very hot.

The exhaust back pressure can be measured using a gauge calibrated in inches of mercury. Most exhaust systems require the back pressure be less than 2.5" Hg as measured just before entering the muffler or catalyst.

Task E6 Inspect, test, service, and replace O_2 sensor, EGR valve and associated controls and wiring.

Oxygen sensors (O_2) are placed into the exhaust stream to determine the amount of oxygen that is left in the exhaust gases after combustion has taken place. There are two basic types of sensors in use on the CNG engines. One of the sensors is typically mounted in the exhaust pipe 6 to 10" from the outlet of the turbocharger. The second sensor is mounted after the catalyst in the exhaust pipe also. Not all CNG engines will utilize dual O_2 sensors but all will have at least 1. O_2 sensors exist in a harsh environment. They can be exposed to temperatures as high as 1375° F. The O_2 sensor has to be at a minimum temperature of 650° F to operate. Most O_2 sensors are now electrically heated to bring them up to this temperature more quickly, allowing the engine to go to closed loop sooner and enabling the ECM to control emissions sooner.

The O_2 sensor is basically a small voltage generator or up to 0.9 vdc. The sensor analyzes the oxygen content in the exhaust stream, compares it to the content in the outside ambient air and produces a voltage. A lean condition will produce a low voltage (<0.5 vdc) and a rich condition will produce a higher voltage (>0.5 vdc). The outside air can be sensed through a small hole at the base of the bulb or through the wires connected to the bulb. The O_2 sensor bulb is very sensitive to contamination such as water, oil, or grease. Because some of the sensors draw air through the wires, no grease should be placed at the connector as this will renders the sensor inoperative. The threads of the sensor have to be lubricated with an anti-seize that contains no lead in it. Sensors are rendered useless very quickly when contaminated with lead, antifreeze or excessive engine oil consumption.

As an O_2 sensor ages, it will usually deteriorate to the lean side or lower voltage emitted thus causing the ECM to believe the engine is running lean. This will cause the ECM in turn to add more fuel and timing. Spark knocking or detonation can occur and causing drivability issues. To test the sensor, the diagnostic software must be connected to the ECM and pointed to the AFR reading. A know good sensor has to be inserted in a spare adaptor port. Start the engine and at idle record the reading from the original sensor. Stall the engine and record the AFR reading when under a load. Turn the engine off. Unplug the original sensor and plug the spare sensor into the connector. Start the engine and record the new readings as previously done. If the readings vary more than ± 0.1 in the AFR between the sensors, the original sensor has started to drift out of range and needs to be replaced.

The Exhaust Gas Recirculation (EGR) system is the latest adaptation to help reduce emissions even further than an electronically controlled CNG engine with a catalyst. Oxides of nitrogen (NOx) are the smog-forming agent that is emitted when combustion temperatures exceed 2600° F.

Exhaust gas is re-circulated into the combustion chamber to introduce a gas that is inert, meaning it has less oxygen, to cause the fuel and air mixture to burn cooler. By adding more fuel into the cylinder, maintaining the same amount of oxygen, cooling the incoming air with the aid of the CAC and introducing the inert exhaust gas along with the fuel air mixture, higher cylinder pressures are created maintaining the horsepower output with less NOx molecules. Exhaust gases are cooled by using a cooler that has coolant circulated through it to lower the temperature of the gases. The engine requires a cooling system that has a higher heat rejection capability because of the heat of the exhaust that is transferred through the cooler into the coolant. The engine has an EGR control valve mounted after the cooler that can control the amount of gas re-introduced into the air intake through a mixer and measurement sensor. The EGR control valve is controlled by the ECM. As the ECM senses engine load, the EGR control valve will begin to open as commanded to allow cooled exhaust gases to be introduced into the combustion chamber. The mixer has a sensor integrated into it to monitor the flow of exhaust gases through the EGR system. The ECM monitors the electrical circuitry and also is capable of logging trouble codes if the EGR valve or sensor becomes defective. The manufacturer's diagnostics steps should be followed to repair the EGR system.

Task E7 Inspect and repair exhaust brake system.

An exhaust brake is a valve that is plumbed in the exhaust system to restrict the flow of exhaust out of the engine. Although not largely used on CNG-powered transit buses, they will be discussed. Exhaust brakes on CNG engines work in the same manner as for the diesel engines but require special valving and signals to enable them to work. By restricting the exhaust flow out of the exhaust system, the cylinder pressures rise and react against the travel of the piston on the compression as well as the exhaust strokes, absorbing energy from the piston, crankshaft and eventually the rear wheels through the drive train when the accelerator is released and helping the vehicle to slow down without using the vehicle's brakes, reducing brake component wear. Diesel engines have a much higher rate of retardation due to the higher compression ratios, 16 to 17:1 and higher cylinder pressures. CNG engines with a 10:1 compression ratio as well as lower cylinder pressures will not be able to slow the vehicle as well as a diesel engine powered vehicle.

For this reason, a transmission hydraulic brake retarder is employed on CNG-powered vehicles. Maximum braking efficiency up to the engine's rated horsepower can be obtained through a hydraulic retarder. Exhaust braking in a CNG engine is minimal in comparison.

An exhaust brake installed on a CNG engine will require fuel shutoff valves mounted between a CNG cylinder and the mixer, an opening and closing valve mounted on an air bypass passage provided in the intake system, and a controller for closing the fuel shutoff valve(s) and the opening and closing valve prior to operation of an exhaust brake valve. Also, the exhaust brake controller for an engine incorporates the means for turning off the fuel supply on the basis of an exhaust brake actuating signal when the accelerator pedal returns to the idle position, an oxygen concentration sensor provided on the exhaust pipe, and a controller for moving the actuator of the exhaust shutter after it is detected on the basis of a signal from the oxygen concentration sensor that oxygen in an exhaust gas has reached a concentration out of an inflammable range.

Search the Internet for "exhaust brake system".

F. Fuel System Inspection, Diagnosis, and Repair (12 Questions)

Disclaimer: The study guide is to be used for the purpose of preparing for the H1 Transit Bus CNG ASE test. The study guide should not be used to train or certify Technicians. Rather, the guide's purpose is to steer the Technician toward the areas of expertise required to become more proficient and safe when working with Compressed Natural Gas. The tasks and their descriptions are not manufacturer specific and therefore are not exhaustive in describing all areas of troubleshooting, repairs and components. The Technician must be aware that almost every day new products or procedures are developed to ensure the systems become more user friendly.

F1. High-Pressure System (8 Questions)

Task F1.1 Inspect fuel storage cylinders for physical damage, delamination, leaks; inspect mountings, shields, and hardware; repair or replace as required.

NGV fuel tanks are used to store compressed natural gas in natural gas fueled vehicles. Two basic types of cylinders will be discussed: Type III manufactured by Structural Composite Industries (SCI) and Type IV cylinders manufactured by Lincoln Composites (LC). A type III cylinder is a fuel container having a load bearing metal liner reinforced by an external composite matrix. The matrix comprises multiple layers of high strength fibers bonded together with resin. The fibers are applied in hoop (circumferential) and helical (axial) directions to cover the entire liner except for the cylinder ports. A Type IV cylinder is a fuel container having a non-load bearing liner (High Density Polyethylene (HDPE) liner), aluminum bosses for accepting valves and/or pressure relief devices (PRD), and a high strength composite wall consisting of carbon and fiberglass reinforcements in an epoxy resin. All NGV fuel tanks are manufactured under the Department of Transportation (DOT), American National Standards Institute (ANSI), Compressed Gas Association (CGA) and other standards to name a few.

Fuel cylinders are required to be inspected by a trained inspector: every 36 months minimum, when the vehicle has been involved in a fire, subjected to impact, exposed to excessive heat, or believed to have been potentially damaged by any means. Other reasons for inspection include: if the tank exhibits unusual behavior, defined as emitting a natural gas odor, unexpected loss of gas pressure, rattling, looseness, or unusual snapping or hissing sounds. If a cylinder(s) is transferred to another vehicle or if the tank installation is altered significantly, inspection of the cylinders will also be required. Tanks that have been involved in an incident which may have the potential of Level 2 or Level 3 damage should be depressurized prior to inspection. As part of the record keeping, the serial number of each tank, the expiration date the tank needs to be removed from service, the date and identification of the Technician who performed the inspection, and if any defects where found, what the disposition of the defects were.

Lincoln Composites gives the following warning: failing to perform inspections on a regular basis, or promptly in the case of a potentially damaging incident or unusual behavior, may result in a serious accident causing severe damage and injury.

Fuel cylinders can be mounted under the floor, in the upper rear HVAC area, or on the roof of transit buses. Cylinders mounted under the floor are more prone from damage by road debris, speed bumps, and curbs. Roof-mounted cylinders are prone to damage from tree limbs and overhead obstructions. The under floor cylinders can be equipped with shields that help to protect the cylinder from small debris. Shields will also help identify if any large object struck the shields and potentially damage the cylinders. Any dents on the shields should direct a Technician to perform an inspection of the area before allowing the vehicle back into service. The above roof cylinders are placed in an enclosure as well that is easily inspected for damage. Any sign of damage to the enclosure should be an indication that a visual inspection should be conducted prior to placing the vehicle back in service.

There are 4 basic levels of damage a Technician will have to take into consideration. Level 1 damage is considered inconsequential to the safe operation of the tank with no further action required. Level 2a damage involves loose fibers that can be removed and edges of a cut or gouge may be tapered by hand filing or sanding and painting the area with polyurethane paint. Level 2b damage requires the Technician to remove the tank from service. The tank manufacturer will need to be contacted for evaluation and for possible re-work procedures before returning back in service or for disposition. Level 3 damage requires the tank to be removed from service, condemned and destroyed. Level 3 damage cannot be repaired. The manufacturer's standards and specifications for determining the levels of damage should be consulted prior to the beginning of the inspection process to ensure the Technician is knowledgeable for the type of cylinder to be inspected.

The cylinders should be cleaned with cool water and a mild soap solution to allow the Technician to view all aspects of the cylinders, lines, mounts, frames and valves. Steam or high-pressure water should not be used as the upper limits of the temperature to cause a PRD to vent could be reached. High-pressure water could potentially cause the epoxy to be affected.

Prior to inspection, the Technician should have the following basic inspection tools to perform a comprehensive inspection as recommended by the Natural Gas Vehicle Association and the CNG fuel cylinder manufacturers:

1. Mirror with extending handle
2. Measuring device, metal or plastic ruler at least 6" in length
3. 10 ft. tape measure
4. Liquid leak detector and plain water to rinse the leak detector solution from the fittings and hardware after testing for leaks
5. A small metal object for tap testing on Type IV cylinder walls only or as applicable (a large coin works well)
6. Cloth rags or paper towels
7. Flashlight/drop light
8. Depth gauge
9. The manufacturer's specification and inspection procedures handbook readily available.
10. The previous documents relating to the vehicle's tanks being inspected.

The Technician performing cylinder inspections needs to be able to identify the different types of damage, as noted below, a cylinder can suffer as described by the Natural Gas Vehicle Association and fuel cylinder manufacturers.

- **Abrasion:** Damage to a CNG cylinder caused by rubbing away of container material by friction and sometimes referred to as scuffs.
- **Impact:** A strong blow to the cylinder that does not cut the composite surface. Surface damage may consist of dents or hairline cracks in the resin opaque or frost colored in appearance. Separation between the fiber layers of the composite material when damage occurs below the surface.
- **Chemical Attack:** Appears as deterioration of the epoxy resin surrounding the fibers. It can appear as cracking, crazing or discoloration of the epoxy resin and or fibers.
- **Delamination:** Gaps between the layers of the composite fiber. This type of damage can be caused by large force applied at a right angle to the composite surface.
- **Cut Damage:** Damage caused by a sharp object impacting with the composite surface.
- **Heat or Fire Damage:** Shown by discoloration, burning or melting of the cylinder and/or composite material, paint, labels or valve materials. Burnt composite usually has no resin left impregnated within it, and the fibers in the burned can be loose, easily broken and rubbed off.
- **Fuel Leakage:** Any leakage from the cylinder, liner, neck or dome area and its associated valves is not acceptable and must be resolved. If the liner is found to be the source of the leak, the cylinder must be taken out of service and destroyed.
- **Stress Corrosion Cracking (SCC):** Cracks in the composite material, usually at a right angle to the fiber direction. SCC can appear as a family of cracks or a single crack. SCC can be caused by exposure to harsh chemicals such as corrosive soaps and battery electrolyte to name a few.
- **Missing or Illegible Labels:** Replacement labels can be obtained from the manufacturer with the applicable cylinder serial number and manufacture date.

The manufacturer's specifications must always be consulted to determine damage level and the disposition of the tank.

The surface of the tanks should not be in contact or in close proximity with other objects that could cut, gouge, or wear the surface of the tank. A minimum of ½" all around clearance is recommended. Additional clearance may be needed in areas near the exhaust system and areas where the vehicle can flex while in service. An 8 inch minimum ground clearance is recommended for under vehicle applications.

The mounting brackets, frames and associated hardware need to be inspected at the same time of the inspection to ensure their integrity and safety. If for any reason a cylinder needs to be defueled and vented, local, state and federal regulations must be followed. Defueling and venting of cylinders causes static electricity. The vehicle must be grounded to earth using the required cables and ground points specified by the manufacturer's service manual and procedures. The cylinders must never be purged using compressed air. Only nitrogen or water can be used for purging the cylinders. Condemned cylinders must be purged with either water or nitrogen and rendered un-useable by drilling 2 ea ½" holes at one end of the cylinder preferably aligned with the manufacturer's label and with a steel marker inscribed with the word condemned.

CNG Cylinders are to not ever be used for anything else other than natural gas.

The fuel cylinders have threaded ports for the installation of valves. These ports are designed to allow the use of an O-ring installed on the neck of the shutoff valve for sealing purposes. If any leak is detected at the threaded port, the tank should be defueled, the valve removed and the port and valve threads inspected for any defects. The O-ring should be replaced. Use only an approved lubricant such as Dow Corning 55 on the O-ring, re-install the valve, refuel cylinder and check for leaks.

Task F1.2 Inspect and interpret cylinder labels for cylinder compliance. Determine needed action.

The fuel tanks or cylinders come labeled from the manufacturer. Whenever cylinders are being inspected, the labels have to be inspected to ensure they are legible and all the necessary information is on each label. In addition, SCI and LC affix 2 or more labels on each tank. With the 2 labels 180 ° apart, this helps to ensure that one label will be visible no matter what position the tank is installed. The information necessary on each label is as follows:

- Service pressure of 3600 psig at 70° F
- Cylinder type (I, II, III, IV)
- Cylinder serial number
- Manufacturer part number
- Test date (MM/YY)
- Service temperature range
- FMVSS304 marking or NGV 2 marking Cylinder inspection frequency information
- CNG Only statement
- Expiration date
- Manufacturer Contact information
- The statement "For use only with container manufacturer's approved pressure relief devices and valves"

If for any reason a label becomes illegible or falls off, the manufacturer can be contacted with the tank serial number and a new label can be obtained and secured back on the tank. Tanks with a damaged or illegible label should not be allowed to return back in service until the label is replaced with a new decal from the tank manufacturer or until the tank can be repositioned to allow one of the labels to be visually inspected.

Task F1.3 Inspect storage cylinder valves, solenoids, pressure relief device (PRD), fittings, piping, clamps, check valves, and vent tubing; repair or replace as required.

Manual shutoff valves, solenoids, check valves, pressure relief devices (PRD) and related tubing/fittings/clamps are at the heart of the fuel storage system. These components allow the fuel to be stored safely and to be transferred from the cylinders (tanks) to the filtering, pressure regulators and eventually to the engine's sensors and valves.

Each fuel shutoff valve is equipped a with thermally activated quick response PRD. The valves provide the Technicians the ability to isolate individual cylinders from the rest of the system for the purpose of performing maintenance on the fittings and lines without the need to defuel the cylinders. At no time should a pressurized cylinder be disconnected or moved. The fuel shutoff valve is mounted in the end of the fuel cylinder. Shutoff valves can be separate of the electronic solenoid or incorporated into solenoid to be as a unit. The stand alone valves usually have a knob that can be operated to turn the fuel either on or off. The electronic solenoid has a shutoff valve with a screw that is mounted directly on the end of the solenoid to allow turning the fuel on or off. If the knob or the screw do not operate easily or fail to operate when trying to turn the fuel on or off, the valve (for the stand alone) and the entire solenoid (for the electronic solenoid) have to be replaced.

Solenoids are electromagnet valves that are normally closed. The valves are usually powered by 12 volts (22 watts). As a rule of thumb, one 12 volt solenoid draws less than 2 amps when operating correctly. Each solenoid will have its own fuse protection but some manufacturer's have up to 4 solenoids on one fused output. The advantage to having only one solenoid on one fuse is that in the event of one shorted to ground solenoid causing the fuse to blow no other solenoid would be affected and not reducing the fuel capacity by more than one cylinder. Solenoids have to be tested at normal preventive cycles to ensure that each one is capable of turning the fuel on and off. Some vehicles have test switch panels. Others require the power connectors to be removed at each solenoid.

A Pressure Relief Device (PRD) has no field serviceable parts. Every fuel cylinder has a minimum of 1 PRD. Longer and larger cylinders can be connected with a manifold with multiple PRD. The PRD is designed to relieve pressure out of the fuel system in the event the device is exposed to temperatures above that which could be potentially unsafe for the cylinder whether the manual valve is open or closed. Any time a PRD shows signs of partial or full actuation, the cylinder associated with the PRD should immediately be defueled. A visual inspection for signs of exposure to excess heat damage should be conducted and appropriate action taken. At each inspection cycle, the Technician should use an approved leak detection method and check the PRD vent line at the roof area for any sign the Eutectic material of the PRD could be beginning to fail. The vent line caps should be in place to prevent dirt or debris from falling in the line. Each manufacturer has their own approved PRD that can be used with their system. When replacing a PRD, the Technician must replace it with the same PRD type and blow-off temperature. Before installing a PRD, visually inspect the PRD to ensure there are no visible signs of damage or evidence or mishandling. PRD must be free of scratches, dings, and/or blemishes. The manufacturer's specifications must be consulted for correct torque settings. Torque seal must be applied to the PRD and fitting to enable visual inspection to see if there is any movement after the installation. In vehicles that have enclosed fuel cylinders, the PRD must vent above the roof line through a high-pressure line and oriented in a manner as to not entrap fluids. The line must be clamped every 12 inches for lengths over 24 inches to prevent whipping in the event of activation.

Fuel check valves are used to allow fueling the vehicle without having to use the tank's fuel solenoid during the operation. At the end of fueling process, the check valve will prevent the fuel from escaping back into the main fuel line. The check valves should be visually inspected for signs of damage, evidence of mishandling, free of scratches, dings and blemishes. Check valves should only be replaced with the same, manufacturer approved check valve.

High-pressure lines should be constructed of manufacturer's approved material, usually a stainless steel material, seamless and tempered. The lines must be inspected to ensure they are installed in a manner to prevent damage to the plumbing when the vehicle flexes or the tanks expand under pressure.

Task F1.4 Inspect fuel lines, clamps and fittings, fill receptacles, ¼ turn valve, and high-pressure gauge; repair or replace as required.

Whenever servicing the fuel lines, the fuel must be depleted in the system. Turn off all the hand valves at the fuel tanks. Start the engine until it stalls. The high pressure must read 0 psi before any attempt is made to service the fuel lines. Fuel lines of the CNG system are usually made of cold finished, bright, annealed seamless Type 316 stainless steel tubing. Furthermore the manufacturer's recommend a minimum wall thickness for the diameter of the tubing such as 0.035" wall thickness for ¼" tubing, 0.065" wall thickness for ½" tubing, and 0.095" wall thickness for ¾" tubing. The lines serve " as a pathway for the natural gas going into storage and back out for eventual use. The fuel lines just like any other CNG high-pressure component are designed to operate at 3600 psi with optimal flow to and from the cylinders. The lines allow for expansion and contraction through their strength and installation design. Most lines will have an expansion loop to prevent the lines from failing due to flexing. Tubing must be free of scratches or abrasions and must not be out-of-round. Tubing with scratches may lead to connection leaks or failure of the tubing. Out-of-round tubing will prevent the fitting ferrule from making a balanced contact and penetration on the tubing and also will lead to fuel leaks. The clamps should be inspected to ensure the integrity of the insulators and that they properly secured. Any defects should be corrected.

All fittings should be tested for leaks with an electronic leak detector or an approved leak detection solution such as Snoop or Hoke. If any leaks are found, they must be corrected. Never attempt to correct a leak without first depleting the fuel from the system. The fitting seat, ferrules, and tubing should be inspected for scratches, nicks, burrs, and integrity.

The fill receptacle must be checked to ensure it is not worn or damaged and is not leaking from the check valve or the O-ring at the fill manifold. The rubber protector cap prevents dirt or other debris from accumulating on the receptacle and being pushed into the fuel system while refueling. The cap must be attached to the valve and installed when not refueling the vehicle.

The ¼ turn valve is usually located at the curbside rear of the vehicle. Its location must be identified with a decal labeled *CNG Shut-Off Valve*. The ¼ turn valve handle should be red and only capable of turning ¼ of a turn. When in the vertical position, the fuel should be turned on and when in the horizontal position the fuel turned off. The ¼ turn valve is used to turn off the fuel when servicing the engine components or the fuel filters. Test the valve by turning the valve off and run the engine until it stalls. Check the gauge to make sure it reads 0 psi. Wait several minutes and re-check the gauge to ensure it still reads zero. If the valve is defective, replace it according the manufacturer's recommendations.

The high-pressure fuel gauge is located downstream from the ¼ turn valve. The gauge usually reads from 0 to 5000 psi and glycerin filled. Check the gauge for accuracy and that the glycerin has not leaked out or the glass is not broken and is legible. Replace if defective. To replace the gauge, the fuel must be depleted. Apply approved thread sealant or an O-ring as needed. Torque the gauge to 12–15 lbs/ft.

Task F1.5 Inspect high-pressure regulator, fuel filter, and coolant lines for leaks, proper mounting or damage; repair or replace as required.

The CNG fuel high-pressure regulator, usually manufactured by ITT ConoFlow, reduces the high fuel pressure (3600 psi) down to a lower pressure range from 90 to 135 psi. The different engine manufacturers have different low-pressure requirements; therefore, the regulators are not interchangeable from one engine brand to another. Coolant from the engine is routed through the high-pressure regulator to reheat the fuel and should be checked for external coolant leaks. An electronic leak detector should be used to check the surge tank for the presence of natural gas in case the high-pressure regulator seals malfunction. The diagnostic software should be used to check the high-pressure regulator for proper operation. The ECM can log high and low-pressure codes as well to direct the Technician in case of intermittent malfunction. A coalescing type fuel filter is mounted between the ¼ turn valve and the high-pressure regulator to remove dirt or contamination from the fuel. The high-pressure filter, the housing and associated hardware should be checked for leaks, defects and proper mounting. If any defects are found, the high-pressure regulator or associated hardware should be replaced or repaired as needed.

The high-pressure regulator is equipped with a PRD and a tube that goes to the roof line. If the PRD fails or starts to leak, the high-pressure regulator needs to be replaced as a unit. The line should be removed and the PRD fitting checked with an approved leak detection solution.

Task F1.6 Service high-pressure regulator, filter, coolant lines, hoses, and high-pressure tubing as required.

The high-pressure regulator is a non–field-serviceable item. The regulator should be replaced as a unit. The coolant lines should be checked to ensure they are not porous and the hose connections are tight. Some manufacturers have incorporated a screen mesh into the coolant line coming from the coolant booster pump to the high-pressure regulator to prevent contaminants from circulating through the high-pressure regulator heat exchanger. If coolant fails to circulate through the high-pressure regulator, the screen should be considered as a potential problem area. Most manufacturers incorporate shut off valves for the coolant to allow the servicing of the coolant lines as well as the high-pressure regulator.

An approved electronic methane leak detector or solution such as Snoop or Hoke should be used to check for fuel leaks from the fittings and associated tubing. The PRD should be checked by disconnecting the small line and placing the leak detector at the fitting. Any fuel leaks should be corrected before allowing the vehicle back in service. Coolant leaks should be repaired as needed. At regular preventive maintenance intervals, the regulated pressure should be observed with the diagnostic software to ensure the regulator is capable to deliver the minimum and not exceed the maximum pressures when the engine is at idle and under load. If the high-pressure regulator fails to maintain approved pressures, no attempt should be made to adjust it; the regulator needs to be replaced and not repaired or adjusted.

Task F1.7 Defuel and vent cylinders; remove and replace cylinders.

Defueling of CNG cylinders may be necessary to service one or more cylinder, valves, fittings, tubing, or PRD to assure the system is safe to work with and require depressurization. It is good practice to deplete the fuel in the cylinders, whenever possible, to shorten the amount of time needed to remove the fuel.

Only qualified personnel may depressurize CNG cylinders to make sure the cylinders are handled safely. All federal, state, and local fire regulations and procedures must be followed by the Technician performing the defueling and venting. At no time is it acceptable to loosen fittings, manipulate tubing, and remove cylinders when the system is still pressurized.

Cylinders that will not require defueling can be isolated by turning the manual hand valve to the off position while the engine is not running. The exhaust system should be cool to the touch. The discharge port should be connected to earth ground such as a water main pipe or a steel rod pushed at least 8 ft into the ground connected with a minimum 3 gauge wire to dissipate the static charge that can build up during defueling. Venting should be done outdoors in an area where no smoking is allowed. Allow the system to vent until ambient pressure and temperature are achieved.

The cylinder(s) or associated hardware(s) can be serviced when the system is no longer venting. Cylinders should be removed using a lifting sling securely attached to both ends to allow balancing. It is recommended that steel devices such as a chain not be used to lift the cylinders because the epoxy could be damaged by the metal.

Prior to installing a cylinder on a vehicle, the cylinder must be inspected to ensure it is in compliance with the inspection requirements. The label should be legible all required information such as the date of certification, when the tank expiration will occur, no visible damage that needs further inspecting or testing is obvious. When installing cylinders back in the vehicle, the reverse procedure of removal is used. The Technician must follow the manufacturer's recommendations concerning the securement of the cylinders, installing the fuel lines and fittings, and the refueling and leak testing before allowing the vehicle back in service.

Task F1.8 Refuel system; check for leaks.

Before refueling the fuel system, it must be checked for the correct installation of all components, tubing and fittings. A thorough inspection must be performed before any fuel or pressure is introduced into the cylinder(s). Once the Technician is satisfied the system is safe for filling, it can be filled to about 500 psi. Check for signs of leaks, noises or frosting on the lines or valves. If there any signs of leakage, they should be corrected before continuing the filling process. At no time are the fittings to be loosened when there is any pressure in the fuel system. When satisfied the system is safe to continue fueling, bring the pressure to 1500 psi and repeat the leakage test. Finish the fueling process and recheck for leaks and any unusual condition. If there are any leaks detected, it may be necessary to defuel the cylinder(s) to correct the defect(s).

F2. Low-Pressure System (2 Questions)

Task F2.1 Inspect low-pressure filters, regulator, coolant lines, switch(es), sensor gauge, and tubing; repair or replace as required.

The low-pressure fuel filter as its names implies is in the lower pressure area of the fuel system, downstream of the ¼ turn fuel shutoff, high-pressure fuel filter and ITT fuel high-pressure regulator.

The low-pressure fuel filter is of a coalescing type to remove oil and contaminants down to 1 micron in size that were able to pass through the primary high-pressure filter.

On the Detroit Diesel engine, fuel travels from the secondary (low-pressure filter) to the low-pressure fuel shutoff solenoid which is equipped with a pressure gauge for monitoring and diagnosing fuel problems, to the low-pressure regulator. The line connecting the fuel shut off solenoid to the low-pressure regulator is critical that it should not be longer than 18" in length to prevent backfiring during engine shutdown. If the line is too long, the fuel will not be completely depleted at shutdown and allow the residual fuel to continue into the engine and be ignited by the hot surfaces resulting in a backfire. The low-pressure regulator senses a boost signal through a balance line from fuel the mixer to the low-pressure regulator that will allow more or less fuel to flow depending on the amount of boost developed. The low-pressure regulator has coolant routed through it to further help in keeping the fuel warm enough to prevent the engine from stumbling or running rough. At the inlet of the low-pressure regulator, a fuel pressure sensor (FPS) is teed into the passage. In the event the fuel pressure becomes too high or too low, the FPS will send a signal to the ECM that will log a DTC which will assist the Technician in diagnosing the fuel pressure related problem. The low-pressure regulator routes the fuel through a Pulse Stepper Valve (PSV) that can further assist in adjusting the fuel mixture by responding to a signal from the ECM to either allow more or less fuel as determined by the O_2 sensor feedback on whether the fuel mixture is too lean or rich. Next, the fuel is routed to the mixer where it is mixed with the passing air, through the throttle valve and into the intake manifold to be delivered to the cylinders.

The Cummins engines do not utilize a low-pressure fuel regulator. Instead, the fuel travels from the ITT fuel high-pressure regulator to the secondary fuel filter and on to the fuel shut off valve mounted directly on the engine. When the fuel shut off valve is commanded on by the ECM, fuel travels to the fuel control valve which is biased through a sensing line that directs boost pressure to one end of the fuel control valve (FCV) to increase or decrease fuel as boost pressure signals are sensed. The fuel flows to the ECM controlled FCV, gas mass flow (GMF) sensor and to the mixer through a screen that helps to align the molecules of gas for better mixing with the air passing through the mixer.

When the engine is in the closed loop operation, the O_2 sensor will give feed feedback to ECM about the oxygen content in the exhaust stream and the ECM will direct either the PSV for the Detroit Diesel or the FCV for the Cummins engine to control the fuel, emissions, and performance of the engine.

Task F2.2 Service low-pressure coolant lines, filters, gaskets, seals, and O-rings as required.

Safety should always be the **Number One Priority** whenever working on or around a CNG system. The low-pressure system, although not as highly pressurized as the high-pressure system, can injure a Technician if not properly serviced. The ¼ turn valve needs to be turned off and the engine run until it shuts off due to lack of fuel. The gauges, both the high and low sides, have to be inspected to ensure they both read 0 psi. The secondary or low-pressure fuel filter housing can now be removed for servicing. All applicable laws and regulations have to be followed when disposing of the filter elements. Some filter housings are reusable with just the element needing replacement. If equipped with O-rings, they need to be installed correctly and according to the vehicle manufacturer's recommendations and torque specifications. Some require lubing of the O-ring with Dow Corning 55 or comparable lube. Other engine manufacturer's secondary fuel filters are one piece and completely discarded and replaced with new ones. Again the manufacturer's requirements have to be followed as to whether or not the gasket needs to be lubed as well as the tightening instructions.

Once the filters are replaced the use of an electronic methane leak detector or an approved leak detection solution needs to be used to ensure there are no leaks. Turn the fuel on gradually. This should be enough pressure to detect the presence of any leaks. Once the Technician is assured there are no leaks the engine can be started and run for a few minutes and then turned back off and the fuel filter rechecked for any signs of leaks.

The low-pressure regulator is a component that has both fuel and coolant passages and gaskets. The regulator should be inspected regularly for any signs of coolant leaks as well as fuel leaks. If there are any leaks detected they should be corrected as needed.

F3. Fuel Controls (2 Questions)

Task F3.1 Inspect fuel shutoff solenoids, throttle actuator, fuel metering devices, control valves, and fuel mixer; repair or replace as required.

Inspection of the fuel system involves many valves, actuators and metering devices. A thorough knowledge of the fuel system used on each type of vehicle and their respective systems will assist the Technician in performing the necessary inspection. When the engine is turned off, the fuel solenoid valves have to be able turn the fuel off without any leakage. Some vehicles are equipped with a high-pressure fuel shutoff solenoid at each fuel cylinder while others have one solenoid to turn off all the high fuel pressure. If equipped with only one solenoid, the power connector can be disconnected with the engine running. The engine should shut off and the high-pressure gauge should read 0 psi and remain there. If the pressure creeps, then the solenoid is defective and needs replacing. The same operation can be performed for vehicles that are equipped with solenoids at each tank by removing all the connectors with the engine running and making sure the engine shuts off and the high-pressure gauge reads 0 psi and remains there. Some manufacturers have incorporated a test switch panel at the rear of the vehicle that can control the power to the solenoids and allow testing without having to climb on the roof of the vehicle.

Fuel metering devices are electronically controlled and need servicing only when a fault code is set and the Technician is directed to service the valve due to a malfunction. The connectors should be inspected to make sure they are connected correctly and the locks on the connectors are properly set. The throttle actuator or plate is electronically controlled and monitored. It is considered to be both an input and output device. The input is in the form of a pulse width modulated signal to operate the actuator. The output is a signal from the position sensor to report to the ECM its position for comparison. The throttle plate also receives a switched battery voltage to operate the electric motor that positions the valve in the correct position after receiving the input signal from the ECM. The connector and lock should be inspected to ensure they are intact and with no corrosion. Occasionally the throttle actuator will become defective. Excessive backfiring can cause the throttle valve to become bent, allowing air to pass through and not allowing the engine to idle correctly. This can also occur if there are any air leaks beyond the throttle actuator such as missing caps or loose bolts. With the engine running, air leaks can be found by listening for air escaping from the piping or intake manifold and they should repaired as needed.

The fuel mixer is the point that the intake air and the metered fuel are introduced together. It has no moving parts and this makes it very reliable. The mixer should be visually inspected for damage from contamination or abuse from mishandling or improper installation. The fuel mixer cannot be serviced and therefore should be replaced whenever it is defective.

Task F3.2 Inspect engine control module (ECM/ECU), sensors, wiring harness(es), connector(s), and other modules; repair or replace as required.

The ECM is inspected for proper mounting to the engine or other area. The wiring and connectors are checked to make sure that no components are rubbing or touching them. As the ignition switch is turned on, the Technician should observe the CEL and SEL indicators in the dash. They should light up for five seconds and then go off. If the CEL and SEL do not come on for five seconds, this would indicate to the Technician the ECM has not turned on. If the ECM is not receiving battery power, in which case the SEL and CEL lights not turning on, possible causes to be investigated may include blown fuses, an ignition signal that has not been received, or a defective ECM . If any DTCs are set, then either the CEL or SEL or both indicators will turn back on as needed after the self-check. The diagnostic software and interface should be connected to check for any DTC and appropriate action taken as needed. Once the codes have been repaired, they should be cleared. When an ECM is replaced, the Technician should insert the connectors in a manner that will not allow the pins to be accidentally bent and that leaves the seals properly positioned. The connector securement screws should be torqued to the specified setting. If the connectors, pins, or sockets need replacement, the approved crimpers should be used to ensure a good crimp will be established.

When diagnostic codes for low or high voltage from the sensors are set, the Technician should follow the troubleshooting steps established to check the integrity of the wiring and pins. Whenever working with connectors is necessary, especially at the ECM and ICM, the Technician should be extremely careful when plugging the connectors to prevent the pins and sockets from being damaged. Only the proper male pins should be plugged into the female sockets to prevent the tension springs from being damaged and later causing poor connections.

G. Starting and Charging System Diagnosis and Repair (8 Questions)

Task G1 Perform battery state-of-charge test; determine needed service.

Disconnect the battery cables and the terminals from the battery posts. Install a voltmeter and a battery load tester to the posts of the battery to be tested. Remove the surface charge from recently charged batteries by applying a 300-amperes load across the battery for 15 seconds with the load tester. Do not remove the surface charge from batteries that have been in storage. Turn the load off and wait four to five minutes for the battery to recover. A fully charged battery will have an open circuit voltage of 12.6 volts. An open circuit voltage of 12.4 indicates that a battery is less than fully charged and could need to be recharged before load testing. An open circuit voltage of 12 volts means the battery is nearly completely discharged.

Determining State of Charge

12.75 & above 100%	Charged
12.60 to 12.74 85 to 100%	Charged
12.40 to 12.59 75 to 85%	Charged
12.20 to 12.39 50 to 75%	Charged
12.00 to 12.19 25 to 50%	Charged
12.00 & below fully discharged	

An open circuit voltage test will not determine if the battery will be able to perform the work needed to start the engine or operate electrical components. The open circuit test can help to determine whether or not a battery needs to be charged before load testing.

Task G2 Perform battery load and capacitance tests; test battery cables for shorts, opens, and high resistance. Determine needed service; repair as required.

The battery discharge rate for a capacity test is usually ½ of the cold cranking amp rating of the battery. The battery is loaded for 15 seconds to eliminate surface charge if the battery was recently charged (within the past two hours). Wait four to five minutes to allow the battery to recover, and then apply ½ of the cold cranking amps on the load tester again. Battery voltage must remain above 9.6 volts with the battery temperature at 70° F. or above.

In recent years, a new style of tester, known as a capacitance tester, has been introduced in the field of testing batteries. The tester uses low-frequency pulses in the form of an A/C voltage. The tester requires no amperage to be depleted from the battery. The tester is connected to the battery posts utilizing special lead adaptors connected to the steel studs on top of a battery. The adaptors are used to prevent inaccurate readings when the low frequencies pass through steel. The Technician follows an on screen menu to continue the test. The test takes approximately eight to ten seconds. The tester returns the values of the voltage of the battery and its state of capacitance in cold cranking amps. The tester will also test for the battery impedance, a form of resistance and display whether the battery passed or failed the test. If the battery fails either test, the battery should be discarded. If the test result is questionable, the battery can be charged and retested.

Test the batteries for a voltage drain to ground by placing the DVOM selector switch in the volt DC position. Place the positive test lead of the meter on the positive post of the battery. Touch the negative lead of the DVOM anywhere around the battery top to test for voltage without touching ground. Any reading above 0.1 volts is problematic because it allows the battery to self-discharge and should be resolved. Usually the battery, the terminals, and the posts need to be wiped clean with a mild baking soda and water solution and a sponge to neutralize the electrolyte. If the cables show signs of corrosion or swelling at the connections, the cables should be replaced. The voltage drop method can be used to check the resistance of the cables. More than 0.5 volts drop in the positive side of the starter and 0.3 volts drop in the negative or ground side of the starter is considered too high and should be checked to find and repair the defect.

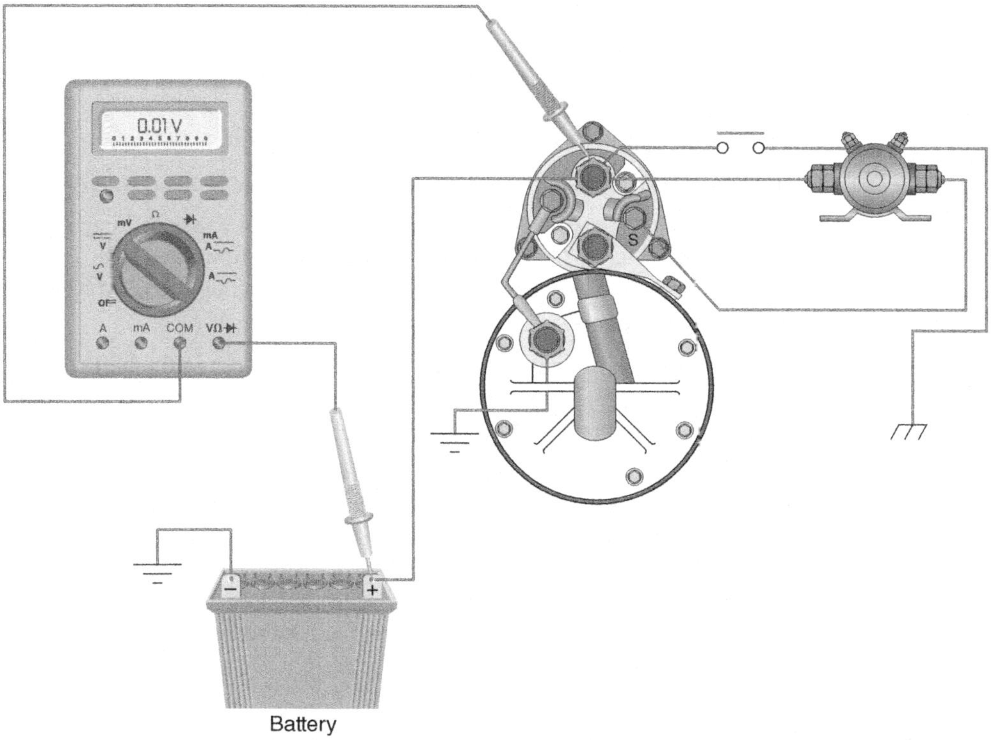

In the above diagram, the positive lead of the voltmeter is connected on the positive post of the battery and the negative lead is connected to the starter solenoid positive battery stud. One way to determine the best way to connect a meter is to think about the most positive part of the circuit you are going to test, in this case the battery positive post, so the positive lead is connected at the positive battery post. The meter can be placed on the V dc scale. Once connected, the circuit needs to work, in other words to crank the engine over without starting. The engine not starting is necessary to allow a long enough cranking demand to be placed on the connections and cables to get an accurate reading. For a CNG engine, fuel should be turned off at the ¼ turn valve and depleted. Crank the engine for three to four seconds while observing the meter for the highest and the maintained reading. In the diagram above, the meter reading is 0.01V dc which is very good. The reading can be performed on the positive cables and connections as well as the ground side of the circuit. Many times the grounds are overlooked, leading to incorrect diagnosis and unnecessary part replacement and repairs.

Task G3 Charge battery using slow or fast charge method as appropriate.

An open circuit voltage test is used to determine a battery's state of charge. Follow these steps to test voltage. If the battery has just been charged or has recently been in for service, the surface charge must be removed before an accurate voltage measurement can be made. To remove the surface charge, crank the engine for 15 seconds. Do not allow the engine to start. If the battery is out of the vehicle, a load test of half the rated cold-cranking amps for 15 seconds will remove the surface charge.

After loading the battery by either method, let the battery rest for one to two minutes. Connect a voltmeter across the battery terminals and observe the reading. If the reading is 12.6 volts or more, the battery should be fully charged. If the voltage reading is below 12.6 volts, the battery probably needs to be charged.

A built-in hydrometer provides a guide for battery testing and charging. In testing, the green dot means the battery is sufficiently charged for testing. The hydrometer reading is not an indication of a good, bad, or fully charged battery. If the colored dot is not visible and the hydrometer has a dark appearance, it means the battery must be charged before the test procedure is performed. Green is an indication of at least a 65 percent charge. Dark is an indication of a less than 65 percent charge. Light yellow or clear indicates a low electrolyte level. If the colored dot is clear or light yellow, this means that the electrolyte level is below the hydrometer. This may be caused by prolonged charging, a broken case, tipping, or excessive venting. In this case first tap the battery, and if the clear or light yellow remains replace the battery.

If a side terminal battery or threaded-top terminal battery is in the vehicle, connect the charger leads to the studs or nuts at the battery terminals. If a side terminal or threaded-top terminal battery is disconnected from the vehicle electrical system for any reason, install appropriate terminal adapters.

These rules apply to the sealed-type battery-charging:

- Do not charge a battery if the built-in hydrometer is clear or light yellow. Replace the battery.
- Charge rates between 3 and 50 amperes are satisfactory as long as spewing of the electrolyte does not occur or the battery temperature does not exceed 125° F (52° C). If spewing occurs or the temperature exceeds 125° F (52 C), the charging rate must be reduced or temporarily halted to permit cooling.
- A battery is sufficiently charged when the green dot in the built-in hydrometer is visible. No further charging is required. Tap the hydrometer lightly on top at hourly intervals during charging and see if the green dot reappears.

Battery charging consists of a charge current measured in amperes for a period in hours. A 25-ampere charging rate for two hours would be equivalent to a 50-ampere hour charge to the battery. In most cases, batteries whose load test values are less than 200 amperes will have the green dot visible after at least a 50-ampere hour charge. Most batteries whose load test values are greater than 200 amperes will have the green dot visible after at least a 75-ampere hour charge. In the event that the green dot does not appear after this amount of charging, continue charging for another 50 or 75 ampere hours. If the green dot still does not appear, replace the battery.

Task G4 Start vehicle using jumper cables, a booster battery, or an auxiliary power supply.

If the vehicle will not start due to discharged batteries, it can often be started by using energy from other batteries or another vehicle—a procedure called "jump-starting." There are basically two types of transit bus starting systems; they can be broken down to 12 volt and 24 volt. The correct starting voltage must be identified before any connections battery-to-battery connections can be established. Failure to check the system voltage could lead to engine or vehicle component damage to both vehicles when trying to jump-start the vehicle. If the Technician is not familiar with the vehicle, looking at the tag on the starter will help to guide the Technician. An examination of the battery compartment will help in identification because manufacturers usually install decals to identify the battery connections and identification. Red cables are usually 24 volts. White or blue cables should indicate the 12 volt system. Most NGVs contain four or more batteries to attain both 12 and 24 volts. 2 batteries are connected in parallel for 12 volts and 2 batteries are connected in series to achieve 24 volts. In addition, most NGVs contain a special plug to allow jump-starting. This special plug is located in the engine compartment. Position the vehicle with the good batteries so the jumper cables will reach, but never allow the vehicles to touch. Also, be sure the cables do not have loose or missing insulation.

In both vehicles:

- Apply the parking brake. Ensure the transmission is in neutral.
- Turn off the ignition (engine control switch) and all lamps and accessories. Check that the other vehicle the ignition is off also.
- Plug one end of the special jumper cable into one of the vehicles. Plug the other end to the other vehicle.
- Start the engine of the vehicle with the charged batteries and turn the fast idle switch on to bring the engine to 1,000 rpm.
- Allow two to three minutes for the dead batteries to build up some charge. Crank the vehicle with the discharged batteries to start. Do not crank for more than 30 seconds without allowing the starter to cool down for at least two minutes in between cranking periods.
- Once the engine has started, remove the jumper cables.

If the vehicles do not have the special jumper adapter, jumpers with clamps will need to be used. Polarity must be observed very closely to prevent damage from occurring.

After positioning the vehicle with the good batteries so the cables will reach the other vehicle, in both vehicles:

- Apply the parking brake. Ensure the transmission is in neutral.
- Turn off the ignition (engine control switch) and all lamps and accessories. Check the other vehicle the ignition is off also.
- Making sure the cable clamps do not touch any other metal parts, clamp one end of the first jumper cable to the positive (+) terminal on the battery, and the other end to the positive terminal on the other battery. Never connect + to −.
- Clamp one end of the second cable to the negative (−) terminal of the good (charged) battery. Make the final connection to the frame rail, chassis or to the other end to the positive terminal on the other battery. Never connect + to −.
- Start the engine of the vehicle with the charged batteries and turn the fast idle switch on to bring the engine to 1000 rpm.
- Allow 2 to 3 minutes for the dead batteries to build up some charge. Crank the vehicle with the discharged batteries to start. Do not crank for more than 30 seconds without allowing the starter to cool down for at least 2 minutes in between cranking periods.
- Once the engine has started, remove the jumper cables in the reverse order they were connected.

Auxiliary power sources:

When jump-starting from another vehicle is not successful, it is possible to use a portable generator designed for this use. However, care should be exercised when attempting to jump-start an NGV engine with this type of generator because sophisticated and dedicated electronic management systems are sensitive to voltage spikes. Most electronically managed engines are designed to withstand short term voltage overloads, but the penalty for excessive overloads can be very costly. Other electronic modules on the bus like the ABS, traction control, and automatic temperature control systems may also suffer damage from voltage spikes. Consideration should be given to charging the batteries or moving the vehicle into the shop whenever possible.

Task G5 Inspect, clean, repair, or replace batteries, battery cables, disconnects, and clamps.

The external parts of a battery can be cleaned with a baking soda and water solution. Always wear hand and eye protection when servicing batteries and their components. If the built-in hydrometer indicates light yellow or clear, the electrolyte level is low, and the battery should be replaced. The low electrolyte level may be caused by a high voltage regulator setting that causes overcharging. When disconnecting the battery cables, always disconnect the negative cable first and the positive cable last. The order is reversed when the cables are reinstalled. This will help protect against accidental contact with the positive terminal and ground, causing sparks and an unsafe condition.

Spray the cable clamps with a protective coating to prevent corrosion. A very light coating of grease or petroleum jelly will also prevent corrosion. Protective pads that clip under the clamp around the terminals to prevent corrosion are also available. Other spray protection products are commercially available to help control corrosion.

When replacing a battery, it is important to check the charging system. A defective alternator or regulator could cause a discharged or overcharged condition, resulting in a damaged battery. The cables should be inspected and the terminals cleaned or replaced when needed as part of a good PM routine.

Task G6 Inspect, test, and reinstall or replace starter relays, safety switch(s), and solenoids; test engine prelube system.

Most heavy duty starters have a solenoid connected on top to energize the starter drive gear and deliver power to the motor that will rotate the engine. Due to the large current required for the solenoid to energize, a remote starter relay (sometimes called a mag switch) is added to the circuitry to protect the starter buttons and switches from damage from the high current. This relay is usually mounted on the frame or firewall near the starter assembly.

To test the remote relay, a good quality DVOM is recommended. Test the voltage from the battery at the power terminal 30. The voltage should be the same as the voltage required for the starter solenoid to operate. Test the ground resistance at terminal 86; the resistance should be 0 ohms. Attempt to start the engine by depressing the start button and notice the voltage at terminal 85; the voltage should be the same as the rated voltage of the relay. If the mag switch energizes, voltage should be transmitted from terminal 30 to terminal 87, in turn causing the starter motor solenoid to energize.

It is important to remember that most vehicles today incorporate a neutral safety switch to prevent the starter from engaging when the transmission is in gear, especially vehicles equipped with automatic transmissions. The neutral switch usually delivers a ground signal to the remote starter relay. The switch can be tested for continuity using a DVOM. Check to make sure there is 0 ohms in neutral and infinity when the transmission is in gear between the return side of the switch and ground. The manufacturer's specifications should always be consulted when in doubt.

Electronically controlled transmissions such as the Allison World Transmission are provided a neutral signal from the ECU of the transmission. It is always important to keep in mind, if the transmission ECU is not being powered up due to an electrical malfunction in its circuitry, which could be reason for the starter or the multiplexing system to not receive the necessary signal to allow the starter to energize.

Whenever servicing the components, after a determination of the repairs needed, always disconnect the ground cable at the batteries to render the system safe.

Task G7 Perform alternator amperage and voltage output tests; perform relay terminal output voltage test; perform charging circuit voltage drop tests; determine needed repairs. Perform repairs as required.

Testing the alternator output voltage and amperage is easily performed. First, the charging system has to be identified. Some vehicles have a 12 volt charging system, while others have a 24 volt charging system and others have a combination of 12 and 24 volt systems. If a vehicle has a combination system, it can be comprised of 12 and 24 volt alternators or a 24 volt alternator with an equalizer for the 12 volt charging capability. For the 12 volt system, the charging rate is normally 14 volts plus or minus .2 volts. The 24 volt system charging rate is normally 28 volts plus or minus . 2 volts. If a battery load tester is available, connect it across the batteries to be tested, observing correct polarity. Start the engine and increase to a fast idle (about 1500 rpms). Observe the unloaded output voltage; for the 12 volt system it should be 14 volts and for a 24 volt system it should be 28 volts. Attach an amp clamp around the alternator output wire. With the engine at high idle, use the load tester to reduce battery voltage to obtain the highest amperage reading without causing the voltage to fall below 12 or 24 volts, depending on the system voltage. This procedure forces the alternator to increase its output to near its maximum. The amp clam should record output within 5 percent of the alternator's capacity. If a load tester is not available, turning on as many electrical loads as possible may be used as a substitute. If an undercharge is noted, some of the causes are loose or glazed belts, damaged or corroded wires and connectors, a defective battery, a defective AC generator, a defective diode trio or rectifier bridge assembly, or a faulty voltage regulator. Follow the relevant diagnostic procedures to find the specific fault.

Some AC generators such as the Delco Remy Genetron 50DN have a relay terminal at the rear of the generator. This relay terminal is utilized to trigger accessories such as air-conditioning or other systems. In the event the charging system fails, these systems will be shut off by the lack of output from the relay terminal. The terminal normally outputs about 50 percent of the normal charging rate of the generator. Place an accurate voltmeter across the relay terminal and use a good ground. Read and record the voltage. The voltage should be about 7 volts on a 12 volt generator and 14 volts for a 24 volt generator.

To test the voltage drop of a charging circuit, an accurate digital volt ohm meter (DVOM) must be utilized. Connect the red lead to the output terminal of the generator. The negative lead must be connected to the positive lead of the 12 or 24 volt batteries of the system to be tested. Start the engine and run at high idle, around 1000 rpm. Turn on as many accessories as possible to make the generator charge at a high rate. Remember the charging circuit has to be working to near capacity to ensure an accurate reading of voltage drop. The voltage drop should not exceed .1 volt per connection between the generator and the battery (ies). The manufacturer's wiring schematics should be consulted to determine the connection points to assess the system.

Causes of excessive voltage drops are: loose or corroded connections, defective or too small cables, and too many connections. Many times voltage drop points can be identified because excessive heat is generated at the point of trouble. An infrared pyrometer is a tool that can be used to identify higher than normal heat in a circuit that can be the source of the voltage drop.

Task G8 Perform starter current draw and circuit voltage drop tests; determine needed repairs. Perform repairs as required.

It is valuable know how much battery voltage is dropped in the starting circuit overcoming resistance. This may be accomplished by voltage drop testing. Voltage drop is the amount of electrical pressure that is lost or consumed in pushing current through the circuit. Voltage drop occurs when current passes through load components and resistances. Most of the voltage should be dropped across the starter and not across the cables and wiring that conduct voltage and control the starter. Voltage drop testing to measure resistance has several big advantages over using an ohmmeter. First, it is very accurate. You can find very small resistances where voltage is lost (dropped) passing through connections, relays, switches, fuses, wiring, and other voltage-conveying components in the circuit. Another advantage is that it is a dynamic test, performed with the circuit live and under load. It is not necessary to isolate the component being tested from the circuit by disconnecting or removing it. Simply connect the voltmeter in parallel with the portion of the circuit being tested, being careful to observe polarity.

For example, to test for resistance caused by battery terminal corrosion, you would place one voltmeter lead on the post and the other one on the cable, and then observe the meter while cranking the starter. A good connection will show that only about 0.1 V is dropped. This is for only one connection. When the voltage drop across all of the positive battery cable is measured, the reading should be less than 0.5 V. You can test the positive battery cable by placing one lead on the battery post and the other lead on the solenoid's battery cable stud, thus testing the entire cable. A similar test with one lead on the negative battery post and the other on the starter frame will test the entire ground side of the circuit. There should be less than 0.3 V dropped across the ground circuit. Most of the battery voltage should be dropped across the starter motor. About 0.3 V is dropped across the starter solenoid and the remainder is consumed elsewhere in the circuit.

Voltage drop testing is also useful to find unwanted resistance in the charging system as discussed previously. The tests are similar.

H. Ignition Systems Diagnosis and Repair (8 Questions)

Task H1 Diagnose ignition-related problems such as no-starting, engine misfire, poor drivability, spark knock, power loss, and poor mileage on vehicles with distributorless ignition systems; determine needed repairs.

The NGV engine utilizes the power triangle principle of fuel, air, and heat for energy. The ignition system of modern CNG engines is of the distributorless type and computer controlled. The first step is to establish the ECM is capable or energizing. The Technician should check the Stop Engine Light and Check Engine Light to make sure they both turn on for 5 seconds and then turn off. This tells the Technician the ECM has initialized. If there are active codes, the light(s) can turn back on after that point.

When the engine cranks but fails to start, a point of diagnostics must be established. If there are no active code lights, check for power to the coils. The Cummins engine's coils receive power through the Igniter control module, up to 300 volts. The Detroit Diesel Series 50G coil power comes directly from a fused, switched battery power and the coils should be receiving battery voltage at pin A of the coil whenever the key switch is turned on.

If the fuse is good and the coils are getting voltage, the DTCs should be checked in the ECM to try to determine the defect that is causing the no-start condition. Using the software, the fuel delivery pressures should be checked to ensure the fuel solenoids are activating to deliver fuel to the engine.

If the engine starts but runs poorly or misses, the spark plugs, coils, or boots or related components should be inspected for their condition. Worn or damaged spark plugs should be replaced. Special attention should be placed on the electrodes and their gap. The Detroit Diesel Series 50G engine model TK and the Cummins engines spark plugs are gapped to 0.015", the Detroit Diesel Series 50G MK engine model spark plugs are gapped to 0.011". Cummins requires the gap be checked but no regapping is allowed; the plug should be replaced if out of specification. When checking the gap, a wire gauge is recommended to prevent the precious metals from being damaged by the sharp edges such as a feeler gauge.

The spark plug resistance can be checked using an ohmmeter. Cummins' acceptable resistance range is between 9K to 17K Ω. The outside ceramic insulator needs to be inspected for spark flashover, black dots, or traces from spark etching. Any sign of flashover requires the spark plug(s) be replaced. Flashover can be minimized by using isopropyl alcohol to clean the complete exterior of the spark plugs before they are installed in the engine.

A lean fuel condition can cause an engine to miss under load. The diagnostic software can help to diagnose this symptom. The Air Fuel Ratio (AFR) for the Detroit Diesel is a good indicator and is received directly from the Oxygen Sensor (O_2). The closer to 1.0 the AFR reading is, the richer the condition. Numbers above the 1.5 range are considered a lean condition. The Cummins Insite software will display the Fuel Deviation Compensation as a + or −. A lean condition will be indicated when a + fuel compensation is being applied.

Spark knock is usually a condition caused by high cylinder temperatures due to overfueling, tight valves, high boost pressures, high intake air temperature. Overfueling could be caused by a sticking fuel control valve, a defective or out of calibration O_2 sensor, high air inlet restriction causing a lack of air, and many other causes. Tight valves require adjustment to prevent damage. High boost pressures are a symptom of a sticking waste gate. High intake air temperatures are indicative of a malfunctioning CAC or the engine fan not turning on or not turning fast enough to cool the air passing through the CAC. The diagnostic software can further assist and guide the Technician in performing the necessary repairs.

Task H2 Check for possible ignition system related diagnostic trouble codes (DTC); verify performance parameters.

The diagnostic software and the appropriate interface or a DDR must be connected to the diagnostic port to allow the Technician to check the ECM for diagnostic trouble codes. If there are any ignition related defects, the ECM should log a DTC related to the problem(s) such as for the coils, ignition control module (ICM), or for any knock codes that exist. The appropriate corrective action should be undertaken. Codes for the coils indicate the power circuit wiring and fuse should be checked as well as the integrity of the insulation of the secondary circuit. The Insite software allows the Technician to view the voltage secondary circuit of the coils to see if a low or high firing voltage is required to fire the spark plug while the engine is running. High voltages would indicate a possible lean condition, a defective spark plug, or high resistance between the coil and the plug. A low voltage could indicate that the spark is jumping to ground without passing through the plug or the spark plug gap is closed. Resetting of the AFR table before and after any corrective action is taken will ensure the engine will be able to learn to operate at its most efficient range without delay. Cummins requires the AFR table to be reset using the Insite diagnostic software after replacing the gas mass flow (GMF) sensor, the O_2 sensor, and the fuel control valve. After the repairs are completed, the Technician should road test the vehicle.

Task H3 Inspect, test, repair, or replace ignition primary circuit wiring and components.

All wiring, connectors, and pins for the primary ignition circuit must be in good condition. Any loose pins, corrosion, bare wires can be a source of malfunction. The Detroit Diesel Series 50G engines utilize ignition switched battery voltage on the coil harness wire no. 440 to the coil's + pin to energize the power to the coils and the ICM controls the coil's ground, the − pin, from a timing signal sent by the ECM to ignite air/fuel mixture. The Cummins engines L-Gas Plus and the ISLG models use a signal from the ECM to ICM to energize the primary ignition circuit of the coils with as much as 300 volts. The coils will then deliver the 5 to 20 KV required, jumping the spark plug gap with Multiple Spark Discharges (MSD) and igniting the compressed air/fuel mixture when commanded. The troubleshooting trees in conjunction with the diagnostic software will allow the Technician to test for voltage to the coil(s), resistance caused by loose pins and sockets or corrosion, open circuit, or wiring shorts with a DVOM. In general, the continuity measurement of the wiring is usually $\leq$10 Ω end to end, with an insulating capability $\geq$100K Ω pin to pin and to ground. The engine manufacturer's recommended practices of repairing the primary ignition circuit have to be followed to prevent drivability issues or damage to the engine.

Task H4 Inspect and test spark plugs, boots, tubes and seals, and wiring; check and adjust spark plug gap.

The spark plugs as well as the spark plug boots/extenders, or spark plug wires, are at the heart of the ignition system and are all part of the ignition secondary system. They are subjected to high voltages in the magnitude of 5 to 30 KV in the normal operating range. They need routine inspection whenever they are being serviced, such as when replacing the spark plugs. The electrodes have to be parallel to each other. They cannot be eroded or damaged. Contamination from the ceramic insulator and the rubber boots needs to be removed with the use of isopropyl alcohol or the components replaced to prevent the spark from flashing over to ground and causing drivability issues as well as DTCs that will result in operator complaints. Spark plug boots/extenders need to be checked for cracks or pinholes that could lead to spark flashover and drivability issues. The spark plug wires have to be checked for continuity with an ohmmeter. The maximum resistance is 10K Ω per foot of wire. Detroit Diesel recommends the secondary ignition components be installed clean and dry. Cummins pre the spark plug boots with a high-quality insulating dielectric compound from the factory. The dielectric compound should be prevented from being applied to the tip of the plug that comes in contact with the coils conductor.

The spark plugs are to be replaced on a routine schedule. Manufacturers have developed diagnostic charts to identify the conditions that affect the spark plug life and durability. Proper spark plug gapping is crucial for the proper operation and life of the spark plug as well as cleanliness when handling the plug. The Detroit Diesel manual requires the plug be gapped according to the type of engine model such as the Series 50G TK or MK models. The TK model requires a gap of 0.015" whereas the MK model requires a gap of 0.011" and the gap can be adjusted. Cummins engines spark plugs come pregapped and Cummins recommends the plugs not be reapped. The reason that Cummins does not want a plug to be regapped is that whenever a feeler gauge is placed between the two electrodes, the precious metals tend to be removed, and this causes the plug life to be reduced and sometimes the plugs to be rendered totally useless. Some Technicians find that referring to a used plug that is gapped correctly as a comparison sample is helpful.

The Technician must use a clean spark plug socket when installing plugs back into the cylinder head. They cannot be contaminated with any grease, oil, or dirt. The engine manufacturers recommend the use of a magnetic socket to hold the plug in the socket and to prevent it from dropping into the adaptor tube and damaging the electrode or closing the gap. The use of a socket with a rubber adaptor to hold the plug is not recommended because the rubber can become contaminated and the contamination is then transferred onto the ceramic of the spark plug and can lead to the spark traveling through the dirt and to ground, causing a misfire.

Task H5 Inspect, test, or replace ignition system secondary circuit wiring and components.

The Cummins 8.9L LGAS-Plus and ISLG and the Detroit Diesel Series 50G CNG engines' ignition secondary circuits are primarily composed of the spark plug boots or extenders and the spark plugs connected to the coils mounted on top of the spark plugs they fire. The spark plug boots and the spark plugs must be free of any contamination to prevent spark flashing to ground. Isopropyl alcohol is recommended to clean all of the secondary components. If any of the components are damaged or have signs of tears or burn holes they have to be replaced. Cummins requires the replacement of the spark plug boots when the spark plugs are replaced. The boots come prelubed with a high-quality insulating dielectric compound and in sufficient quantity. The compound should be prevented from coming in contact with the plug's conductor to prevent the compound from wicking up the coil's extender wire and eventually onto the coil's conductor, preventing good continuity, creating high resistance, and misfiring of the cylinder. The Cummins L10-G engine model utilizes spark plug wires that are connected to a coil pack. The resistance of the wires has to range from 500 to 1500 Ω.

The Detroit Diesel Series 50G requires the ignition's secondary circuit components be installed clean without the use of any compounds or lubricants. The boots need to be inspected for cracks, tears, or burn marks that can lead the spark to leak to ground and cause a misfire situation. Failure to follow this guideline can lead to spark flashover and the login of DTCs.

Task H6 Inspect, test, and replace ignition coil(s).

Whenever the coils are removed from the engine, they should be checked for cracks, contamination from water or engine oil, and traces of spark jump, and replaced as necessary. An ohmmeter can be used to check the resistance of the primary circuit of the Detroit Diesel Series 50G and should have a value of less than 1 Ω, usually ≤.55 Ω. Because the secondary side of the coil is solid state, it cannot be checked without the use of highly sophisticated tools. One method of determining if a coil is defective is to swap it with one from an adjacent cylinder after checking the spark plug and checking to see if the problem followed it to the new cylinder location. If a coil becomes cracked, the result is usually a misfire under a load when the spark intensity is at its greatest. If the coils are improperly aligned and forced onto the spark plug this can also cause the coil to become damaged, cracked, or distorted over time. For the Cummins L-Gas Plus and the ISL G coils, an ohmmeter will be connected at the spark plug connection and to the pins of the primary circuit. The resistance measurement must be less than 20K Ω. If greater than 20K Ω the coil is considered to be defective and must be replaced.

The coil (COP) needs to be checked on its underside and along the body for cracks. When the coils are removed on a Cummins engine, the spark plug boot should be replaced as a rule. The boots contain a special grade of dielectric compound to prevent spark flashover to the ceramic and also to keep the boot from seizing onto the spark plug, causing the boot to break off and making it difficult to remove the plug. If any contaminants are found on the coil, the coil should be cleaned with isopropyl alcohol. Regular dielectric compound should not be applied as the voltage breakdown is much lower at high temperatures than the compound applied from the manufacturer.

The Detroit Diesel coils, plugs, and extenders should be installed dry and clean to prevent spark flashover. The plug extenders should be inspected for signs of burning or deterioration that could lead to the spark jumping to ground.

Task H7 Inspect, test, and replace ignition system pick-up sensor or triggering device.

The CNG engine is electronically controlled. The engine utilizes sensors to determine the firing order and its position due to not having a mechanical type of distributor. Detroit Diesel calls its sensors Timing Reference Sensor (TRS) and Synchronous Position Sensor (SRS); Cummins calls its sensors Camshaft Speed/Position Sensor and Crankshaft Speed/Position Sensor. These sensors are responsible for transmitting engine speed and position to the ECM to determine the speed of the crankshaft and its position in relationship to the camshaft to fire the coils. A minimum gap between the sensor and the pulse wheel or reluctor must be maintained to prevent damage to the sensors. A maximum gap also must be maintained to assure a good signal from the target. The sensors are powered by 5 volts DC from the ECM. A signal is generated when a target, in the form of a tooth, passes in front of the sensor. The location and design of these sensors vary by manufacturer. When the key switch is turned on, the ECM monitors the sensors and is capable of returning faults for low, high voltage. If too many or too few targets are seen by the ECM during an engine revolution, it will return different DTCs for the appropriate symptom. The manufacturer's troubleshooting trees will assist the Technician in the steps necessary for diagnosing the defects associated with each DTC.

Task H8 Inspect, test, and replace ignition control module or controller.

Ignition control modules (ICMs) are used to amplify the ground signal to allow the COP to fire. If an engine has a misfiring cylinder, the ICM could be the cause. If the ECM does not log a diagnostic code, the coil can be swapped with an adjacent cylinder's coil and observed to see if the problem is still with the original cylinder or if it migrated with the coil swap. The manufacturer of the engine will usually ask that all pins, connectors, wiring, voltage and resistance be checked as well as the fuses before an ICM is replaced.

On the Cummins L GAS-Plus and the ISL G engines, the ICM amplifies the coil signal to a voltage as high as 300 volts to drive the coils to multiple spark discharges (MSDs). Because of the lean burn technology and the large spark plug gap employed by these engines, it is necessary to have multiple sparks during one burn cycle to ensure complete combustion.

5 Sample Test for Practice

Sample Test

Please note the letter and number in parentheses following each question. They match the task in Section 4 that discusses the relevant subject matter. You may want to refer to the overview using the cross-referencing key to help with questions posing problems for you.

1. Technician A says that if there are any exhaust leaks in the engine compartment, they must be repaired to prevent the possibility of exhaust from entering the passenger compartment. Technician B says the leak only needs to be repaired if the vehicle is equipped with a rear-roof-mounted HVAC. Which Technician is correct?
   A. A only
   B. B only
   C. Both A and B
   D. Neither A nor B (E5)

2. Technician A says that when an oil pump is disassembled, all mated surfaces must be checked for wear. Technician B says that because oil pump gears receive so much wear, the oil pump should routinely be replaced during an overhaul. Which Technician is correct?
   A. A only
   B. B only
   C. Both A and B
   D. Neither A nor B (D2)

3. Blue smoke emission from the tailpipe can be caused by all of the following EXCEPT
   A. Worn valve guides.
   B. Worn valve seals.
   C. Broken spark plug center electrode.
   D. Worn piston rings. (A5)

4. Which of the following tools is LEAST-Likely to reveal an out-of-round cylinder sleeve?
   A. An inside micrometer
   B. A telescoping gauge
   C. A feeler gauge
   D. A dial bore gauge (C3)

5. Technician A says the low-pressure fuel regulator is the component that regulates the fuel pressure from the high of 3600 down to around 100 psi. Technician B says the high-pressure fuel regulator is the component that regulates the high fuel pressure from 3600 down to around 100 psi. Which Technician is correct?
   A. A only
   B. B only
   C. Both A and B
   D. Neither A nor B (F2.1)

6. An NGV engine develops knock codes on hot days only. Technician A says the air intake temperature could be too high, causing auto ignition of the fuel. Technician B says the IAT sensor could be defective and reading the temperature of the air as being too low. This results in the ECM not commanding the cooling fan to turn on soon enough, causing the temperatures in the cylinders to be too high and auto ignition of the fuel.
   A. A only
   B. B only
   C. Both A and B
   D. Neither A nor B (E3)

7. The LEAST- Likely cause of a wiring harness failure is
   A. Too large a fuse protection installed for the wire diameter.
   B. A loose ground connection.
   C. Securing the harness with a insulated clamp.
   D. Using a zip tie to secure the harness to the frame. (A3)

8. An engine will not crank. Technician A says that low fuel pressure can be the cause. Technician B says that some modern NGV engines use programmable logic controls and ladder logic and diagrams must be available to be able to diagnose these vehicles.
   A. A only
   B. B only
   C. Both A and B
   D. Neither A nor B (A11)

9. To check a cylinder block deck for warpage, Technician A says to use a straightedge and feeler gauge. Technician B says to measure the deck around each cylinder using a dial indicator. Which Technician is correct?
   A. A only
   B. B only
   C. Both A and B
   D. Neither A nor B (C2)

10. A Technician is starting a fuel cylinder inspection. Technician A says the tank manufacturer's inspection guidelines are only needed if a questionable defect is observed. Technician B says the inspection tools are only needed if a questionable defect is observed. Which Technician is correct?
    A. A only
    B. B only
    C. Both A and B
    D. Neither A nor B (F1.1)

11. The vanes of a turbocharger compressor wheel are excessively worn. Technician A says that this could be a result of a loose or damaged hose between the air filter and the turbo. Technician B says that the aftercooler is not properly cooling the air before it enters the turbocharger, causing the compressor blades to become fatigued. Which Technician is correct?
    A. A only
    B. B only
    C. Both A and B
    D. Neither A nor B (E2)

12. Technician A says that NGV engine air cleaners will last longer due to the clean burning of Natural Gas. Technician B says that a manometer or magnehelic gauge is the only method of accurately measuring and recording if an air cleaner needs replacement. Which Technician is correct?
    A. A only
    B. B only
    C. Both A and B
    D. Neither A nor B (A7)

13. A customer has brought in a NGV that has an engine oil leak. Upon inspection, the lower half of the engine is covered in oil. Technician A says that the engine should be cleaned before an accurate diagnosis can be given. Technician B says that if the lower portion of the engine is leaking, it is Most-Likely the oil pan gasket. Which Technician is correct?
    A. A only
    B. B only
    C. Both A and B
    D. Neither A nor B (A2)

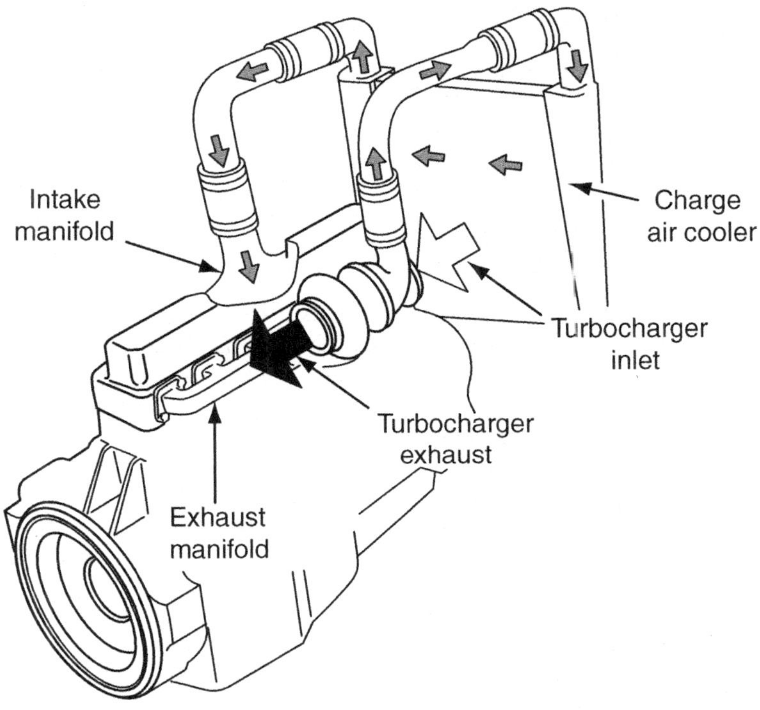

14. Referencing the figure above, what cools the boost air?
    A. Engine coolant in the intercooler
    B. Ram air through the charge air cooler
    C. Filtered intake air from the turbocharger
    D. Exhaust gas routed through the charge air cooler (E4)

15. Which of the following is LEAST-Likely to produce a loss of power in a CNG engine at full power?
    A. High exhaust back pressure
    B. High manifold boost pressure
    C. Oil and dirt in the air tubes or cooling fins of the aftercooler
    D. Waste gate stuck open (A12)

16. To inspect a cylinder head for cracks, all of the following may be necessary EXCEPT
    A. Using fine grit sandpaper to remove excess gasket material from the head to allow inspecting for visible cracks between cylinders and ports.
    B. Checking the torque of head bolts.
    C. Using magnetic crack detection to find small cracks.
    D. Removal of excess carbon buildup between the valves.
        A. A only
        B. B only
        C. Both A and B
        D. Neither A nor B (B3)

17. An NGV engine is being serviced. The Technician notices the engine oil dipstick is missing. A new dipstick is installed. After testing the engine performance, the Technician notices the dipstick has started to move out of the tube again. Technician A says that excessive crankcase pressure should be suspected and can be measured with a manometer. Technician B says the dipstick lock is defective and the dipstick should be wired down into place to prevent it from being pushed up and potentially lost again. Which Technician is correct?
    A. A only
    B. B only
    C. Both A and B
    D. Neither A nor B (A10)

18. A water pump mounting hole in the cylinder block has stripped threads. Technician A says that the block has to be replaced. Technician B says it may be possible to install a helicoil. Which Technician is correct?
    A. A only
    B. B only
    C. Both A and B
    D. Neither A nor B (B2)

19. Servicing of the fuel system is being discussed. Technician A says no special precautions need to be taken when draining the CNG fuel filters because the filters are equipped with a drain valve. Technician B says that when the engine is turned off, the engine depletes all of the fuel in the fuel filters; therefore the engine only needs to be turned off before changing the fuel filters. Which Technician is correct?
    A. A only
    B. B only
    C. Both A and B
    D. Neither A nor B (A6)

20. Spark plug tube adaptors are being discussed. Technician A says that before a spark plug tube adaptor can be removed from the engine only the radiator needs to be drained. Technician B says the cylinder head also needs to be drained. Which Technician is correct?
    A. A only
    B. B only
    C. Both A and B
    D. Neither A nor B (B5)

21. Which of the following would Most-Likely indicate an oil cooler failure?
    A. Low oil pressure
    B. High oil pressure
    C. Milky gray sludge in radiator
    D. High engine oil level (D4)

22. Technician A says bent valve train pushrods and tubes should be replaced, not straightened. Technician B says pushrods or tubes must be checked for ball socket integrity. Which Technician is correct?
    A. A only
    B. B only
    C. Both A and B
    D. Neither A nor B (B8)

23. Technician A says to clean and blow out the seal groove in a rocker housing cover with compressed air. Technician B says that when using silicone for the valve cover gasket, a light coat of oil in the seal groove makes the gasket easier to install. Which Technician is correct?
    A. A only
    B. B only
    C. Both A and B
    D. Neither A nor B (C1)

24. The certification label on a cylinder cannot be read because of its position between the roof and the tank. Technician A says the tank needs to be condemned. Technician B says the tank is acceptable if it can be repositioned so the label can be read.
    A. A only
    B. B only
    C. Both A and B
    D. Neither A nor B (F1.2)

25. Technician A says that when honing a cylinder liner, the hone should be rhythmically stroked up and down to provide a good crosshatch pattern. Technician B says once the honing has been completed, the liner should be washed with a soap solution to remove stone grit. Which Technician is correct?
    A. A only
    B. B only
    C. Both A and B
    D. Neither A nor B (C4)

26. An electric oil pressure gauge indicates zero pressure, but the engine is well lubricated. Which of the following is Most-Likely at fault?
    A. The oil passage to the sensor is open.
    B. The wrong engine oil was installed.
    C. The wiring to the sensor is open.
    D. The wrong sensor was installed. (D1)

27. While checking the oil level in the engine, the Technician notices evidence of coolant in the engine oil. All of the following could cause this EXCEPT
    A. A crack in the exhaust valve seat.
    B. A crack in the cylinder liner.
    C. An internal leak in the water pump of a gear train driven water pump.
    D. A leak in the head gasket. (B1)

28. When inspecting an oil filter housing or its mounting, which of the following is LEAST-Likely to be a procedure?
    A. Visually check for cracks
    B. Visually check gasket surface for nicks
    C. Magnaflux the housing to detect small cracks
    D. Visually inspect housing passageways for obstructions (D3)

29. An engine drops a valve 20 hours after an overhaul. Technician A says the probable cause was a missing or a defective valve seal. Technician B says the probable cause was improperly installed keepers. Which Technician is correct?
    A. A only
    B. B only
    C. Both A and B
    D. Neither A nor B (B7)

30. Technician A says that on a turbocharged CNG engine, the turbocharger is lubricated by engine oil and the center section is cooled by coolant. Technician B says that if the turbocharger shaft bearing fails, the engine oil and filters should be changed. Which Technician is correct?
    A. A only
    B. B only
    C. Both A and B
    D. Neither A nor B (D5)

31. A NGV engine is being tested for lack of power. Technician A says that a boost pressure leak could be the cause. Technician B says a stuck open waste gate could be the cause. Which Technician is correct?
    A. A only
    B. B only
    C. Both A and B
    D. Neither A nor B (A8)

32. A CNG engine develops too much manifold boost at high engine loads. Technician A says that the air filter is punctured. Technician B says that the wastegate could be malfunctioning. Which Technician is correct?
    A. A only
    B. B only
    C. Both A and B
    D. Neither A nor B (E1)

33. Technician A says that a customer's concerns must be listened to closely. Technician B says to always start with the easiest applicable test to perform when diagnosing an NGV vehicle. Which Technician is correct?
    A. A only
    B. B only
    C. Both A and B
    D. Neither A nor B (A1)

34. An NGV vehicle is brought to the shop with lack of power. A DTC is logged for high exhaust back pressure. Technician A says the waste gate could be stuck closed, causing too much pressure to be introduced into the exhaust system from overboosting. Technician B says the catalyst should be checked for high restriction using a magnehelic gauge calibrated in inches of water and the reading should not exceed 2.4" of water. Which Technician is correct?
    A. A only
    B. B only
    C. Both A and B
    D. Neither A nor B (A9)

35. The fuel shutoff solenoid, also known as the Title 13 valve, is being checked to make sure it is functioning correctly. Technician A says it can be checked by disconnecting the power connector from it while the engine is running causing the engine to shut off when the fuel is depleted. Technician B says it can also be checked by turning off the ignition switch while the engine is running. Which Technician is correct?
    A. A only
    B. B only
    C. Both A and B
    D. Neither A nor B (F3.1)

36. Exhaust brakes utilize all of the following to slow a vehicle down EXCEPT
    A. A restriction of the flow of exhaust gases.
    B. Back pressure in the intake system of the engine.
    C. The compression stroke of the pistons pushing against the pressurized exhaust gases.
    D. The exhaust stroke trying to push the spent exhaust gases out of the exhaust. (E7)

37. A CNG transit bus is being diagnosed for a no-start condition at the layover zone. The engine cranks but will not run. Technician A says the CEL and SEL indicators should be checked to make sure the ECM is powering up when the ignition switch is turned on. Technician B says the fuel ¼ turn shutoff valve needs to be checked before anything else because the Technician that serviced the vehicle may have turned the fuel off. Which Technician is correct?
    A. A only
    B. B only
    C. Both A and B
    D. Neither A nor B (F3.2)

38. A manual ¼ turn shutoff valve is being discussed. Technician A says the handle should be horizontal when the fuel is turned off. Technician B says the handle should be vertical when the fuel is turned on. Which Technician is correct?
    A. A only
    B. B only
    C. Both A and B
    D. Neither A nor B (F1.3)

39. Technician A says that valve crossheads should be Magnafluxed to check for cracks. Technician B says the valve crossheads should be checked for squareness. Which Technician is correct?
    A. A only
    B. B only
    C. Both A and B
    D. Neither A nor B (B6)

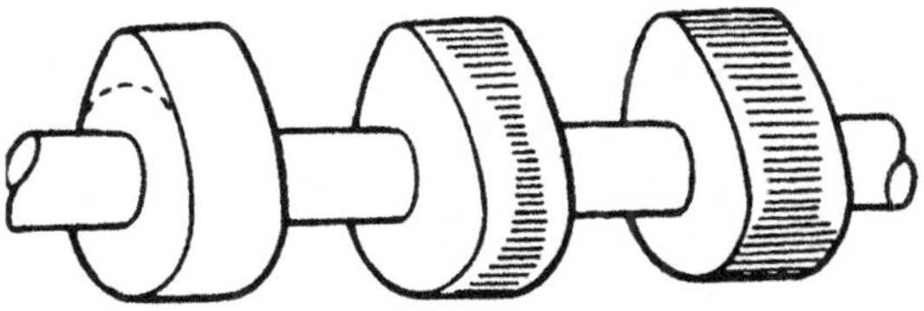

40. Two checks of the heel-to-toe measurement on the camshaft in the figure above provide different readings, but they are still within specifications. Technician A says to check for grooving of the cam lobe. Technician B says that providing the readings are within specifications, no further checks are required. Which Technician is correct?
    A. A only
    B. B only
    C. Both A and B
    D. Neither A nor B (C7)

41. A loud squealing sound is heard randomly especially going around corners or when de-accelerating. Technician A says the A/C compressor belts should be checked. Technician B says that if the vehicle is equipped with a belt driven hydraulic cooling fan, the belts should be inspected for proper tension and to ensure that they are not glazed. Which Technician is correct?
    A. A only
    B. B only
    C. Both A and B
    D. Neither A nor B (A4)

42. Which of the following is LEAST-Likely to be revealed during inspection and testing of valve springs?
    A. Springs that produce harmonic distortion
    B. Springs with uneven free height
    C. Cracks in the spring coil
    D. Springs that are not square (B4)

43. A vibration that has an occurrence rate equal to engine speed is LEAST-Likely to be caused by
    A. An out-of-balance turbocharger.
    B. A bent connecting rod.
    C. An out-of-balance crankshaft.
    D. A dented vibration damper. (A13)

44. The LEAST-Likely cause of cam follower roller pitting is
    A. Shutting the engine off without a five-minute cool-down period.
    B. Driving the engine immediately after start-up with no idle period to build up oil pressure.
    C. Dirty oil.
    D. High engine mileage. (B9)

45. Surges in coolant level could be caused by all of the following EXCEPT
    A. A loose water pump belt.
    B. A defective spark plug.
    C. A restricted radiator.
    D. A blown head gasket. (A14)

46. Technician A says that an EGR cooler uses coolant to cool the exhaust gases before they are introduced back into the cylinder to help reduce the formation of NOx. Technician B says that an engine with EGR requires a cooling system with a higher heat rejection than a non-EGR engine. Which Technician is correct?
    A. A only
    B. B only
    C. Both A and B
    D. Neither A nor B (E6)

47. Troubleshooting has indicated that a thermister-type coolant temperature sensor is not returning a signal voltage. Which of the following should be done first?
    A. Check the reference voltage (V-Ref) value
    B. Consult the manufacturer's service literature
    C. Simulate V-Ref to check for signal voltage
    D. Check thermister internal resistance using an ohmmeter (D8)

48. A CNG vehicle's fuel tanks are being refueled. Technician A says the tanks need to be fueled in increments and leak checked at each increment. Technician B says that frosting at the fuel solenoids is normal during refueling. Which Technician is correct?
    A. A only
    B. B only
    C. Both A and B
    D. Neither A nor B (F1.8)

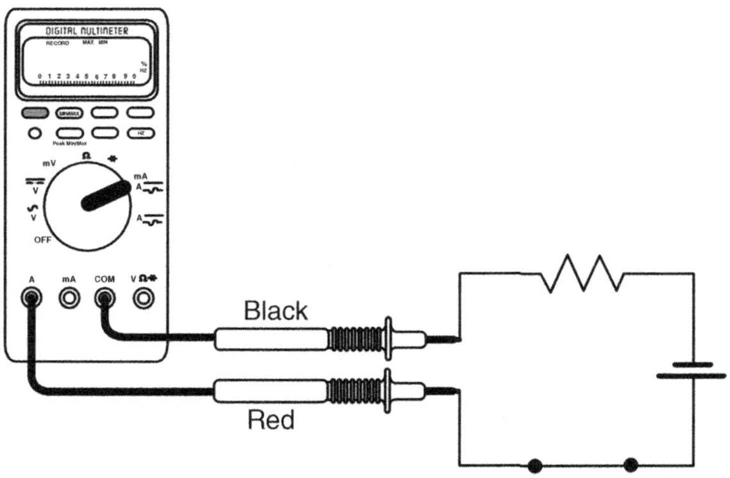

49. In the figure above, what is about to be measured in the circuit shown?
    A. Current flow
    B. Voltage drop
    C. Circuit resistance
    D. Wattage consumed (G7)

50. The LEAST-Likely condition to cause a cylinder to misfire is
    A. Handling the spark plug with bare fingers.
    B. A contaminated spark plug tube.
    C. A cracked coil extension tube.
    D. Cleaning the spark plug with isopropyl alcohol. (H1)

51. Technician A says that petroleum jelly can be used on battery cable clamps to prevent corrosion. Technician B says that it is permissible to use protective pads and a little grease to prevent corrosion on battery cable clamps. Which Technician is correct?
    A. A only
    B. B only
    C. Both A and B
    D. Neither A nor B (G5)

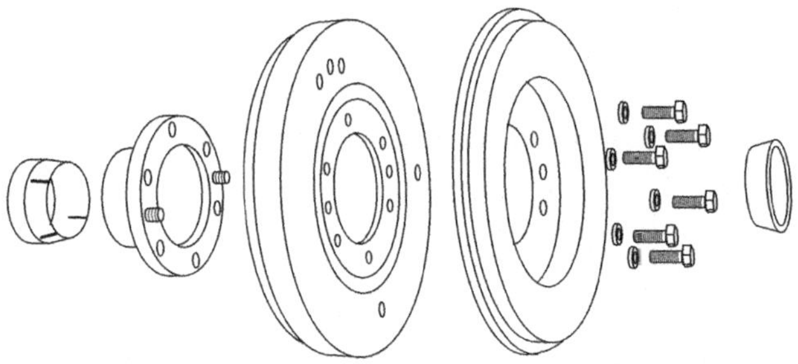

52. Which of the following properly describes the component in the figure above?
    A. Flywheel assembly
    B. Vibration damper assembly
    C. Timing gear assembly
    D. Water pump assembly (C16)

53. Technician A says that a radiator pressure cap with a 9 stamped on it should open to release pressure at between 8 psi (55 kPa) to 10 psi (69 kPa). Technician B says that the same cap should open at about −0.6 psi (4.3 kPa) differential (vacuum) pressure. Which Technician is correct?
    A. A only
    B. B only
    C. Both A and B
    D. Neither A nor B (D13)

54. All of the following will cause uneven wear on the main bearings or stress fractures on the crankshaft EXCEPT
    A. Full load operation in excess of eight hours.
    B. An unbalanced vibration damper.
    C. An incorrectly installed flywheel housing.
    D. Loose torque converter bolts. (C14)

55. Technician A says the piston ring end gap should be checked on the piston. Technician B says the piston ring end gap is checked with only the rings in the cylinder. Which Technician is correct?
    A. A only
    B. B only
    C. Both A and B
    D. Neither A nor B (C13)

56. Which of the following SAE acronyms is used to identify a major chassis electronic system with processing capability such as the engine or transmission?
    A. MID—message identifier
    B. SID—subsystem identifier
    C. PID—parameter identifier
    D. FMI—fault mode indicator (A16)

57. Technician A says that all gears in the timing gear train must be inspected for tooth wear, including the crankshaft gear. Technician B says that a slight roll or lip on each gear tooth is acceptable, because gears will normally mate this way during the first 500 miles of operation. Which Technician is correct?
    A. A only
    B. B only
    C. Both A and B
    D. Neither A nor B (C10)

58. CNG high-pressure fuel lines are being discussed. Technician A says the lines are made of tempered high-strength steel. Technician B says that if there is a scratch on the line where a fitting is going to be placed, the scratch needs to be buffed out with a ScothBrite pad to prevent a leak from the scratch. Which Technician is correct?
    A. A only
    B. B only
    C. Both A and B
    D. Neither A nor B (F1.4)

59. Technician A says back-flushing the radiator of NGV transit buses should be accomplished annually. Technician B says air locks usually occur when the cooling system is filled with the bleeder valves open. Which Technician is correct?
    A. A only
    B. B only
    C. Both A and B
    D. Neither A nor B (D10)

60. The ITT high-pressure regulator is being discussed. Technician A says that the high-pressure regulator setting can be adjusted by the Technician in the field. Technician B says coolant is routed through the high-pressure regulator. Which Technician is correct?
    A. A only
    B. B only
    C. Both A and B
    D. Neither A nor B (F1.5)

Voltage and Temperature Chart

Temperature	Minimum voltage
21°C (70°F & above)	9.6
10°C (50°F)	9.4
−1°C (30°F)	9.1
−10°C (15°F)	8.8
−18°C (0°F)	8.5

61. As shown in the chart above, what would be the minimum allowable voltage during a battery load test if the temperature is 50° F (10° C)?
    A. 9.6
    B. 9.4
    C. 9.1
    D. 9.8 (G2)

62. A wire from the ICM to the coil is being checked, end to end, to make sure that it is good. A resistance reading of .9 Ω is found. Technician A says the wire or pins are defective and needs to be repaired. Technician B says the wire is ok because the maximum reading is up to 10 Ω. Which Technician is correct?
    A. A only
    B. B only
    C. Both A and B
    D. Neither A nor B (H3)

63. Technician A says that the belt tensioner does not need to be replaced unless the tensioning spring is broken. Technician B says that worn pulleys can cause the belt to prematurely wear out. Which Technician is correct?
    A. A only
    B. B only
    C. Both A and B
    D. Neither A nor B (A19)

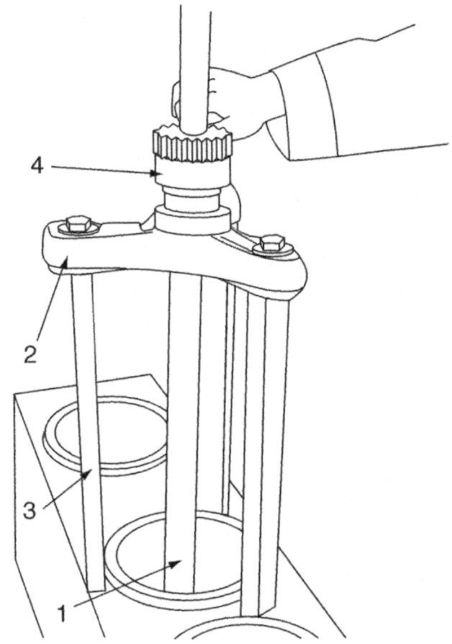

64. As shown in the figure above, Technician A says that puller adapter plates or shoes must fit snugly in sleeves to prevent cocking in the block. Technician B says that if puller adapter plates are larger than the sleeve OD, they are better because they cannot slip. Which Technician is correct?
    A. A only
    B. B only
    C. Both A and B
    D. Neither A nor B (C5)

65. The Cummins COP coils are being discussed. Technician A says that dielectric compound must be applied to the spark plug boots before installing onto the spark plug. Technician B says the boots come prelubricated from the manufacturer if they need to be lubricated with dielectric and must be installed as is from the container. Which Technician is correct?
    A. A only
    B. B only
    C. Both A and B
    D. Neither A nor B (H6)

66. Technician A says that a flywheel ring gear is heat-shrink-fitted to the flywheel. Technician B says that if ring gear teeth are damaged, the flywheel assembly must be replaced. Which Technician is correct?
    A. A only
    B. B only
    C. Both A and B
    D. Neither A nor B (C18)

67. Technician A says to replace the fan if the blades are cracked. Technician B says bent fan blades can cause engine vibration. Which Technician is correct?
    A. A only
    B. B only
    C. Both A and B
    D. Neither A nor B (D14)

68. To determine battery potential, which measuring instrument is used?
    A. Refractometer
    B. Voltmeter
    C. Ohmmeter
    D. Ammeter (G1)

69. The firing order of an inline, six-cylinder, four-stroke cycle CNG engine is 153624. Which of the following would be correctly described as companion cylinders?
    A. 1 and 5
    B. 4 and 5
    C. 3 and 6
    D. 5 and 2 (B10)

70. A vehicle comes to the shop with a complaint of screeching belts. Technician A says it probably is just a matter of readjusting the belt tension properly. Technician B says the belts should be checked to see if they have become glazed due to prolonged slipping and becoming overheated. Which Technician is correct?
    A. A only
    B. B only
    C. Both A and B
    D. Neither A nor B (D7)

71. A CNG engine misfire in one cylinder is being discussed. Technician A says the spark plug condition and gap should be checked before condemning the ICM. Technician B says the coil of the misfiring cylinder should be swapped with a known good cylinder's coil to see if the problem follows the suspected coil. Which Technician is correct?
    A. A only
    B. B only
    C. Both A and B
    D. Neither A nor B (H8)

72. When using jumper cables to start a vehicle with a dead battery, always
    A. Connect (+) positive to (−) negative to create a series-aiding circuit.
    B. Connect (+) positive to (+) positive and (−) negative to (−) negative to create a parallel circuit.
    C. Connect the (−) negative cable first and the (+) positive cable last.
    D. Electronically ground both vehicles before making any connections. (G4)

73. Technician A says that if the pressure in the radiator is too great, the pressure cap will relieve the excess pressure and fluid to an expansion tank. Technician B says that most radiator caps have a release pressure of 15 psi or lower. Which Technician is correct?
    A. A only
    B. B only
    C. Both A and B
    D. Neither A nor B (D15)

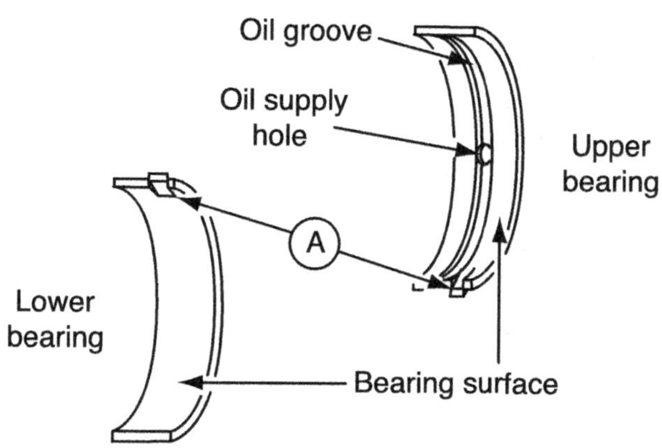

74. The notches in the bearing shell at points A in the figure above are used
    A. To direct the oil evenly around the bearing.
    B. To indicate a defect in the bearing shell
    C. To indicate this is an oversized bearing shell
    D. To position the bearing shell in the bearing bore (C9)

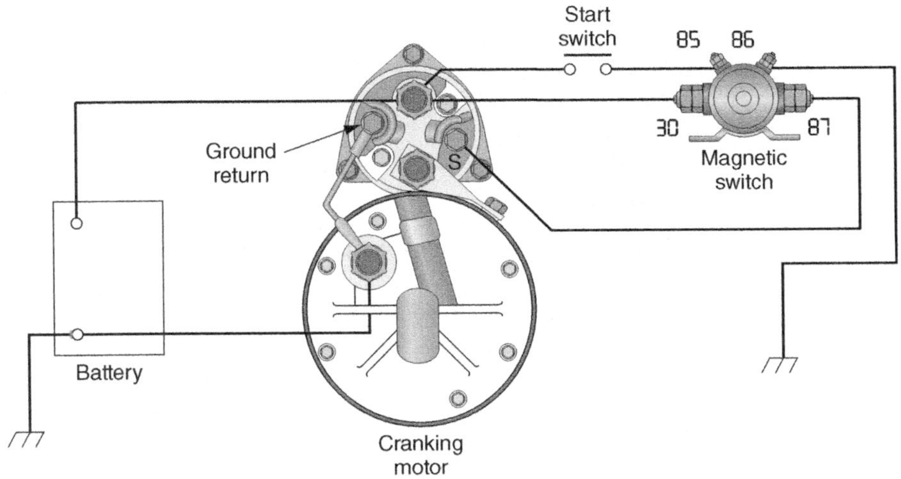

75. A remote start switch is being tested in the figure above. The Most-Likely positions of the DVOM test leads are
   A. Terminal 85 to 30.
   B. Terminal 86 to 87.
   C. Terminal 86 to 30.
   D. Terminal 85 to chassis ground. (G6)

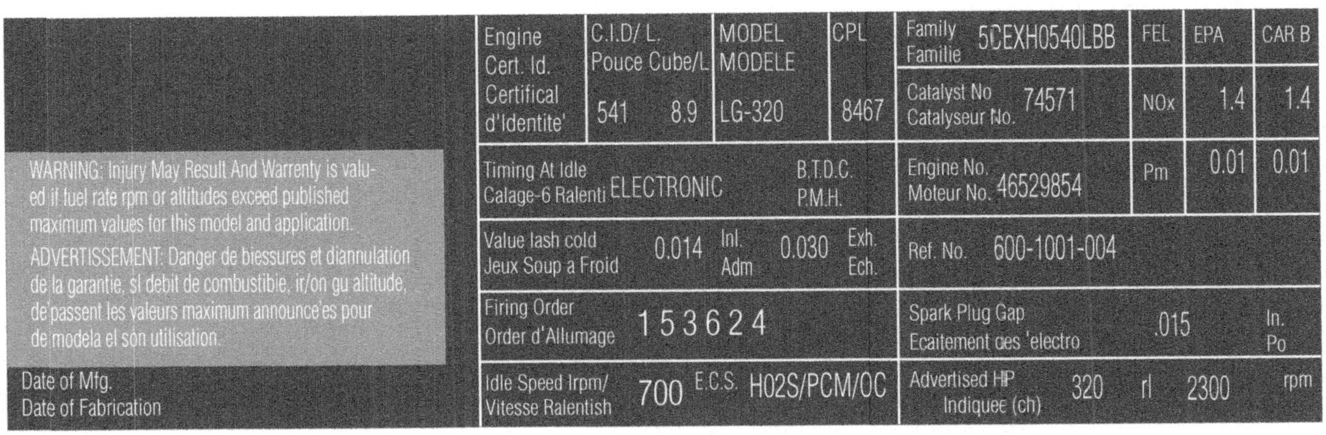

76. All of the following information will be found on the engine certification and emission decals EXCEPT
   A. Valve lash.
   B. Timing specifications.
   C. Oil pressure specifications.
   D. Emission certification level, date and certifying agencies. (A17)

77. Technician A says worn camshaft bushings can cause low oil pressure. Technician B says camshaft bushings can be reinstalled after being removed with the correct driver. Which Technician is correct?
   A. A only
   B. B only
   C. Both A and B
   D. Neither A nor B (C6)

78. Technician A says that CNG engine oil will become black with use. Technician B says that CNG engine oil is rated 15W-40 GEO with special additives for particular engine requirements. Which Technician is correct?
   A. A only
   B. B only
   C. Both A and B
   D. Neither A nor B (D6)

79. All of the following actions should be followed when replacing the secondary fuel filter EXCEPT
   A. Turn off the ¼ turn fuel valve to isolate the fuel filter and pressure regulating components from the rest of the system.
   B. Run the engine until it shuts off after the fuel is depleted.
   C. Check the high- and low-pressure gauges to make sure they both read 0 psi.
   D. Lubricate the O-ring with engine oil before installing the housing on the filter head. (F2.2)

80. Spark plugs are being discussed. Technician A says to always set the gap a few thousandths of an inch larger than specified to increase the intensity of the spark. A larger gap needs more KV to jump the gap and results in a better burning of the air/fuel mixture. Technician B says the gap must always be set a few thousandths of an inch smaller than specified to increase the life of a spark plug. As the electrodes wear, the gap will eventually increase to specifications increasing the life of the plugs. Which Technician is correct?
   A. A only
   B. B only
   C. Both A and B
   D. Neither A nor B (H4)

81. When checking for excess resistance in the starting circuit, Technician A says to use an ohmmeter to check the resistance between (−) negative cable and ground while cranking the starter. Technician B says to check the voltage drop between battery positive cable and the starter terminal stud while the starter is cranking. Which Technician is correct?
   A. A only
   B. B only
   C. Both A and B
   D. Neither A nor B (G8)

82. The LEAST-Likely reason to defuel a CNG cylinder is
   A. To replace a PRD.
   B. To replace a manual shutoff valve.
   C. To perform a cylinder inspection.
   D. To replace the fuel cylinder. (F1.7)

83. A CNG engine has spun a crankshaft bearing and the crankshaft journal is scored. Technician A says that the crankshaft should be reconditioned by metalizing and machining to original size. Technician B says the journal could be repaired by grinding all the main journals and installing undersize bearings. Which Technician is correct?
   A. A only
   B. B only
   C. Both A and B
   D. Neither A nor B (C8)

84. A CNG engine fails to reach operating temperature. Technician A says that a leaking thermostat seal could be the cause. Technician B says that a loose fan drive belt could be the cause. Which Technician is correct?
   A. A only
   B. B only
   C. Both A and B
   D. Neither A nor B (D9)

85. Which of the following would be the best source for the most current service information?
    A. Technical service bulletins
    B. Service manuals
    C. Parts manuals
    D. Operators manuals (A18)

86. Spark plugs are being installed on a Cummins CNG engine. Technician A says the boots come prelubricated with the required type and amount of dielectric compound from the manufacturer and should be installed as they come from the container. Technician B says the spark plug outer ceramic area needs to be cleaned with alcohol and a clean towel to prevent the boots from sticking to the spark plugs preventing the boots from being difficult to remove at the next service interval. Which Technician is correct?
    A. A only
    B. B only
    C. Both A and B
    D. Neither A nor B (H5)

87. A CNG engine logs a DTC for low oil pressure at idle only. The LEAST-Likely cause is
    A. Partially stuck-open oil pressure regulator valve.
    B. Engine oil level too high.
    C. Diluted or contaminated engine oil.
    D. Oil becoming too hot due to a clogged oil cooler. (A15)

88. A camshaft fails a visual inspection of the cam lobes due to pitting. Technician A says damaged camshaft lobe surfaces can be reground to factory specifications. Technician B says flat spots on the lobes can be silver-soldered and machined to original specifications. Which Technician is correct?
    A. A only
    B. B only
    C. Both A and B
    D. Neither A nor B (B11)

89. Cracks are present in the piston skirt. Technician A says to check the connecting rod and bottom of the cylinder liner for damage caused by improperly installed connecting rods. Technician B says to check for insufficient piston-to-cylinder liner clearance. Which Technician is correct?
    A. A only
    B. B only
    C. Both A and B
    D. Neither A nor B (C11)

90. A CNG engine's ignition system is being discussed. Technician A says the triggering devices for the ignition system are always placed on the flywheel housing to monitor the position of the flywheel to determine the position of the crankshaft. Technician B says the ignition system's triggering devices are housed in a distributor. Which Technician is correct?
    A. A only
    B. B only
    C. Both A and B
    D. Neither A nor B (H7)

91. Technician A says not to charge a lead-acid battery if the built-in hydrometer is clear or yellow. Technician B says that the battery is charged when the built-in hydrometer eye is green. Which Technician is correct?
    A. A only
    B. B only
    C. Both A and B
    D. Neither A nor B (G3)

92. The LEAST-Likely cause of stuck rings in the ring grooves of a CNG engine is
    A. Overfueling.
    B. Overheating.
    C. Insufficient lubrication.
    D. Cold running engine due to a stuck-open thermostat. (C12)

93. If a CNG engine develops low compression in a cylinder, the LEAST-Likely cause is
    A. Carbon buildup in the combustion chamber.
    B. Worn exhaust valves and/or seats.
    C. Worn compression rings.
    D. A cracked piston. (A20)

94. The Technician is troubleshooting a CNG engine with a complaint of a misfiring cylinder. The LEAST-Likely action to be taken is
    A. Reset the AFR table before continuing the diagnostics of the engine.
    B. Scan for Diagnostic Trouble Codes with the appropriate software.
    C. Replace the spark plugs and road test the vehicle.
    D. Speak with the operator and look up the past repair history of the vehicle to determine what repairs were performed previously. (H2)

95. Which of the following would be the correct tool to check flywheel housing alignment?
    A. An alignment machine
    B. A micrometer
    C. A dial indicator
    D. A straightedge and feeler gauge (C17)

96. Which of the following is LEAST-Likely to cause a cooling system failure?
    A. A water pump drive belt having two cracks within any two-inch section.
    B. An occasional drop of coolant from the pump weep hole of the water pump seal.
    C. A loose clamp on a heater hose.
    D. Rust or scum floating in the radiator. (D12)

97. A vehicle is brought to the shop for servicing with a complaint of an intermittent trace gas detection sensor light coming on. Technician A says an electronic methane leak detector should be used in the area to make sure there is no leak. Technician B says that if no leaks are detected or found at the zone of the trace light, the ITT high-pressure regulator PRD line should be disconnected and tested to ensure the PRD is not starting to leak causing the trace light on the gas detection panel to illuminate. Which Technician is correct?
    A. A only
    B. B only
    C. Both A and B
    D. Neither A nor B (F1.6)

98. When checking the coolant condition with an SCA test strip, the Technician finds that the coolant is extremely overconditioned. What should the Technician do?
    A. Add more antifreeze to increase the SCA.
    B. Continue to run the engine until the next PMI.
    C. Drain the entire coolant system and add the proper SCA mixture.
    D. Run the vehicle with no SCA additives until the next PMI. (D11)

99. Technician A says that end gaps on piston rings must be offset to limit blowby. Technician B says that when installing a piston and connecting rod, studs should be covered to prevent damage to the crankshaft. Which Technician is correct?
    A. A only
    B. B only
    C. Both A and B
    D. Neither A nor B (C15)

6 Additional Test Questions for Practice

Additional Test Questions

Please note the letter and number in parentheses following each question. They match the task in Section 4 that discusses the relevant subject matter. You may want to refer to the overview using the cross-referencing key to help with questions posing problems for you.

1. All of the following symptoms are indicators of high air inlet restrictions EXCEPT
   A. Slow acceleration
   B. Low power
   C. Collapsing inlet rubber boots
   D. Blue smoke (A7)

2. Technician A says that excessive crankcase pressure can be caused from a plugged crankcase breather. Technician B says that all CNG engines use EGR to control emissions and will have high crankcase pressure due to the higher combustion chamber pressures causing more gases to pass by the compression rings. Which Technician is correct?
   A. A only
   B. B only
   C. Both A and B
   D. Neither A nor B (A10)

3. A cooling system is pressurized with a pressure tester to locate a coolant leak. After 15 minutes, the test gauge has dropped from 9 to 5 psi, and there are no visible signs of coolant leaks in the engine compartment. Technician A says the engine may have a leaking head gasket. Technician B says that the heater core may be leaking. Which Technician is correct?
   A. A only
   B. B only
   C. Both A and B
   D. Neither A nor B (A2)

4. A transit bus is equipped with a logic-controlled driver's display panel. All the lights are erratic including the low oil pressure indicator light. The ECM has not turned on the check engine lamp and all of the accessories are still functioning, including the HVAC. Which of the following is the Most-Likely cause?
   A. A faulty dash multiplexing controller signal
   B. A faulty alternator
   C. A faulty oil pressure sensor
   D. A faulty ECM (D1)

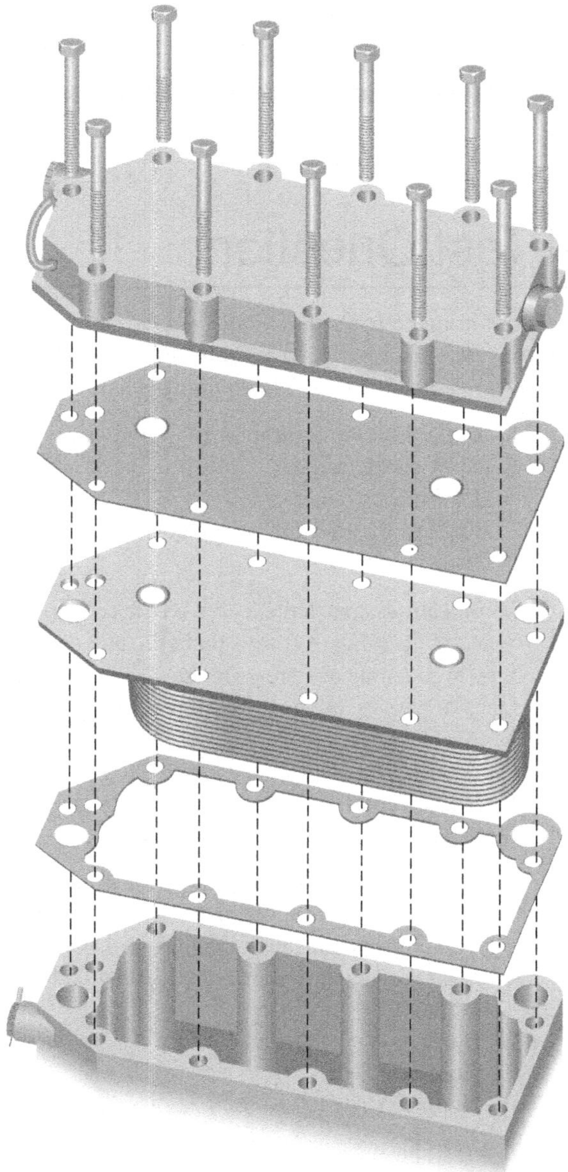

5. The figure above shows what part of the lubrication system?
   A. Oil filter
   B. Oil cooler
   C. Filter housing
   D. Oil pump (D4)

6. A customer brings a vehicle that has drivability issues. There are no indicator lights on the dash but the operator complains of a warning buzzer. Technician A says that the ignition switch should be switched off and turned back on while observing the CEL and SEL indicators to ensure they work. Technician B says the DDR or diagnostic software will allow the Technician to power up the CEL and SEL indicators through the ECM to test them. Which Technician is correct?
   A. A only
   B. B only
   C. Both A and B
   D. Neither A nor B (A1)

7. On a CNG engine with a charge air cooler, the Technician notices a soot-type material in the intake manifold. What should the Technician check first?
   A. CAC
   B. Air intake piping
   C. Intake manifold
   D. Head gasket (E2)
8. Technician A says the ECM does not have any control over the temperature of the air as it passes through CAC of an electronically controlled CNG engine. Technician B says the coolant is what cools the air passing through the CAC.
   A. A only
   B. B only
   C. Both A and B
   D. Neither A nor B (E4)
9. Technician A says oil leaks are always caused by gasket material failure. Technician B says that oil leaks can sometimes be located by adding a dye to engine oil that is visible under ultraviolet light. Which Technician is correct?
   A. A only
   B. B only
   C. Both A and B
   D. Neither A nor B (A2)
10. Which of the following is LEAST-Likely to provide satisfactory results in an attempt to identify an engine noise?
   A. Listening to the exhaust to determine which cylinder is causing the noise
   B. Using a stethoscope
   C. Manually disconnecting the power at each coil for coil-over plugs or removing the individual spark plug wires
   D. Performing an electronic cylinder cut-out test using the diagnostic software (A4)
11. The Most-Likely method to determine when a paper-type air cleaner element needs to be replaced is
   A. The use of the indicator mounted on the housing or intake piping.
   B. Poor engine performance.
   C. The use of a magnehelic gauge or a water manometer.
   D. Oil change intervals. (A7)
12. A turbo charger is being replaced. The compressor wheel blades are damaged. The LEAST-Likely cause of this condition is
   A. Torn air cleaner element.
   B. Loose V-band at the turbine housing.
   C. Disconnected air intake hose.
   D. Mishandling during the installation of the turbocharger. (E3)
13. During a visual inspection, a fuel cylinder is found to be missing the proper labeling. Technician A says the cylinder needs to be defueled and condemned. Technician B says the manufacturer of the cylinder needs to be contacted with the serial number of the tank to obtain a new label to be installed on the tank. Which Technician is correct?
   A. A only
   B. B only
   C. Both A and B
   D. Neither A nor B (F1.1)

14. A NGV engine is being diagnosed with the diagnostic software. The engine is off with the key switch in the on position. The air intake manifold pressure is displayed as 14.7 psi. Technician A says this is a normal condition because the reading is absolute pressure and therefore displays atmospheric pressure plus or minus the manifold pressure. Technician B says the intake manifold pressure sensor is defective because the gauge should be displaying the pressure as 0 until the engine starts to run. Which Technician is correct?
    A. A only
    B. B only
    C. Both A and B
    D. Neither A nor B (A8)

15. A center-cooled turbocharger has catastrophically failed damaging the cooling portion allowing the coolant to leak out. Technician A says the CAC needs to be cleaned out to ensure that no oil or coolant is left inside that could damage the engine when it is restarted. Technician B says the engine oil and the filters should be replaced to remove as much of the coolant or metal particles as possible from the engine's lubrication system. Which Technician is correct?
    A. A only
    B. B only
    C. Both A and B
    D. Neither A nor B (E4)

16. Technician A says that a worn CNG engine lube pump may cause low oil pressure. Technician B says that CNG engine oil pumps are positive displacement design. Which Technician is correct?
    A. A only
    B. B only
    C. Both A and B
    D. Neither A nor B (D2)

17. Technician A says that fuel filters used in CNG systems are made of pleated paper to remove oil and contaminants that could damage engine sensors or control valves. Technician B says the fuel filters used in CNG systems are a coalescing type to remove any oil and contaminants that could damage engine sensors or control valves. Which Technician is correct?
    A. A only
    B. B only
    C. Both A and B
    D. Neither A nor B (F2.1)

18. An NGV is brought to the shop. An odor of mercaptan is detected. Technician A says the odor is normal after refueling. Technician B says a special dye can be added to the fuel to detect the leak as used for the HVAC system. Which Technician is correct?
    A. A only
    B. B only
    C. Both A and B
    D. Neither A nor B (A2)

19. Technician A says to use a straightedge across the exhaust manifold surfaces to align separate cylinder heads on the engine block. Technician B says to blow out head bolt holes to remove any foreign material and to ensure the bolts do not bottom out prematurely and cause improper clamping of the components. Which Technician is correct?
    A. A only
    B. B only
    C. Both A and B
    D. Neither A nor B (B2)

20. On a vehicle with an electrical oil pressure gauge, all of the following will cause inaccurate readings EXCEPT
    A. Low voltage to the gauge.
    B. A short in the wire from the transducer.
    C. Low oil pressure.
    D. A defective transducer. (D1)

21. Dielectric compounds are being discussed. Technician A says the external ceramic of spark plugs of CNG engines should be cleaned with isopropyl alcohol and a clean towel before being installed in the engine. Technician B says the CNG engine spark plug boots come prelubricated from the factory and are ready to be installed from the container. Which Technician is correct?
    A. A only
    B. B only
    C. Both A and B
    D. Neither A nor B (H1)

22. An engine produces a deep-sounding thumping noise that is especially noticeable at slow rpm and high engine loads. The Most-Likely cause is
    A. A knock caused by worn crankshaft main bearings.
    B. A bottom-end knock caused by a failed rod big end journal.
    C. A knock caused by a worn camshaft bearing.
    D. A knock caused by a defective valve rocker arm. (A4)

23. A recently overhauled engine suddenly develops cylinder hydraulic lock. After clearing the liquid from the cylinder, inspecting the heads, and replacing the head gaskets, the problem reoccurs. Technician A says the cylinder liner could have a pinhole in the piston compression ring travel area. Technician B says the cylinder head may have a hairline crack. Which Technician is correct?
    A. A only
    B. B only
    C. Both A and B
    D. Neither A nor B (B3)

24. Technician A says that testing of the oil cooler core is to be done at PM intervals. Technician B says that testing of the oil cooler core is to be done as part of an engine overhaul. Which Technician is correct?
    A. A only
    B. B only
    C. Both A and B
    D. Neither A nor B (D4)

25. The Most-Likely cause of spark plug having high KV is
    A. Too large of a gap between the electrodes of the spark plug.
    B. Tracking on the center electrode of the spark plug.
    C. A defective ICM.
    D. A defective coil. (H2)

26. Technician A says the cylinder head can be resurfaced as much as necessary to bring it back into flatness specifications. Technician B says the cylinder head only warps lengthwise, not side to side. Which Technician is correct?
    A. A only
    B. B only
    C. Both A and B
    D. Neither A nor B (B1)

27. An NGV transit bus with the Dinex I/O logic controls is brought to the shop for not cranking. The starter solenoid is heard clicking. Technician A says the starter is the Most-Likely problem and it should be replaced. Technician B says the fuel door should be checked to make sure it is closed and the safety switch is sending the ground input to the I/O logic controls. Which Technician is correct?
    A. A only
    B. B only
    C. Both A and B
    D. Neither A nor B (A11)

28. A small crack is detected in a wet sleeve around the counterbore flange. Technician A says to use heliarc welding to seal the crack. Technician B says the block needs to have the counterbore recut. Which Technician is correct?
    A. A only
    B. B only
    C. Both A and B
    D. Neither A nor B (C4)

29. A Detroit Diesel Series 50G CNG engine's coil power is being discussed. Technician A says that 12 volt is applied to the coils whenever the ignition is turned on directly from the batteries. Technician B says that 12 volt is applied to the coil by the ICM when the coil has to supply the spark voltage to the plugs. Which Technician is correct?
    A. A only
    B. B only
    C. Both A and B
    D. Neither A nor B (H3)

30. Technician A says that external gear–type engine oil pumps have to be submerged in oil to be able to pump oil. Technician B says internal gear–type oil pumps have to be submerged in oil to be able to pump oil. Which Technician is correct?
    A. A only
    B. B only
    C. Both A and B
    D. Neither A nor B (D2)

31. Upon inspecting the coolant sensor connector, the Technician finds the lock and seal are missing. Technician A says the connector assembly must be replaced with a new one. Technician B says installing a new seal and a zip tie to hold it in place is sufficient. Which Technician is correct?
    A. A only
    B. B only
    C. Both A and B
    D. Neither A nor B (A3)

32. Technician A says before installation, some manufacturers require the head bolts to be cleaned and inspected for erosion or pitting as well as to their lengths. Technician B says that torque-to-yield bolts need to be replaced with new ones because they are not reusable. Which Technician is correct?
    A. A only
    B. B only
    C. Both A and B
    D. Neither A nor B (B2)

33. The Most-Likely material to clean the spark plugs of a CNG engine before installing them is
    A. Contact cleaner.
    B. Brake cleaner.
    C. Isopropyl alcohol.
    D. Water. (H4)

34. Technician A says that when venting a cylinder, compressed air should be used. Technician B says that water can be used. Which Technician is correct?
    A. A only
    B. B only
    C. Both A and B
    D. Neither A nor B (F1.7)

35. An operator comes to the shop with a complaint of an intermittent low coolant signal on the dash. Technician A says the low coolant probe is the defect and should be replaced because when low coolant is sensed by the engine, the low coolant error stays active and will not turn off until the engine is turned off and back on again. Technician B says the coolant level needs to be checked and corrected if needed before making any repairs, and if the coolant is low the cooling system should be pressure tested to determine the problem leak.
    A. A only
    B. B only
    C. Both A and B
    D. Neither A nor B (D1)

36. Which of the following exhaust conditions is the Most-Likely indication of an engine misfire?
    A. Blue-colored smoke from the tailpipe
    B. Excessive rattling of the exhaust system
    C. Cold tailpipe exhaust temperatures
    D. A stumble or miss from the engine (A5)

37. All of the following can cause flashover of the spark plug EXCEPT
    A. An excessive spark plug gap (too wide).
    B. A spark plug contaminated with oil residue from a dirty spark plug socket at installation.
    C. Cleaning the spark plug with isopropyl alcohol before installation.
    D. Handling a spark plug with bare hands during gapping and installation. (H5)

38. Type IV cylinders are being refueled after they were emptied to replace the fuel tank shutoff solenoids. Technician A says small popping sounds coming from the cylinders while refueling them is normal because the liner is settling back into place. Technician B says it could be possible to detect the smell of mercaptan and for an electronic leak detector to detect trace amounts of methane coming from a Type IV tank because of the permeable liner used in these tanks. Which Technician is correct?
    A. A only
    B. B only
    C. Both A and B
    D. Neither A nor B (F1.8)

39. All of the following are causes of catalyst contamination EXCEPT
    A. Coolant leaking into the combustion chamber
    B. Worn oil control rings
    C. Stuck-open waste gate
    D. Worn exhaust valve guide seals (A9)

40. A Crankshaft Position Sensor (CPS) is being discussed. Technician A says the sensor is designed so that once inserted into the block, the gap will be automatically set to the manufacturer's specification. Technician B says the sensor to target must be measured and the sensor must be shimmed to achieve the required gap. Which Technician is correct?
    A. A only
    B. B only
    C. Both A and B
    D. Neither A nor B (H7)

41. The Most-Likely cause of methane in the coolant surge tank is
    A. Defective ¼ turn fuel shutoff valve.
    B. Plugged up secondary fuel filter.
    C. Cracked cylinder head.
    D. A defective ITT fuel high-pressure fuel regulator. (F1.6)

42. An NGV engine cranks but will not start. All the following are possible causes for the no-start condition EXCEPT
    A. No power to the coils due to a defective power fuse.
    B. Fuel ¼ turn valve turned off.
    C. No crank sensor signal received by the ECM.
    D. Stuck-open turbocharger waste gate. (A11)

43. A CNG engine is being diagnosed for a misfiring complaint. Technician A says the coil should be checked for cracks that could cause the spark to jump to ground, causing the misfire. Technician B says the coil should be checked for contamination such as engine oil on the outside because that could cause the spark to jump to ground and cause a misfire. Which Technician is correct?
    A. A only
    B. B only
    C. Both A and B
    D. Neither A nor B (H6)

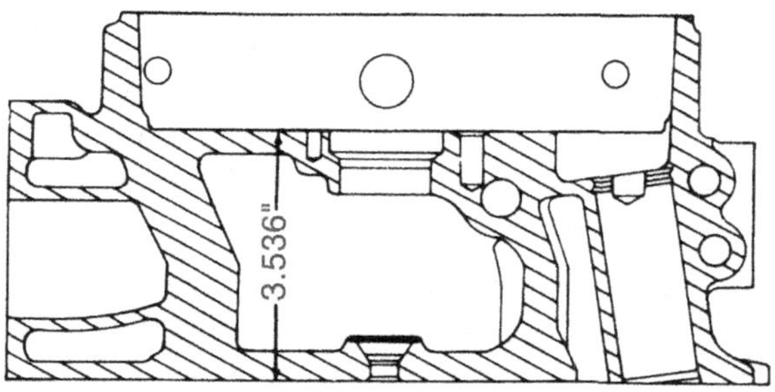

44. The cylinder head deck-to-deck thickness measurement indicated in the figure above is used to
    A. Prevent cylinder head warpage.
    B. Calculate valve guide height.
    C. Determine cylinder warpage.
    D. Determine resurfacing limit. (B3)

45. Technician A says that an NGV with an automatic Allison World Transmission can never be stall tested, only road tested, or a dyno has to be used to simulate a full power condition. Technician B says that an NGV with an automatic Allison World Transmission should only be stalled tested in fourth gear with the DDR or the Allison DOC software clutch enabled test and no longer than 30 seconds to simulate a full power condition. Which Technician is correct?
    A. A only
    B. B only
    C. Both A and B
    D. Neither A nor B (A1)

46. A Cummins ISL G 8.9 engine model is being discussed. Technician A says that the voltage for the primary side of the coil is delivered through an ignition controlled circuit of the vehicle sourced from the batteries. Technician B says that the ICM delivers up to 300 volts to the coil's primary circuit. Which Technician is correct?
    A. A only
    B. B only
    C. Both A and B
    D. Neither A nor B (H8)

47. All of the following statements about the spark plug adaptor tubes are true EXCEPT
    A. The adaptor tubes have to be free of contamination such as oil or dirt.
    B. The adaptor tubes are cooled by coolant surrounding the tube.
    C. The adaptor tubes have O-ring seals to prevent coolant from leaking into the engine oil.
    D. The adaptor tube spark plug threads can be lubricated with antiseize compound if they are not smooth. (B5)

48. Which of the following is LEAST-Likely to be used to check the inside diameter of a valve guide?
    A. Snap gauge
    B. Feeler gauge
    C. Ball gauge
    D. Dial ball gauge (B6)

49. A CNG engine appears to be misfiring. Technician A says to use an infrared pyrometer to find the defective cylinder; a cooler cylinder exhaust port usually indicates a misfiring cylinder. Technician B says the diagnostic software should be used to to determine which cylinder is not firing by canceling out each cylinder individually and listening for the cyliner whose cancelation results in no distinct change in sound. Which Technician is correct?
    A. A only
    B. B only
    C. Both A and B
    D. Neither A nor B (A4)

50. A CNG engine is being discussed. Technician A says the voltage required to jump the spark plug gap ranges from 5 KV to as high as 20 KV. Technician B says the voltage required to jump the spark plug gap is generally around 40 KV. Which Technician is correct?
    A. A only
    B. B only
    C. Both A and B
    D. Neither A nor B (H3)

51. When air-pressure testing an engine block, Technician A says the block needs to be submerged in hot water for about two hours and then the water checked for bubbles. Technician B says cracked cylinder blocks should be replaced. Which Technician is correct?
    A. A only
    B. B only
    C. Both A and B
    D. Neither A nor B (C2)

52. Technician A says cylinder heads are always torqued in sequence according to specifications set by the manufacturer. Technician B says that any pushrods that are bent must be straightened. Which Technician is correct?
    A. A only
    B. B only
    C. Both A and B
    D. Neither A nor B (B7)

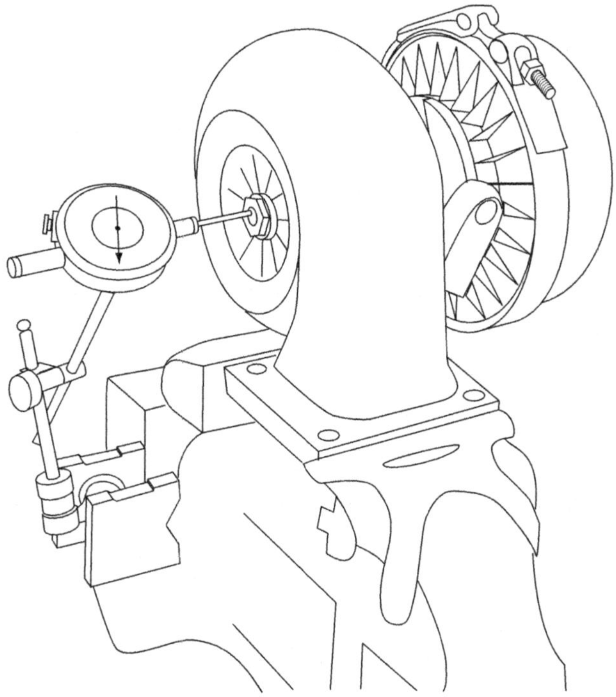

53. In the figure above, the Technician is checking the turbocharger for
    A. Radial runout.
    B. Turbine shaft endplay.
    C. Turbine fin wear.
    D. Compressor fin wear. (E2)

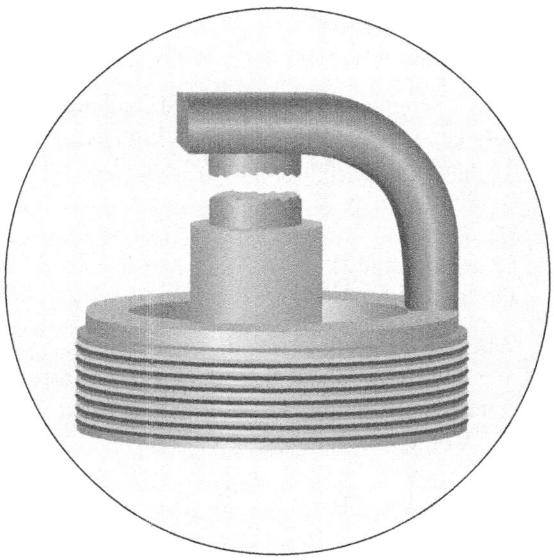

54. The Most-Likely condition of the spark plug electrodes pictured above is
    A. Overheated.
    B. Improperly gapped.
    C. Contaminated with deposits.
    D. Worn through normal usage. (H4)

55. The color of CNG engine exhaust smoke is being discussed. Technician A says the lack of air in a CNG engine will cause blue smoke. Technician B says that because of the smaller carbon chain of methane, CNG engines probably will not produce black smoke. Which Technician is correct?
    A. A only
    B. B only
    C. Both A and B
    D. Neither A nor B (A7)

56. The LEAST-Likely cause of a low fuel pressure DTC being logged when the engine is under a load is
    A. Lack of coolant flow through the heat exchanger of the ITT fuel pressure regulator.
    B. Fuel pressure sensor that has drifted to the low side.
    C. A restricted fuel filter due to contamination.
    D. A defective fuel high-pressure regulator. (F1.5)

57. A misfiring cylinder is diagnosed on a CNG engine to a specific cylinder. The Technician replaces the spark plug on the misfiring cylinder. However, the engine continues to misfire. Which of these could be the cause?
    A. Misadjusted valves on a cylinder
    B. A clogged secondary fuel filter
    C. Low fuel delivery pressure
    D. A boost pressure leak in the air intake piping (A12)

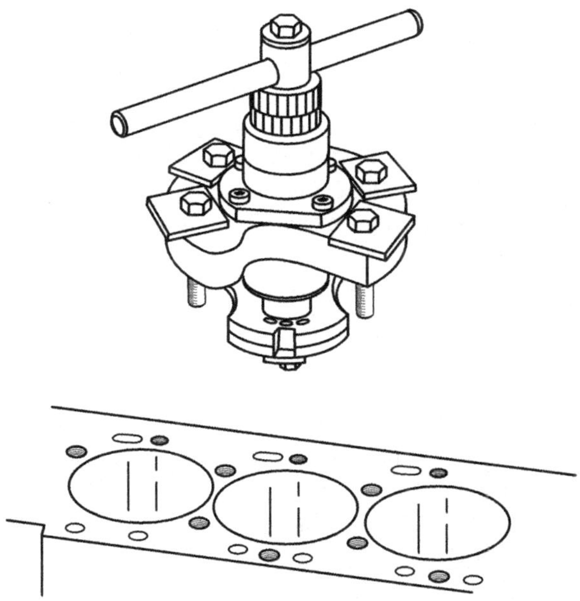

58. What task is being performed in the figure above?
    A. Removing the liner carbon ridge.
    B. Removing the cylinder liner.
    C. Machining the liner crosshatch pattern.
    D. Cutting the liner counterbore. (C3)

59. All of the following are true of high-pressure fuel lines EXCEPT
    A. Usually manufactured using stainless steel material.
    B. Manufactured from welded tubing.
    C. Must be tempered.
    D. Minimum wall thickness requirements for different diameters. (F1.3)

60. An NGV is being inspected in the shop. The rear apparatus box cover is cracked and a piece of it is missing. Technician A says the cover should be replaced with a metal cover because the plastic one is too weak and will fail again. Technician B says a piece of duct tape can be applied to the crack and seal up the missing piece. Which Technician is correct?
   A. A only
   B. B only
   C. Both A and B
   D. Neither A nor B (A3)

61. Technician A says that CNG engine breathers use a mesh-type element to prevent oil from escaping the engine. Technician B says the crankcase breather is piped directly into the air intake system to recycle the crankcase fumes. Which Technician is correct?
   A. A only
   B. B only
   C. Both A and B
   D. Neither A nor B (C1)

62. A turbocharger has failed. The thru-shaft bearings are broken and pieces of the bearings cannot be located. Technician A says the turbocharger drain line should be inspected to see if the missing pieces can be located. Technician B says the engine oil pan should be removed to ensure no broken pieces were allowed to fall into the pan that could be eventually sucked up by the oil pump, even though it is equipped with a pickup-tube screen, locking up or damaging the oil pump and resulting in serious engine damage. Which Technician is correct?
   A. A only
   B. B only
   C. Both A and B
   D. Neither A nor B (D2)

63. Technician A says that when servicing the fuel low-pressure filter of a CNG system, the ¼ turn valve must be turned off and the engine run until the engine shuts off to deplete the fuel from the system and the low-pressure gauge must record and remain at 0 psi before attempting to remove the plug or filter from the system. Technician B says that the fuel filter bowl of a low-pressure filter used on CNG engines can be drained without draining the fuel pressure from the fuel system because the pressure is only approximately 100 psi.
   A. A only
   B. B only
   C. Both A and B
   D. Neither A nor B (F2.1)

64. Before installing the valves into the cylinder head, it is important to check which one of the following?
   A. Valve-to-seat contact
   B. Rocker arm adjustment
   C. Camshaft lift
   D. Head gasket (B7)

65. Technician A says the inlet restriction gauge should always show some restriction whenever the engine is under full load because this signifies the air intake system is sealed and the air has to pass through the air cleaner. Technician B says the air restriction gauge should not go over 20" of water of restriction or this indicates the air cleaner is plugged and needs servicing. Which Technician is correct?
   A. A only
   B. B only
   C. Both A and B
   D. Neither A nor B (E3)

66. The Most-Likely cause of a CNG engine idling too high, above 900 rpm, is
    A. Too much fuel being fed to the engine from a malfunctioning fuel control valve.
    B. A bent throttle plate valve allowing too much air into the intake manifold.
    C. Valves set too tight.
    D. Valves set too loose. (F3.1)

67. All of the following must be checked before installing rocker arms EXCEPT
    A. Rocker arm bushings for wear
    B. Rocker arm shaft for wear
    C. Rocker arm pallet for pitting or scoring
    D. Rocker arm adjustment screws for tightness (B8)

68. What is the LEAST-Likely cause of high exhaust temperature?
    A. Tight exhaust valves
    B. Overfueling due to a defective fuel control valve
    C. High ambient temperature
    D. Clogged air cleaner (A5)

69. Fuel cylinders are condemned by performing the following EXCEPT
    A. Purging with water.
    B. Drilling a minimum of 2, 1/2" diameter holes aligned with the manufacturer's label.
    C. Discarding according to local, state, and federal regulations.
    D. Cylinders can be used as compressed air storage tanks. (F1.1)

70. Technician A says the fuel high-pressure regulator should be checked for coolant leaks whenever the cooling system is pressure tested. Technician B says no coolant is routed through the fuel high-pressure regulator and therefore no coolant leaks can occur. Which Technician is correct?
    A. A only
    B. B only
    C. Both A and B
    D. Neither A nor B (F1.6)

71. Technician A says that an easy way to erase fault codes in CNG engine electronic control modules (ECMs) is to disconnect the vehicle's batteries. Technician B says that when the vehicle's batteries are disconnected, none of the logged codes and audit trails are lost because they are written to EEPROM. Which Technician is correct?
    A. A only
    B. B only
    C. Both A and B
    D. Neither A nor B (A16)

72. Technician A says that when installing a new CNG cylinder, a chain wrapped around the cylinder can be used. Technician B says that a nylon sling with sufficient strength can be used. Which Technician is correct?
    A. A only
    B. B only
    C. Both A and B
    D. Neither A nor B (F1.7)

73. The Most-Likely cause of a strong odor of rotten eggs determined to be mercaptan after fueling of a CNG transit bus is
    A. Normal residual fuel on the fueling receptacle.
    B. The fueling receptacle's check valve could be leaking.
    C. The CNG fuel cylinder is equalizing by venting due to the high pressure in the cylinder after fueling.
    D. A missing rubber cap on the fueling receptacle. (A6)

74. The cylinder counterbores depths can be corrected by the following EXCEPT
    A. Boring the cylinder counterbores to manufacturer's specification.
    B. Resurfacing the cylinder block deck surface to bring the counterbores to specifications.
    C. Grinding the counterbores with a die grinder and a stone.
    D. Utilizing shims under the liner flange to adjust the liner protrusion. (C3)

75. The LEAST-Likely cause of low turbo boost pressure is
    A. Punctured CAC.
    B. Engine overfueling.
    C. Clogged air cleaner element.
    D. Engine power de-rated due to a diagnostic code. (A8)

76. When the oil filter bypass valve opens, what has caused it to do so?
    A. A restricted inlet to the filter
    B. The oil pump is sucking air
    C. A massive oil leak in the lubrication system
    D. A filter element is plugged (D3)

77. A CNG-powered transit bus comes to the shop for repairs. The complaint is that the engine consumes engine oil with no evident blue smoke. Technician A says the exhaust manifold should be removed to access the exhaust ports and check for deposits in the ports caused by the consumption of lube oil. Technician B says that an infrared pyrometer can be used to check the exhaust ports for changes in temperature to determine which cylinder is defective. Which Technician is correct?
    A. A only
    B. B only
    C. Both A and B
    D. Neither A nor B (A5)

78. The turbocharger bearings have failed. Technician A says the oil drain tube needs to be inspected for pieces of the bearings that may have fallen into the drain tube and cause an obstruction for the proper draining of the new turbocharger. Technician B says the bearings may have failed due to not allowing the turbocharger to slow down at shutdown. Which Technician is correct?
    A. A only
    B. B only
    C. Both A and B
    D. Neither A nor B (D5)

79. Technician A says that an NGV engine's CAC uses coolant to remove heat from the boosted air. Technician B says that a CAC can be checked for leaks by plugging the inlet and outlet ports with special caps and pressurizing the charger with a minimum of 75 psi of compressed air. Which Technician is correct?
    A. A only
    B. B only
    C. Both A and B
    D. Neither A nor B (E4)

80. A Cummins L-Gas Plus engine is being diagnosed for high exhaust back pressure. Technician A says the sensor line from the catalyst to the sensor at the top of the engine valve cover could be plugged with contaminants and should be checked before further diagnostics are performed. Technician B says the catalyst is the cause and needs to be replaced. Which Technician is correct?
    A. A only
    B. B only
    C. Both A and B
    D. Neither A nor B (A9)

81. Technician A says that because CNG engine uses methane, a non-leaded fuel, O_2 sensors will not fail because sensors are only contaminated by the use of leaded fuel. Technician B says that contaminants such as deposits from engine oil and coolant consumption in the cylinders are harmful to O_2 sensors. Which Technician is correct?
    A. A only
    B. B only
    C. Both A and B
    D. Neither A nor B (E6)

82. CNG fuel tank labels are being discussed. Technician A says that when a label is missing or illegible, the Technician can make a duplicate copy and glue it back onto the cylinder. Technician B says that a label with the same information from another cylinder can be transferred from another cylinder. Which Technician is correct?
    A. A only
    B. B only
    C. Both A and B
    D. Neither A nor B (F1.2)

83. Technician A says that manufacturers allow for relative movement between the chassis and the engine by using durable rubber couplings in the intake system. Technician B says that if an air intake pipe, tube, or hose is loose or missing a clamp, an inspection should be made of the turbocharger compressor. Which Technician is correct?
    A. A only
    B. B only
    C. Both A and B
    D. Neither A nor B (E1)

84. All of the following are approved methods of servicing high-pressure CNG fuel components EXCEPT
    A. Lowering the line pressure to 500 psi before loosening the fittings to repair a leak.
    B. Using an approved leak detection solution to find a leak.
    C. Close off the ¼ turn valve and running of the engine until it shuts off to deplete all of the fuel from the filters before servicing them.
    D. Venting procedures should only be used in an approved area. (F1.4)

85. When fast filling, less than two hours, a CNG fuel system rated for 3600 psi, Technician A says the fueling station will usually allow an initial charge of 4000 psi to compensate for the heating of the fuel and the friction that results as it is forced through the lines. Technician B says the fuel system pressure should not be higher than 3600 psi because the tanks are only rated for 3600 psi. Which Technician is correct?
    A. A only
    B. B only
    C. Both A and B
    D. Neither A nor B (F1.8)

86. The intake manifold pressure is being discussed. Technician A says that a CNG engine's intake manifold will usually be under a vacuum at idle due to the throttle valve being only partially open and restricting the air and fuel flow. Technician B says the diagnostic software will be able to show the vacuum due to the sensor's capability of reading absolute pressure. Which Technician is correct?
    A. A only
    B. B only
    C. Both A and B
    D. Neither A nor B (A8)

87. A CNG engine has a cylinder that is misfiring intermittently. All of the following are possible causes EXCEPT
    A. An improperly gapped spark plug.
    B. A loose connection at the COP.
    C. The spark plug's electrodes are shorted together.
    D. A spark plug is fouled with oil residue. (H8)

88. A CNG engine is being diagnosed with a DTC for high exhaust temperature. Technician A says that if the engine is being overfueled due to a defective low-pressure regulator or fuel control valve, high exhaust temperatures will result. Technician B says that if the exhaust valves are maladjusted too tight, this can cause high exhaust temperatures. Which Technician is correct?
    A. A only
    B. B only
    C. Both A and B
    D. Neither A nor B (A9)

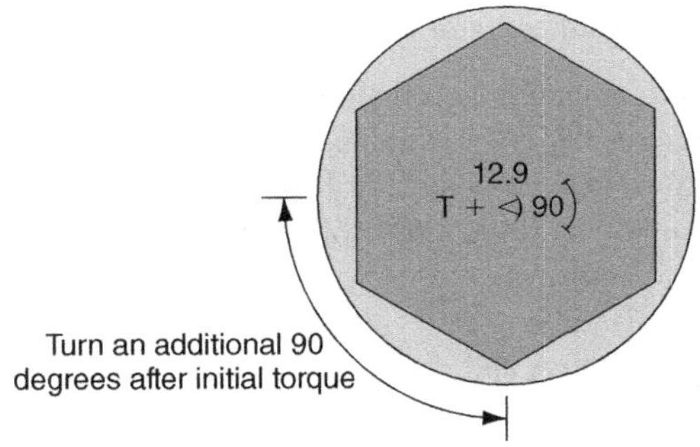

89. Technician A says the markings on the head bolt above show the strength of the bolt. Technician B says the markings on the head bolt above show that it is a torque-turn-type bolt. Which Technician is correct?
    A. A only
    B. B only
    C. Both A and B
    D. Neither A nor B (B2)

90. Spark plug tubes are being replaced in a CNG engine. Technician A says the O-rings should be lubricated with engine oil. Technician B says the O-rings have to be lubricated with engine coolant. Which Technician is correct?
    A. A only
    B. B only
    C. Both A and B
    D. Neither A nor B (B5)

91. Which procedure is LEAST-Likely to be used when performing valve seat maintenance?
    A. Testing seats for looseness with a ball-peen hammer
    B. Using a chisel to remove defective seats
    C. Checking seats with a concentricity indicator
    D. Dressing a valve seat grinding stone (B4)

92. Technician A says the emission labels are only required to be on the engine for delivery purposes. Technician B says the emission label is required to be affixed to the engine at all times. Which Technician is correct?
    A. A only
    B. B only
    C. Both A and B
    D. Neither A nor B (A17)

93. Technician A says that valve crosshead guide pins should be at right angles to the cylinder head milled surface. Technician B says to check crosshead guide pin diameter with a micrometer, and compare to manufacturer's specifications. Which Technician is correct?
    A. A only
    B. B only
    C. Both A and B
    D. Neither A nor B (B6)

94. Technician A says to make sure a silicone gasket is coated with adhesive before installing die cast rocker covers on a cylinder head. Technician B says to always use new gaskets when installing valve covers. Which Technician is correct?
    A. A only
    B. B only
    C. Both A and B
    D. Neither A nor B (C1)

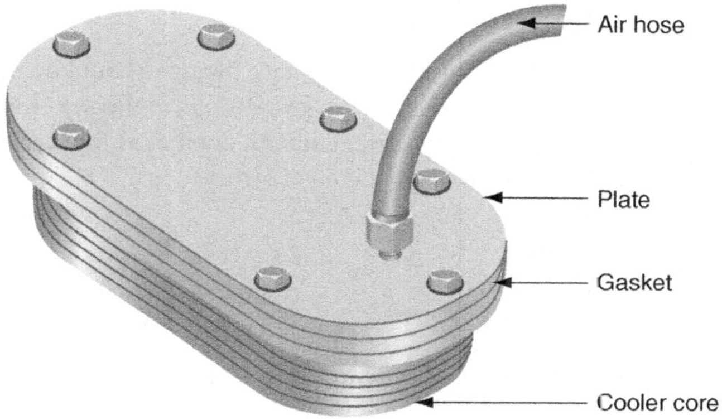

95. The Most-Likely operation being performed above is
    A. Pressure testing of the oil cooler core.
    B. Pressure testing of the coolant core.
    C. Prelubing of the oil cooler core.
    D. Precharging of the oil cooler core. (D4)

96. Technician A says that the parts manuals are not a good guide in re-assembling engine components. Technician B says that the latest publications of the parts manuals will contain the latest information about engine component changes as well as previous versions. Which Technician is correct?
    A. A only
    B. B only
    C. Both A and B
    D. Neither A nor B (A18)

97. All of the following could be a likely cause for a low fuel pressure problem EXCEPT
    A. Clogged fuel filters.
    B. No coolant flowing through the high-pressure regulator causing icing inside the primary fuel filter.
    C. A malfunction low-pressure shutoff solenoid causing it to be only partially open.
    D. A ¼ turn fuel shutoff valve not turned on all the way, restricting fuel flow. (A12)

98. After placing pencil marks around the face of the valve about 1/8" apart and snapping the valve against its seat, all but two lines are found to be visibly broken. Technician A says the valve seat may need to be reground. Technician B says the valve and seat are within limits and will seal when spring pressure is applied. Which Technician is correct?
   A. A only
   B. B only
   C. Both A and B
   D. Neither A nor B (B4)

99. The cleaning of the fire deck of a cylinder head is being discussed. Technician A says that a gasket scraper can be used to remove the deposits from the fire deck area so the head can be inspected for cracks. Technician B says that only fine grit sandpaper or a scuffing pad or similar object can be used to remove the deposits from the cylinder head. Which Technician is correct?
   A. A only
   B. B only
   C. Both A and B
   D. Neither A nor B (B1)

100. Technician A says the spark plug threads should be lubricated with anti-seize compound to ensure easy removal of the spark plugs at the next tune-up. Technician B says the spark plug threads should be lubricated with engine oil to ensure easy removal at the next spark plug change. Which Technician is correct?
   A. A only
   B. B only
   C. Both A and B
   D. Neither A nor B (B5)

101. When testing a cylinder head for potential coolant leakage, Technician A says the coolant needs to be heated to operating temperature. Technician B says the coolant needs to be under pressure. Which Technician is correct?
   A. A only
   B. B only
   C. Both A and B
   D. Neither A nor B (B3)

102. Technician A says a cylinder head should be stored upright on one end to prevent the cylinder head from warping. Technician B says the cylinder head should be stored flat on two blocks of wood to prevent it from warping. Which Technician is correct?
   A. A only
   B. B only
   C. Both A and B
   D. Neither A nor B (B1)

103. Technician A says that some valve crossheads are of the adjustable type and do not usually need adjusting at normal tune-up intervals, only at initial installation. Technician B says that some valve crossheads are of the floating type and do not require any adjustment. Which Technician is correct?
   A. A only
   B. B only
   C. Both A and B
   D. Neither A nor B

(B6)

104. A Detroit Diesel Series 50G comes to the shop with the complaint the engine returns to idle very slowly when the accelerator is released and then dies upon coming to a stop. Technician A says the waste gate is staying stuck close. Technician B says the Pulse Stepper Valve (PSV) may be sticking; and should be checked for proper operation. Which Technician is correct?
    A. A only
    B. B only
    C. Both A and B
    D. Neither A nor B (A12)

105. Which of the following should the Technician check first when installing a cylinder head to the engine block?
    A. Bolt holes for foreign matter and cross threading
    B. Head bolt torque
    C. Rocker pedestal height
    D. Valve bridge adjustment (B7 and B8)

106. A CNG engine is leaking oil from several crankcase seals. Technician A says oil leaks can develop if the crankcase pressure is too high, caused by worn compression piston rings. Technician B says that high crankcase pressure can be caused if the crankcase breather tube becomes plugged or restricted. Which Technician is correct?
    A. A only
    B. B only
    C. Both A and B
    D. Neither A nor B (A10)

107. Technician A says that oil in a head bolt hole in the block can throw off the torque setting of that bolt. Technician B says that a head bolt with damaged threads should be dressed before installing. Which Technician is correct?
    A. A only
    B. B only
    C. Both A and B
    D. Neither A nor B (C2)

108. When repairing DTCs, Technician A says the lower number, active defect codes should be repaired first. Technician B says the DTCs should be recorded prior to deleting them for future references. Which Technician is correct?
    A. A only
    B. B only
    C. Both A and B
    D. Neither A nor B (A16)

109. Technician A says a loss of engine lubrication oil through the crankcase breather is an indication of excess crankcase pressure. Technician B says any measurable crankcase pressure is a sign of trouble. Which Technician is correct?
    A. A only
    B. B only
    C. Both A and B
    D. Neither A nor B (A10)

110. A CNG engine with an in block camshaft has a miss while it is running. The #3 cylinder exhaust valve rocker arm is not actuating. Technician A says the cam lobe could be flat and causing the miss. Technician B says the cam follower roller could be defective and causing the miss. Which Technician is correct?
   A. A only
   B. B only
   C. Both A and B
   D. Neither A nor B (B8)

111. Which of the following tools is LEAST-Likely to reveal an out-of-round cylinder sleeve?
   A. Inside micrometer
   B. Telescoping gauge
   C. Feeler gauge
   D. Dial bore gauge (C4)

112. Technician A says that an intake manifold has to be torqued from the center outward toward the ends. Technician B says the intake manifold needs to be checked with a straightedge to ensure it is not warped beyond the OEM specifications. Which Technician is correct?
   A. A only
   B. B only
   C. Both A and B
   D. Neither A nor B (E3)

113. Technician A says the housing ends of a turbocharger should not be removed from the turbocharger during installation because they come pre-aligned from the factory. Technician B says the housings are designed so they can be rotated to align with the mounting flanges when they are installed on the engine. Which Technician is correct?
   A. A only
   B. B only
   C. Both A and B
   D. Neither A nor B (D5)

114. Technician A says the exterior of a CNG engine's wet cylinder sleeves can be cleaned by solvent immersion. Technician B says the exterior of a CNG engine's wet cylinder sleeves can also be cleaned with glass bead blasting. Which Technician is correct?
   A. A only
   B. B only
   C. Both A and B
   D. Neither A nor B (C4)

115. The Most-Likely cause of a catalyst failure is
   A. High air intake temperature.
   B. Worn spark plugs.
   C. Leak in the pressure side of the air intake system.
   D. Overfueling due to a defective low-pressure regulator overfueling the engine. (E5)

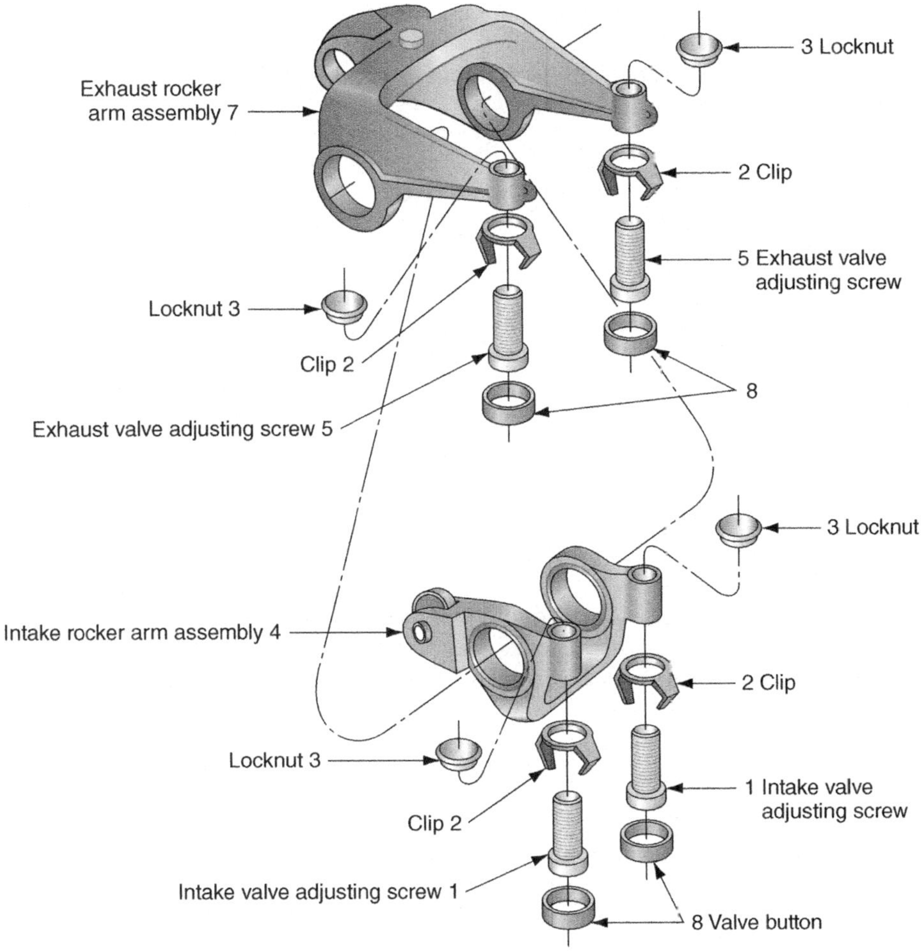

116. A CNG engine is being tuned up when the Technician discovers that item number 8 in the figure above, a valve button, has fallen off. Technician A says the button should be reassembled and the tune-up completed. Technician B says the rocker arm assembly should be replaced. Which Technician is correct?
    A. A only
    B. B only
    C. Both A and B
    D. Neither A nor B (B8)

117. A turbocharger compressor wheel (inlet side) has tell-tale signs of dirt and erosion on the blade tips. Technician A says this is a normal condition on high mileage engines. Technician B says that the erosion of the blade tips is a good indication that some abrasive material such as dust or sand has entered the air system and the air cleaner and associated air intake piping need to be checked for proper installation. Which Technician is correct?
    A. A only
    B. B only
    C. Both A and B
    D. Neither A nor B (E1)

118. Technician A says that drivability issues are always caused by the engine. Technician B says that drivability issues could be a result of a change of operator and driving habits and should be investigated as a cause. Which Technician is correct?
    A. A only
    B. B only
    C. Both A and B
    D. Neither A nor B (A1)

119. A compression test was performed on an engine that has knock codes. The cylinder compression readings are between 215 and 235 psi except one cylinder that is 265 psi with three puffs of the piston.

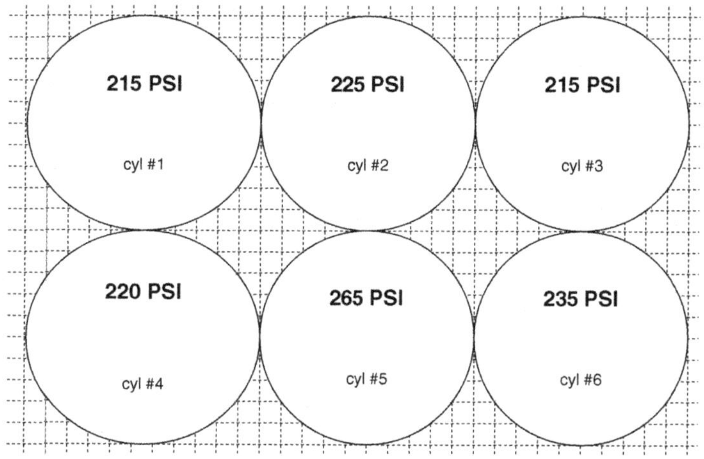

Technician A says the cylinder with the pressure of 265 psi is too high and this could be the cause of the knock codes. Technician B says that as long the compression is above 200 psi the engine is in good shape. Which Technician is correct?

A. A only
B. B only
C. Both A and B
D. Neither A nor B (A20)

120. A CNG engine has one complete rocker arm shaft and a set of rocker arm bushings that failed less than 2000 miles after the cylinder head was replaced. Technician A says the premature failure appears to have been caused by a lack of lubrication and the assembly needs to be checked to make sure the assembly was installed correctly. Technician B says the assembly was probably just defective and only needs to be replaced. Which Technician is correct?

A. A only
B. B only
C. Both A and B
D. Neither A nor B (B9)

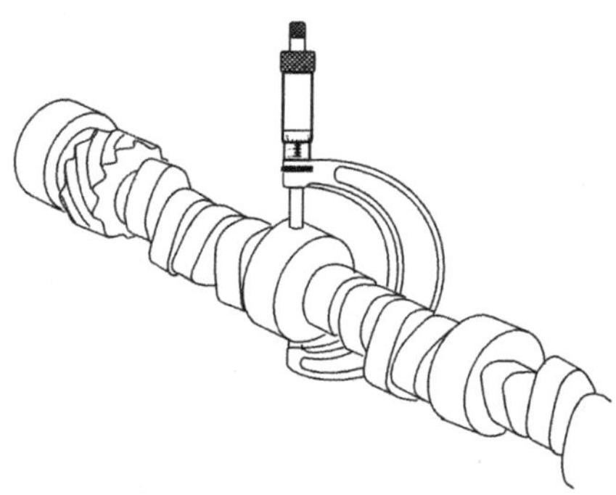

121. What is the Technician measuring in the figure above?

A. Cam lobe
B. Cam bearing journal
C. Cam lift
D. Cam duration (B11)

122. Technician A says camshaft bushings are installed with Loctite to lock them in place and prevent them from moving out of position. Technician B says the bushings have to be indexed with a center punch to keep them from moving out of position. Which Technician is correct?
    A. A only
    B. B only
    C. Both A and B
    D. Neither A nor B (C6)

123. Technician A says to check the bearings and crankshaft before cleaning the crank. Technician B says to measure the crankshaft after cleaning the crank. Which Technician is correct?
    A. A only
    B. B only
    C. Both A and B
    D. Neither A nor B (C8)

124. Technician A says that a feeler gauge should be inserted between the main bearing shell and the crankshaft to check for clearances. Technician B says that a feeler gauge should be inserted between the bearing shell and the engine block to check the main bearing clearances. Which Technician is correct?
    A. A only
    B. B only
    C. Both A and B
    D. Neither A nor B (C9)

125. Technician A says that if the piston cooling nozzle is bent, oil will not be directed correctly on the underside of the piston and cause the piston to overheat and possibly pre-ignition to occur along with knock codes. Technician B says that some pistons are cooled through oil passing up the connecting rod and through the wrist pin to be sprayed on the underside of the piston crown to cool it. Which Technician is correct?
    A. A only
    B. B only
    C. Both A and B
    D. Neither A nor B (C12)

126. All of the following statements are true about turbochargers EXCEPT
    A. Turbochargers can rotate at a high rate of speed to create boost.
    B. Turbochargers of today's CNG engines use a speed sensor to monitor their rotational speed.
    C. CNG engines use waste gates to control the boost created.
    D. The turbocharger's bearings come pre-lubed from the factory and require no further lubrication at installation. (D5)

127. A connecting rod bearing shows signs of more wear on one side of the bearing than the other. Technician A says this is a normal condition for an engine that has high mileage. Technician B says this is an indication the connecting rod could be bent. Which Technician is correct?
    A. A only
    B. B only
    C. Both A and B
    D. Neither A nor B (C14)

128. Technician A says that a fuel cylinder's expiration date can be determined by looking on the cylinder label. Technician B says the label must contain the serial number of the fuel cylinder. Which Technician is correct?
    A. A only
    B. B only
    C. Both A and B
    D. Neither A nor B (F1.2)

129. Technician A says that if the turbocharger coolant and oil lines and hoses become frayed or nicked, they must be replaced, not repaired. Technician B says that turbocharger oil drain tubes must be checked for obstructions when replacing a turbocharger. Which Technician is correct?
   A. A only
   B. B only
   C. Both A and B
   D. Neither A nor B (E2)

130. The fuel fill connection is being discussed. Technician A says a rubber cap is not necessary to be in place because the valve contains a check valve. Technician B says the cap prevents dirt and debris from accumulating on the valve and eventually being pushed into the fuel system when the vehicle is being refueled. Which Technician is correct?
   A. A only
   B. B only
   C. Both A and B
   D. Neither A nor B (F1.4)

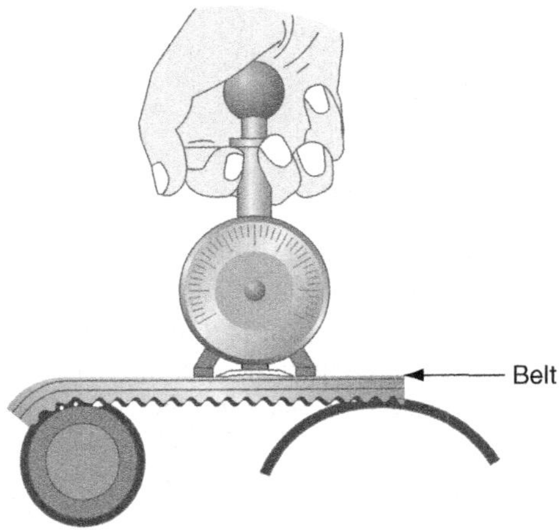

131. In the figure above, what is the technician performing?
   A. Checking belt alignment.
   B. Belt straightening.
   C. Checking the thickness of the belt.
   D. Checking the belt tension. (D7)

132. The LEAST-Likely cause of a flywheel housing TIR being out of specification is
   A. Dirt or contaminants under the flywheel.
   B. Bent mounting dowel pins from improper handling.
   C. Damaged flywheel housing from improper handling.
   D. Improper installation of the crankshaft. (C17)

133. The fuel filters of a CNG transit bus were replaced and their housings torqued to the manufacturer's specifications. Technician A says the fuel can now be turned back on at the ¼ turn valve, and if there is no presence of the mercaptan odor the vehicle is safe for operation. Technician B says the ¼ turn valve can now be turned on. The engine needs to be started and then shut off before the Technician checks for the presence of mercaptan with an electronic leak detector. Which Technician is correct?
   A. A only
   B. B only
   C. Both A and B
   D. Neither A nor B (F2.2)

134. Technician A says that to remove the surface charge from a recently charged battery, apply a 300-ampere load across the terminals for 15 seconds. Technician B says that it is unnecessary to remove the surface charge from batteries that have been in storage. Which Technician is correct?
   A. A only
   B. B only
   C. Both A and B
   D. Neither A nor B (G1)

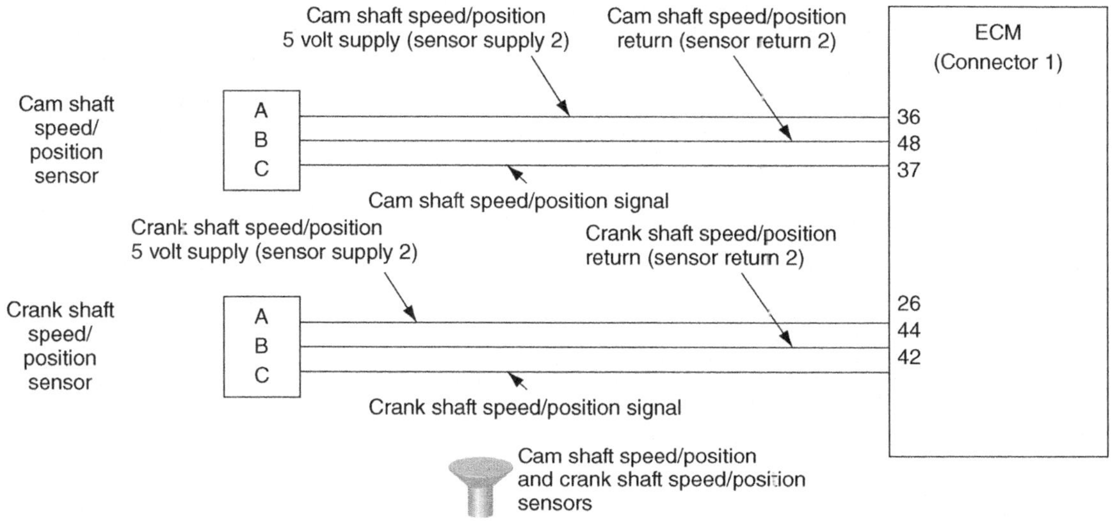

135. Referring to the figure above, the Most-Likely ECM connector and pin to supply the +5 volts to the camshaft and crankshaft speed/position sensors are
   A. Connector 2 pin 36 and 26.
   B. Connector 2 pin 48 and 44.
   C. Connector 1 pin 37 and 42.
   D. Connector 1 pin 36 and 26. (H7)

136. An engine fails to achieve normal operating temperature. Technician A says that a stuck-open thermostat could cause this problem. Technician B says that an overfilled radiator could cause this problem. Which Technician is correct?
   A. A only
   B. B only
   C. Both A and B
   D. Neither A nor B (A14)

137. A CNG valve adjustment procedure is being discussed. Technician A says the valve should set tight, difficult to drag the feeler gauge across the rocker arm and valve stem, to ensure the valves will remain in adjustment for a longer period of time. Technician B says the valves should be adjusted a few thousandths loosely to ensure the valves do not remain open once the engine reaches operating temperature. Which Technician is correct?
   A. A only
   B. B only
   C. Both A and B
   D. Neither A nor B (B10)

138. Technician A says the engine oil should be drained when the oil is at ambient temperature. Technician B says the engine oil needs to be drained after the engine has been off for a minimum of 1 hour to ensure all the oil has returned to the oil pan for complete drainage. Which Technician is correct?
  A. A only
  B. B only
  C. Both A and B
  D. Neither A nor B (D6)

139. After installing new camshaft bushings, the Technician cannot install the camshaft because the bearing-to-bushing clearances are too tight. Technician A says the bushing should be reamed to manufacturer's specifications. Technician B says the bushings should be enlarged with emery cloth to increase the inside diameter of the bushing and permit the camshaft to be installed. Which Technician is correct?
  A. A only
  B. B only
  C. Both A and B
  D. Neither A nor B (C6)

140. Cracks and scoring in the piston skirts are LEAST-Likely to be caused by which of the following?
  A. High crankcase lubrication levels
  B. Improper piston clearance
  C. Overfueling
  D. Engine overheating (C11)

141. Technician A says that CNG engines that have pre- and post-catalyst O_2 sensors are considered to have two-way catalysts. Technician B says that CNG engines that have only a pre-catalyst O_2 sensor are considered to have three-way catalysts. Which Technician is correct?
  A. A only
  B. B only
  C. Both A and B
  D. Neither A nor B (E5)

142. The gas being produced by a battery when it is charging is called
  A. Methane.
  B. Hydrogen.
  C. Hydrogen sulfide.
  D. Carbon monoxide. (G1)

143. The flooded lead-acid batteries of a transit bus are constantly low on electrolyte. Technician A says the voltage regulator could be set too low under charging the batteries. Technician B says the driver could be driving with all the lights on all the time, overworking the batteries. Which Technician is correct?
  A. A only
  B. B only
  C. Both A and B
  D. Neither A nor B (G5)

144. Technician A says the stoichiometric ratio of natural gas is 16.5:1. Technician B says that CNG engines are also capable of operating leaner than the 16.5:1 ratio. Which Technician is correct?
  A. A only
  B. B only
  C. Both A and B
  D. Neither A nor B (H1)

145. The LEAST-Likely cause for a low oil pressure condition is:
    A. A stuck-closed oil pressure relief valve.
    B. A restricted oil pump suction tube.
    C. Worn main bearings.
    D. A clogged oil filter. (A15)

146. Oil pan removal is being discussed. Technician A says the oil should be completely drained before removing the pan. Technician B says that an inspection of the pan after removal can give clues as to the type of defects a Technician might encounter during the repair or rebuild of an engine. Which Technician is correct?
    A. A only
    B. B only
    C. Both A and B
    D. Neither A nor B (C1)

147. Timing gears are being installed on an engine. Technician A says the idler gear can be installed one tooth advanced on the crankshaft. Technician B says the idler can be installed one tooth advanced on the camshaft gear. Which Technician is correct?
    A. A only
    B. B only
    C. Both A and B
    D. Neither A nor B (C10)

148. The results of testing with SCA test strips can indicate all of the following EXCEPT
    A. Coolant exposure to combustion gases.
    B. Improper use of a low silicate antifreeze.
    C. Coolant temperature.
    D. Coolant degradation. (D10)

149. A fuel line is discovered to have a pin hole at the fuel high-pressure regulator area due to having rubbed through as it passes through the bulkhead and requires replacing. Technician A says that a steel brake line can be utilized to replace the defective line. Technician B says the line can be welded and repaired by a certified welder. Which Technician is correct?
    A. A only
    B. B only
    C. Both A and B
    D. Neither A nor B (F1.6)

150. A sealed battery has been charging for more than three hours and the green eye is not showing yet. Technician A says the battery can be gently tilted to see if the eye will change to green because the dot can stick and tilting will free it. Technician B says the battery should be charged another two hours and if the eye does not change to green the battery should be discarded because it will not charge correctly. Which Technician is correct?
    A. A only
    B. B only
    C. Both A and B
    D. Neither A nor B (G3)

151. Technician A says that the fuel low-pressure regulator has coolant passing through it to help maintain adequate fuel temperature and prevent drivability issues. Technician B says the fuel low-pressure regulator does not have coolant passing through it and uses engine compartment heat to maintain adequate fuel temperature. Which Technician is correct?
    A. A only
    B. B only
    C. Both A and B
    D. Neither A nor B (F2.2)

152. Technician A says a spark plug socket with a rubber insert can be used to remove the spark plugs to prevent the porcelain insulator from cracking. Technician B says a magnetic spark plug socket should be used when installing spark plugs. Which Technician is correct?
   A. A only
   B. B only
   C. Both A and B
   D. Neither A nor B (H5)

153. Technician A says the engine of the booster vehicle being needs to be running when connecting up the jumper cables to the dead vehicle's batteries. Technician B says the engine of the booster vehicle can be placed on fast idle to allow the low batteries of the other vehicle to charge up before attempting to start it. Which Technician is correct?
   A. A only
   B. B only
   C. Both A and B
   D. Neither A nor B (G4)

154. An engine cranks slowly and fails to start. Technician A says that the cranking speed may be slow to allow the ECM to turn on the fuel control or low-pressure fuel solenoid. Technician B says to check for a low battery potential or corrosion on the battery terminals and clean or charge as necessary. Which Technician is correct?
   A. A only
   B. B only
   C. Both A and B
   D. Neither A nor B (A11)

155. Technician A says that when a valve of a cylinder starts to open the other valve on the same cylinder is closed and can be adjusted. Technician B says that procedure is incorrect. Only the approved valve overlap method for the companion cylinder can be used when performing valve adjustments. Which Technician is correct?
   A. A only
   B. B only
   C. Both A and B
   D. Neither A nor B (B10)

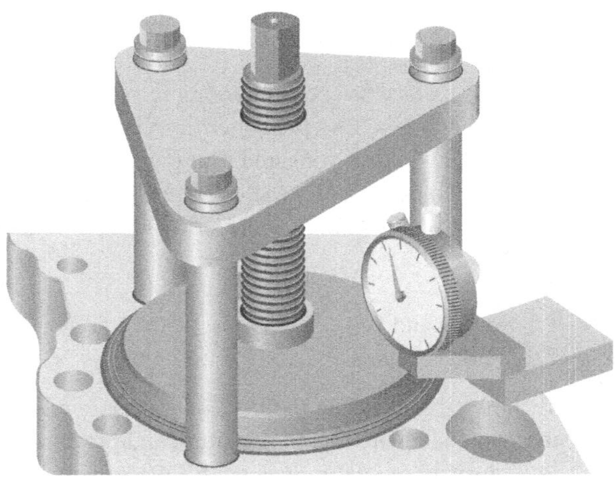

156. In the picture above, what is being measured?
   A. Liner concentricity.
   B. Liner counter bore.
   C. Cylinder block flatness.
   D. Cylinder liner flange protrusion. (C5)

157. Technician A says that a radiator cap in a cooling system that cannot maintain its rating pressure can cause an overheat condition. Technician B says that when the pressure exceeds its rating, the cap relieves the pressure. Which Technician is correct?
    A. A only
    B. B only
    C. Both A and B
    D. Neither A nor B (D13)

158. Technician A says that SCA dropout in the radiator appears as a white powder. Technician B says SCA appears as a black residue in the radiator. Which Technician is correct?
    A. A only
    B. B only
    C. Both A and B
    D. Neither A nor B (D11)

159. Technician A says that a PRD is thermally activated. Technician B says that thermal activation of the PRD usually ranges from 189 to 215° F depending on the manufacturer's requirements. Which Technician is correct?
    A. A only
    B. B only
    C. Both A and B
    D. Neither A nor B (F1.3)

160. A cooling system is being pressure tested. The pressure does not relieve at the predetermined pressure of 9 psi required for the system and does not relieve at 12 psi either when the pressure is further increased. Technician A says the pressure cap could be the problem. Technician B says the vent line could be plugged up and causing the problem. Which Technician is correct?
    A. A only
    B. B only
    C. Both A and B
    D. Neither A nor B (D15)

161. Technician A says if an O_2 sensor becomes slow in responding to variations in the air fuel ratio, this can lead to knock codes. Technician B says O_2 sensors either work or they don't. Which Technician is correct?
    A. A only
    B. B only
    C. Both A and B
    D. Neither A nor B (E6)

162. The Most-Likely component(s) to be replaced along with a full-blocking thermostat and the housing gasket is
    A. The temperature sensor.
    B. The water pump seal.
    C. The thermostat seals pressed into the housing.
    D. The housing plugs. (D9)

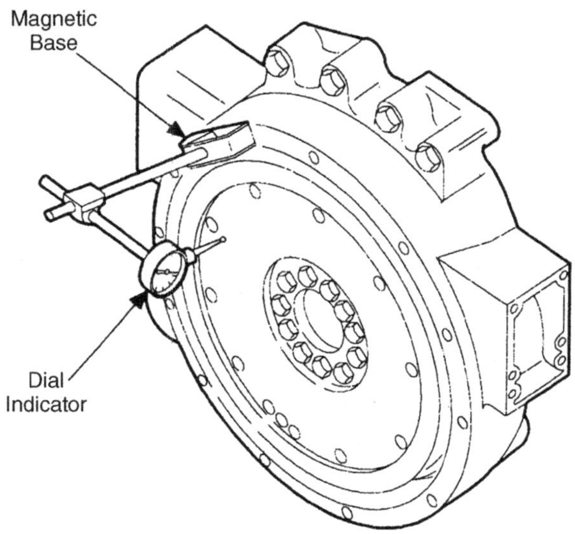

163. Which measurement is being performed in the above figure?
    A. Clutch face surface runout
    B. Pilot bore radial runout
    C. Crankshaft radial runout
    D. Flywheel housing concentricity (C18)

164. A CNG engine is being diagnosed for a DTC that is interpreted as "Throttle Plate Erratic." The LEAST-Likely course of action to be taken to diagnose the problem is
    A. Manually apply force to the throttle valve to force it wide open before turning on the ignition switch to see if the DTC remains.
    B. Check the fuses and the wiring connections to make sure they are tight and clean.
    C. Check to make sure the torquing procedures were followed when the throttle valve was installed and it is not binding.
    D. Check to make sure the ignition voltage supplied to the throttle plate is within range. (F3.1)

165. The cold-cranking amperage (CCA) rating of a battery is an indication of
    A. The amperage a battery can deliver for 30 seconds at 0° F.
    B. The amperage a battery can deliver for 60 seconds at 0° F.
    C. The temperature range a battery is designed to operate within.
    D. The length of time a battery will allow the engine to be operated after the alternator fails. (G2)

166. Technician A says that the fuel filter housings have to be tightened as tight as possible to prevent them from leaking due to the high pressure in the fuel system. Technician B says the manufacturer's specifications have to be followed and are printed on the filter housing or can be found in the service manual. Which Technician is correct?
    A. A only
    B. B only
    C. Both A and B
    D. Neither A nor B (F2.2)

167. Technician A says that when an electrical connector has a broken lock, a zip tie can be used to secure it back in place. Technician B says the connector should be replaced to ensure a weathertight connection so moisture will not be able to intrude into the connector. Which Technician is correct?
    A. A only
    B. B only
    C. Both A and B
    D. Neither A nor B (H3)

168. The LEAST-Likely cause of an engine knock code is
    A. High battery charging voltage above 29.5 volts on a 24 volt system.
    B. Defective belt tensioner.
    C. Poor engine-to-chassis ground.
    D. High intake air temperature. (A19)

169. Technician A says that the camshaft end play needs to be checked and adjusted if necessary before installing the rocker arms and shafts on an overhead cam engine. Technician B says that camshaft end play of an overhead cam engine can only be checked and adjusted after the rocker arms and shafts have been installed. Which Technician is correct?
    A. A only
    B. B only
    C. Both A and B
    D. Neither A nor B (B9)

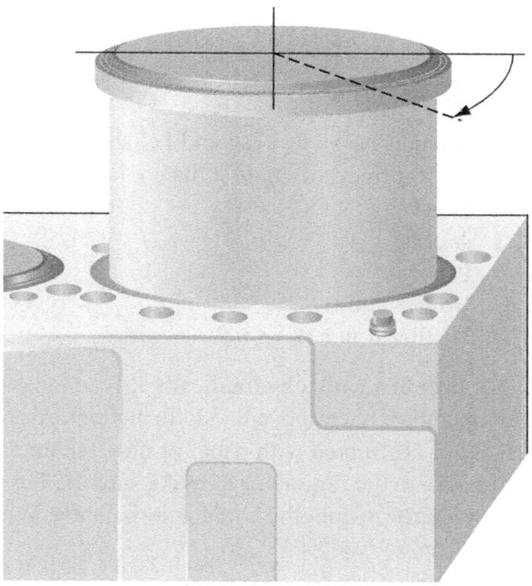

170. Refer to the above figure for this question. Technician A says that the liners are only reusable if the pitting of the liner occurs at a 45° angle to the longitudinal line of the engine block. Technician B says the manual indicates the liners need to be reinstalled at a 45° angle to the original centerline to position the pitting marks away from the original position. Which Technician is correct?
    A. A only
    B. B only
    C. Both A and B
    D. Neither A nor B (C5)

171. Technician A says that the blue indications on a camshaft journal are normal due to the hard facing process of the journals. Technician B says that blue indications are the result of overheating of the journal and the camshaft should be carefully measured and inspected to make sure it is reuseable. Which Technician is correct?
    A. A only
    B. B only
    C. Both A and B
    D. Neither A nor B (C7)

172. The LEAST-Likely acceptable housing concentricity total runout is
    A. 0.005"
    B. 0.015"
    C. 0.009"
    D. 0.002" (C17)

173. A CNG engine is being diagnosed with a low engine oil pressure complaint. Technician A says the oil pump is probably the problem and it needs to be checked for worn components. Technician B says the main bearings could be worn and they should be dropped so they can be checked for wear. Which Technician is correct?
A. A only
B. B only
C. Both A and B
D. Neither A nor B (D3)

174. An NGV engine vehicle is brought to the shop for an overheating complaint. It is equipped with a belt-driven hydraulic pump to drive a hydraulic radiator fan. Technician A says that if the drive belts were loose it would be obvious because the belts would screech. Technician B says that slipping belts do not always screech, it depends on the severity of the slippage and the belt tension should be checked. Which Technician is correct?
A. A only
B. B only
C. Both A and B
D. Neither A nor B (D7)

175. Technician A says that EG and PG should not be mixed. Technician B says that if switching the type of antifreeze used, first the cooling system should be flushed completely. Which Technician is correct?
A. A only
B. B only
C. Both A and B
D. Neither A nor B (D10)

176. A transit bus with a hydraulically driven fan overheats frequently, especially when the ambient temperature rises above 85° F. Technician A says the radiator should be checked to make sure it is not obstructed with paper or dirt. Technician B says the next step is to check the hydraulic pressure to the fan motor to make sure the fan can spin fast enough to cause air flow through the radiator to cool the coolant sufficiently. Which Technician is correct?
A. A only
B. B only
C. Both A and B
D. Neither A nor B (D14)

177. Technician A says that a reading of .8 volts from the O_2 sensor when back probing and monitoring its voltage is an indication of a rich fuel condition. Technician B says that a reading of .2 volts is an indication of a lean fuel condition. Which Technician is correct?
A. A only
B. B only
C. Both A and B
D. Neither A nor B (E6)

178. When a fuel cylinder has reached its expiration date, Technician A says that it can be used as a compressed air tank because the pressures are much lower than that of CNG. Technician B says the cylinder must be destroyed because it cannot be used for any other purpose other than to store CNG as the label indicates. Which Technician is correct?
A. A only
B. B only
C. Both A and B
D. Neither A nor B (F1.2)

179. Technician A says the fuel high-pressure regulator is equipped with a PRD vent line that should be checked for fuel leakage at every preventive maintenance interval. Technician B says if fuel leakage is detected, the regulator assembly should be replaced before the vehicle is released back in service. Which Technician is correct?
   A. A only
   B. B only
   C. Both A and B
   D. Neither A nor B (F1.5)

180. The clear eye indicator on a sealed battery hydrometer is an indication that
   A. The battery is fully charged and ready to be tested.
   B. The battery is partially charged.
   C. The battery is discharged.
   D. The electrolyte level is very low, potentially the top of the plates are exposed, and the battery should be discarded. (G1)

181. The jump-starting of vehicles with different operating voltages is being discussed. Technician A says that a vehicle with a 12 volt system can be used to jump-start another vehicle with a 24 volt starter/system because the 12 volts will not exceed the 24 volts operating system of the vehicle needing to be jumped. Technician B says that a vehicle with a 24 volt system can be used to jump-start another vehicle with a 12 volt starter/system because the 12 volt batteries will act to control the overvoltage condition that could exist when connecting the cables. Which Technician is correct?
   A. A only
   B. B only
   C. Both A and B
   D. Neither A nor B (G4)

182. The LEAST-Likely function that a CNG engine starter solenoid performs is to
   A. Allow the starter drive to slip if the engine requires an excessive amount of power to crank the engine over.
   B. Help to disengage the starter drive gear.
   C. Transfer the power from the battery stud of the solenoid to the starter motor.
   D. Engage the starter drive gear mechanism. (G6)

183. A CNG engine is in the shop for repairs because it will not run. It cranks but will not fire. Upon further checks, the Technician discovers that none of the spark plugs are firing. Technician A says the ICM is defective because all the spark plugs are not firing and it needs to be replaced because the ICM is the component that turns the ground on and off to the coils, causing the spark to occur. Technician B says the ECM is the component that is defective because the ECM is the component that signals the ICM the timing and position of the crankshaft to initiate the firing of the spark plugs. Which Technician is correct?
   A. A only
   B. B only
   C. Both A and B
   D. Neither A nor B (H8)

184. Technician A says if the resistance measurement of a Detroit Diesel Series 50G COP coil between pins 1 and 2 is $\leq .55\ \Omega$, the coil needs to be replaced. Technician B says the resistance reading of a Cummins COP between the spark plug connection and the pins of the connector must be $\leq 20K\ \Omega$ or the coil must be replaced. Which Technician is correct?
   A. A only
   B. B only
   C. Both A and B
   D. Neither A nor B (H6)

185. Technician A says that when performing a starter load test, you should not crank the engine for more than 30 seconds at a time. Technician B says that you must perform this test with a battery charger connected to the battery. Which Technician is correct?
   A. A only
   B. B only
   C. Both A and B
   D. Neither A nor B (G8)

186. The Technician has finished replacing the HOS sensor of a CNG engine and is getting ready to check to make sure the engine is performing well. The Most-Likely next course of action to be undertaken is
   A. Change the spark plugs to make sure they are in good condition.
   B. Adjust the valves.
   C. Replace the fuel filters and air cleaner.
   D. Reset the AFR. (H2)

187. All of the following apply when charging batteries EXCEPT
   A. The battery temperature should not exceed 125° F while the battery is being charged.
   B. The caps of a conventional battery that can be removed to check and adjust the electrolyte level of the cells should be on tightly while the battery is charging.
   C. The battery should only be charged long enough to allow the green hydrometer to rise.
   D. The open circuit voltage of a fully charged battery is 12.4 volts. (G3)

188. Which of the following would Most-Likely result from excessive starting fluid usage during cold start-up?
   A. Engine runaway
   B. Cylinder hydraulic lock
   C. Damage to the fuel system sensors
   D. Damaged piston compression ring lands (C11)

189. A CNG transit bus is brought to the shop for repairs with the complaint that coolant needs to be added every day, though there are no apparent leaks. The LEAST-Likely cause of the need for coolant every day is
   A. Cold coolant leak.
   B. Small crack in the cylinder head.
   C. Loose spark plug adaptor tube in the cylinder head.
   D. Coolant is evaporating away. (D15)

190. Technician A says a fuel cylinder needs to be inspected whenever a CNG vehicle has been involved in an accident that could have damaged the cylinder, mounting brackets, or plumbing. Technician B says a CNG vehicle's fuel cylinders need to be inspected whenever it has been disconnected for servicing. Which Technician is correct?
   A. A only
   B. B only
   C. Both A and B
   D. Neither A nor B (F1.1)

191. Catalysts are being discussed. Technician A says the catalysts being used on CNG engines help to reduce the emission of hydrocarbons. Technician B says the catalysts help to control the emission of oxygen molecules. Which Technician is correct?
   A. A only
   B. B only
   C. Both A and B
   D. Neither A nor B (E5)

192. Technician A says that if extended life coolant (ELC) is purchased as a premix it should not be diluted with water. Technician B says that silicone radiator hoses should have spring-loaded constant torque clamps. Which Technician is correct?
    A. A only
    B. B only
    C. Both A and B
    D. Neither A nor B (D10)

193. When installing a wet cylinder sleeve, all of the following should be checked, EXCEPT
    A. The contact surfaces between the sleeve and the cylinder block.
    B. The O-ring grooves.
    C. The bore of the sleeve for scoring.
    D. The interference fit between the sleeve and block bore. (C5)

194. Technician A says that stuck-open thermostats cause the engine to run colder than normal and this condition can cause the engine oil to wear out prematurely. Technician B says that the engine oil and filters should be replaced prior to placing an engine back in service after the cylinder head needed repairing due to being cracked. Which Technician is correct?
    A. A only
    B. B only
    C. Both A and B
    D. Neither A nor B (D6)

195. Pistons are being discussed. Technician A says that CNG engine pistons can be of a two-piece design. Technician B says that CNG engine pistons can be of a one-piece design. Which Technician is correct?
    A. A only
    B. B only
    C. Both A and B
    D. Neither A nor B (C12)

196. Technician A says the piston rings should be twisted onto the ring lands to prevent from overstressing the rings and damaging them. Technician B says the piston rings should only be spread the required amount for installation to prevent from overstressing the rings and damaging them. Which Technician is correct?
    A. A only
    B. B only
    C. Both A and B
    D. Neither A nor B (C13)

197. An NGV transit bus comes to the shop for a loud clunking noise from the engine compartment when the engine is placed in gear. Technician A says the rear engine mounts are the cause and should be replaced. Technician B says the front engine mounts are the cause and should be replaced. Which Technician is correct?
    A. A only
    B. B only
    C. Both A and B
    D. Neither A nor B (A13)

198. A transit bus NGV engine cooling system is being discussed. Technician A says the pressure cap is the component that controls the pressure in the radiator and the cooling system. Technician B says the water pump increases the coolant flow within the engine as engine speed increases to ensure adequate coolant to all of the internal components of the engine. Which Technician is correct?
A. A only
B. B only
C. Both A and B
D. Neither A nor B (D8)

199. A Technician is trying to communicate with a CNG engine ECM using the diagnostic software. The diagnostic software returns the message that it cannot connect. Technician A says to check the interface module to make sure it is powered up by checking the power light emitting diode (LED) and making sure it is illuminated. Technician B says the proper interface driver and the correct communication port must be selected to allow the software to communicate with the ECM. Which Technician is correct?
A. A only
B. B only
C. Both A and B
D. Neither A nor B (A16)

200. Piston cooling nozzles are being discussed. Technician A says that piston cooling nozzles can only be removed after the pistons and the crankshaft are removed due to limited accessibility. Technician B says the piston cooling nozzles have to be removed before attempting to remove the pistons to prevent from damaging or bending the nozzles. Which Technician is correct?
A. A only
B. B only
C. Both A and B
D. Neither A nor B (C15)

201. Technician A says that some extended life coolants (ELC) require only one SCA treatment in six years. Technician B says that engine cooling fan engagement may be included in transmission hydraulic braking strategy. Which Technician is correct?
A. A only
B. B only
C. Both A and B
D. Neither A nor B (A14)

202. Technician A says that misalignment between pulleys must not exceed 1/16" for each 12" of distance between pulley-running centerlines. Technician B says that belts must not touch the bottom of grooves or protrude more than 3/32" above the top of the groove. Which Technician is correct?
A. A only
B. B only
C. Both A and B
D. Neither A nor B (D12)

203. Technician A says that if a small leak is detected coming from a CNG Swagelok fitting, it can be loosened and retightened until it stops leaking. Technician B says the fitting can only be tightened by an additional ¼ turn without removing the fuel pressure from the lines. Which Technician is correct?
A. A only
B. B only
C. Both A and B
D. Neither A nor B (F1.8)

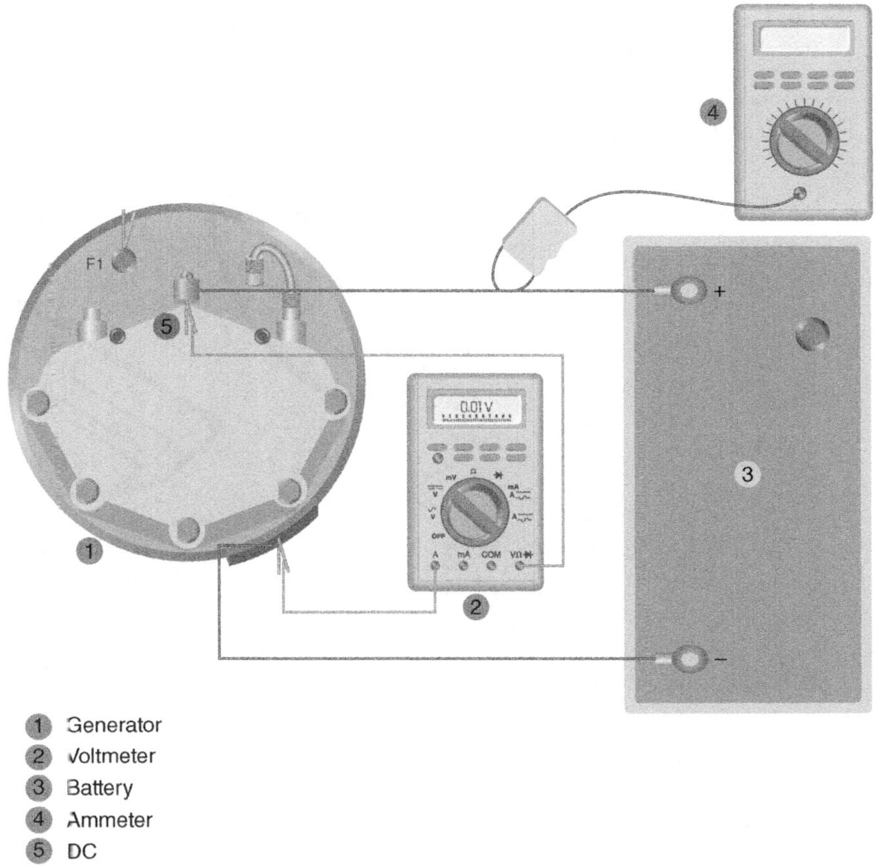

204. In the figure above, the Most-Likely test being performed is
    A. Amperage test.
    B. Voltage test.
    C. Voltage drop test.
    D. Voltage and amperage test. (G7)

205. Technician A says that the angle of the valve and its seat should be ground equal when valve rotators are used on a CNG engine. Technician B says that equal angles of the valve and its seat ensures more metal-to-metal contact to transfer some of the heat from the valve to the valve seat to prevent the valve from burning. Which Technician is correct?
    A. A only
    B. B only
    C. Both A and B
    D. Neither A nor B (B4)

206. While discussing a low-voltage problem at the starter, Technician A says that a corroded battery cable may cause this. Technician B says that a defective starter drive may cause this. Which Technician is correct?
    A. A only
    B. B only
    C. Both A and B
    D. Neither A nor B (G5)

207. In a two 12 V, 8D battery, 12 V electrical system, how are the batteries connected?
   A. Series circuit
   B. Parallel circuit
   C. Series/parallel circuit
   D. Isolation circuit (G8)

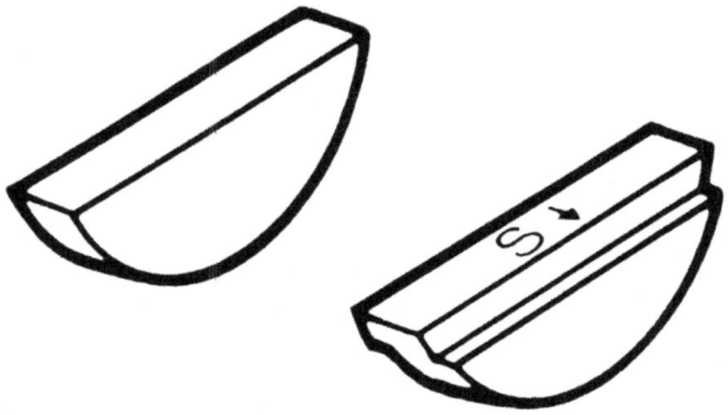

208. The figure above shows keys used on camshaft gears. Technician A says the key on the left is used to offset the timing and it must be replaced with the same type. Technician B says that the camshaft, gear, and key must be replaced because the camshaft Most-Likely jammed and started to shear the key. Which Technician is correct?
   A. A only
   B. B only
   C. Both A and B
   D. Neither A nor B (B11)

209. Which of these would be LEAST-Likely to cause low engine oil pressure?
   A. Worn bearings
   B. Use of oil with a high viscosity index
   C. Restricted suction tube
   D. Worn oil pump (A15)

210. An automatic belt tensioner appears to be weak and is causing the belt to slip when a heavy load is placed on the driven accessory. Technician A says the tensioner should be rotated in a position to create more tension on the belt. Technician B says the tensioner needs to be replaced. Which Technician is correct?
   A. A only
   B. B only
   C. Both A and B
   D. Neither A nor B (D7)

211. Technician A says the service manual will outline the torque specifications for the special fasteners for the components. Technician B says that some torque settings are generic. The size, thread, and grade of the bolt or fastener will determine the torque according to a specific chart. Which Technician is correct?
   A. A only
   B. B only
   C. Both A and B
   D. Neither A nor B (A18)

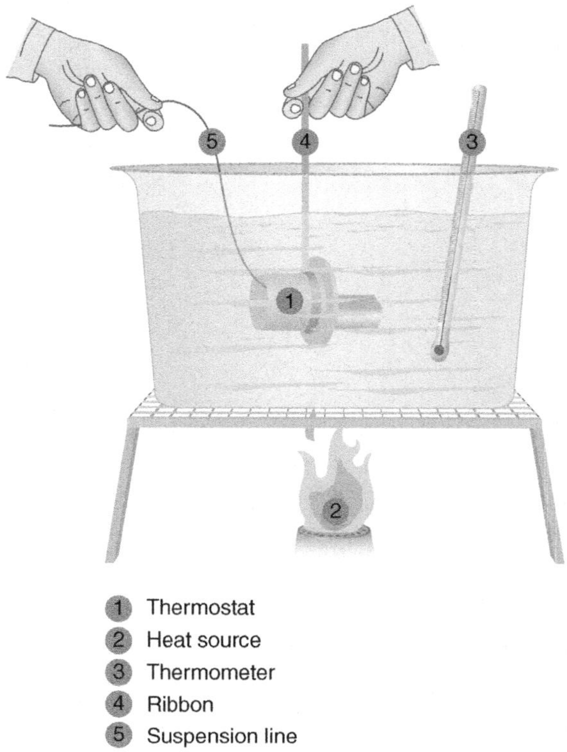

212. In the figure above the Most-Likely test the Technician is conducting is
   A. Determining the opening temperature of the thermostat.
   B. Determining whether the thermostat sticks open after cooling down.
   C. Checking the fully open temperature of the thermostat.
   D. Checking the closing temperature of the thermostat. (D9)

213. All of the following are true EXCEPT
   A. The fuel gauge has to be of a minimum pressure of 5000 psi when the working pressure of the system is 3600 psi.
   B. The PRD vent lines should be secured every 12" for lengths of greater than 24".
   C. CNG high-pressure fuel lines require an expansion loop to prevent stressing the line when temperature differentials occur.
   D. Fuel tanks can be removed from a vehicle with defueling them as long as the fuel shutoff valve can be turned off. (F1.4)

214. The rear two cylinders on an inline six-cylinder engine show erosion on the outside of the liner. Technician A says that this could be caused by low water pump flow. Technician B says that this could be caused by dirty oil. Which Technician is correct?
   A. A only
   B. B only
   C. Both A and B
   D. Neither A nor B (D12)

215. Technician A says that some engine ECM cases are grounded directly to the engine block or chassis through the mounting bolts. Technician B says that some ECMs are mounted using nonconductive isolators to prevent the ECM case from grounding and from being subjected to outside interferences. Which Technician is correct?
   A. A only
   B. B only
   C. Both A and B
   D. Neither A nor B (F3.2)

216. A thermostatically controlled fan, as shown in the figure above, may rely on all of the following to regulate fan operation EXCEPT
   A. A coil spring.
   B. Air temperature.
   C. Radiator temperature.
   D. Air pressure.

217. A PRD is being discussed. Technician A says the vent line from a PRD must be above the roof line to prevent fuel from accumulating in the roof area. Technician B says the PRD's eutectic material can be replaced in the field. Which Technician is correct?
   A. A only
   B. B only
   C. Both A and B
   D. Neither A nor B (F1.3)

218. A connection rod bearing shell is being checked to determine if it is reusable. Technician A says a ball-end micrometer can be used to measure the bearing shell thickness to determine if the bearing is within the manufacturer's specifications. Technician B says a dial caliper can be used to measure the bearing shell thickness to determine if the bearing is within the manufacturer's specifications. Which Technician is correct?
   A. A only
   B. B only
   C. Both A and B
   D. Neither A nor B (C14)

219. Emission labels are being discussed. Technician A says the emission certification labels are not required for CNG-powered transit buses and are only required for over the road CNG coaches. Technician B says that if an emission certification label is missing or illegible, a photocopy of another identical engine's label is an acceptable form of replacement. Which Technician is correct?
   A. A only
   B. B only
   C. Both A and B
   D. Neither A nor B (A17)

220. A CNG engine is being diagnosed with a low oil pressure light that comes on at idle but goes out when the engine is revved up. Technician A says the engine oil pump is defective and it should be replaced. Technician B says the main bearings are probably worn and need to be replaced.
 A. A only
 B. B only
 C. Both A and B
 D. Neither A nor B (A15)

221. One of the camshaft bearings fails to pass visual inspection due to a large nick in the bearing's surface. Technician A says you can ream out a defect in the camshaft bearing surface. Technician B says you cannot replace only one cam bearing; the complete set must be replaced. Which Technician is correct?
 A. A only
 B. B only
 C. Both A and B
 D. Neither A nor B (B11)

222. A new camshaft gear is being installed on a camshaft. Technician A says to check the keyway of the camshaft to make sure it is not cracked or worn. Technician B says to check the key and replace it if it shows signs of fretting or wear. Which Technician is correct?
 A. A only
 B. B only
 C. Both A and B
 D. Neither A nor B (C7)

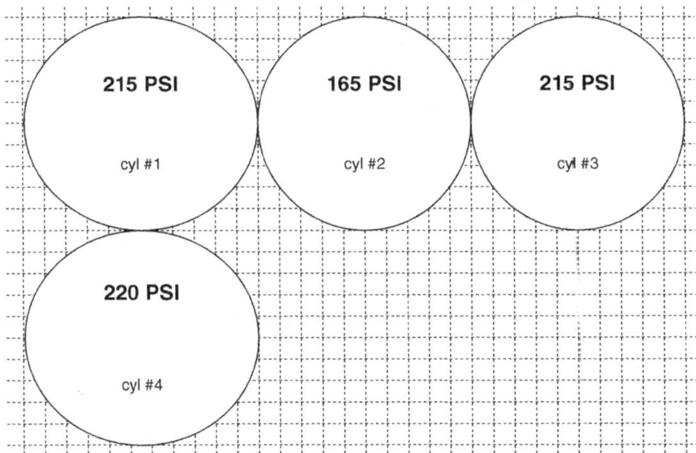

223. A compression test is being performed on a CNG engine. The indicated compression pressure is 220 to 215 psi except one cylinder that has a reading of 165 psi with three puffs of the piston Technician A says to put 2 oz. of engine oil into the cylinder and repeat the test to see if the pressure comes up to determine if the problem is with the compression rings or the valves. Technician B says that in order to isolate the compression rings as defective, a leak-down test can be performed. Which Technician is correct?
 A. A only
 B. B only
 C. Both A and B
 D. Neither A nor B (A20)

224. Technician A says that the starter ring gear is usually pinned or bolted to the flywheel assembly. Technician B says the ring gear is welded to the flywheel. Which Technician is correct?
   A. A only
   B. B only
   C. Both A and B
   D. Neither A nor B (C18)

225. Technician A says that after a water pump seal has been replaced, the engine should be run for a couple of minutes without any coolant to allow the ceramic sealing surfaces to seat with each other. Technician B says that when a water pump seal is replaced, the ceramic seals should be checked for scratches that could have occurred due to sand or grit contamination in the cooling system, which is evidence that the coolant filter may not be working correctly. Which Technician is correct?
   A. A only
   B. B only
   C. Both A and B
   D. Neither A nor B (D12)

226. Technician A says that when a fuel high-pressure regulator is replaced, the fuel fitting O-rings need to be replaced along with the regulator. Technician B says the connections need to be checked for leaks with a soapy water solution or Snoop before releasing the vehicle back in service. Which Technician is correct?
   A. A only
   B. B only
   C. Both A and B
   D. Neither A nor B (F1.5)

227. Oil traces on the underside of a COP are usually an indication of
   A. An oil leak in the engine rocker cover gasket.
   B. Use of the wrong type of coil.
   C. A potential crack in the coil.
   D. Normal condition. (H6)

228. When removing a starter motor from a vehicle, what should the Technician remove first?
   A. The starter bolts
   B. The starter solenoid
   C. The battery ground cable
   D. The starter magnetic switch (G6)

229. Engine fuel pre-ignition is being discussed. Technician A says that high intake air temperature caused by a malfunctioning fan can cause pre-ignition of the fuel and air mixture and knocking. Technician B says that a high turbo boost pressure can cause pre-ignition of the fuel air mixture and knocking as well. Which Technician is correct?
   A. A only
   B. B only
   C. Both A and B
   D. Neither A nor B (H1)

230. The accessories such as the driver's fan and the HVAC of a transit bus that is equipped with a 24 volt Delco-Remy 50 DN Alternator are not functioning. Technician A says the Relay (R) terminal on the 24 volt alternator should be checked for ½ of the charging voltage of the alternator. Technician B says the fuse for the HVAC system should be checked to make sure it is not burned out. Which Technician is correct?
   A. A only
   B. B only
   C. Both A and B
   D. Neither A nor B (G7)

231. Technician A says the roller cam followers need to be installed in the engine dry. Technician B says the roller cam followers need to be coated with bearing grease before being installed in the engine. Which Technician is correct?
   A. A only
   B. B only
   C. Both A and B
   D. Neither A nor B (B9)

232. Technician A says that if a tensioner has lost the ability to control belt tension due to a weak spring, a shorter belt can be installed to increase the tension. Technician B says the tensioner needs to be replaced. Which Technician is correct?
   A. A only
   B. B only
   C. Both A and B
   D. Neither A nor B (A19)

233. When installing the wrist pin retaining snap rings in an articulating piston assembly, what is the most commonly recommended position of the snap ring gap?
   A. Downward
   B. Upward
   C. Horizontal
   D. Does not matter (C11)

234. Technician A says the oil filter gasket should be lubed with clean engine oil prior to mounting it to the engine. Technician B says that the oil filter housing bypass valve needs to be removed and inspected every time a filter is replaced. Which Technician is correct?
   A. A only
   B. B only
   C. Both A and B
   D. Neither A nor B (D3)

235. The LEAST-Likely reference that will contain part number information about preventive maintenance components such as oil filters, acceptable lubricants, and preventive maintenance schedules is
   A. A parts manual.
   B. A service manual.
   C. A manufacturer's electronic reference CD.
   D. An operator's manual. (A18)

236. When performing failure analysis on a set of crankshaft main bearings, one is observed to be heat discolored on the backing (block) side. Which of the following would be Most-Likely to create this condition?
   A. Excessive big end play
   B. Insufficient bearing crush
   C. Belled crankshaft main journal
   D. Degraded engine lube (C9)

237. The piston shown in the figure came from an inline six-cylinder CNG engine. All the other pistons showed normal wear. This was the only piston showing damage. Which of the following is the Most-Likely cause?
   A. Partially worn exhaust cam lobe
   B. Restricted exhaust system
   C. Restricted air cleaner
   D. Overfueling condition of the engine (C15)

238. All of the following are correct procedures if a pressure cap is found to be defective EXCEPT
   A. Replace the cap.
   B. Test the coolant.
   C. Check the radiator hoses.
   D. Adjust or replace the spring in the cap. (D13)

239. Technician A says that fluid or viscous type damper can be checked for proper operation by operating the engine for approximately 30 minutes at different loads and speeds. After 30 minutes the engine is turned off and the temperature of the damper is measured. A damper that reads ambient temperature is an indication that it is not functioning correctly. Technician B says that before a damper is installed back on the crankshaft, the mounting bolt holes have to be checked for signs of elongations. If any of the bolt holes are elongated the damper needs to be replaced. Which Technician is correct?
   A. A only
   B. B only
   C. Both A and B
   D. Neither A nor B (C16)

240. When performing a load test on a fully charged battery, the Technician finds that the battery voltage drops below manufacturer's specifications. The Technician should do which one of the following?
   A. Recharge the battery and return it to service.
   B. Replace the battery.
   C. Check the alternator belt to make sure it is not slipping.
   D. Replace the voltage regulator. (G2)

241. The Most-Likely resistance reading of a CNG spark plug wire is
   A. 50 to 150 Ω.
   B. 500 to 1.5K Ω.
   C. 5K to 15K Ω.
   D. 500 to 1.5K Ω per foot of wire. (H5)

242. The LEAST-Likely cause of a misfiring cylinder is:
    A. Defective coil.
    B. Too large of a spark plug gap.
    C. Loose valve adjustment.
    D. Cracked piston. (H2)

243. When installing a new starter relay or mag switch because it tested as defective, all of the following steps are required EXCEPT
    A. Disconnect the battery to the chassis ground connection first.
    B. Disconnect the starter positive cable at the starter solenoid.
    C. Remove and install the new starter relay and wiring.
    D. Reinstall the battery to chassis ground connection. (G6)

244. A CNG transit bus with a standard 12 V charging/cranking circuit has to be jump-started. Technician A says that the battery-to-battery connections should be made with the boost vehicle engine switched off. Technician B says the positive jumper cable clamps should be connected last and disconnected first. Which Technician is correct?
    A. A only
    B. B only
    C. Both A and B
    D. Neither A nor B (G4)

245. Technician A says that flywheel housing concentricity needs to be checked when the engine is equipped with a flywheel. Technician B says the flywheel housing concentricity needs to be checked when the engine is equipped with flexplates. Which Technician is correct?
    A. A only
    B. B only
    C. Both A and B
    D. Neither A nor B (C17)

246. The ECM of an NGV engine can be programmed to perform each one of the following when an overheating condition exists EXCEPT
    A. Illuminate a warning light and immediately turn off the engine.
    B. Illuminate a warning light and gradually ramp down the power of the engine until it shuts down or the overheating condition no longer exists.
    C. Illuminate a warning light and allow the engine to continue to run.
    D. Illuminate a warning light and give the driver up to 30 seconds before the engine is shut down. (D8)

247. Technician A says the valve adjustments can be performed when the engine is hot if the mechanic removes 0.002" from the adjustment to compensate for the expansion of the engine parts (e.g., 0.031" − 0.002" = 0.029"). Technician B says the engine temperature needs to be lower than 175° F. Which Technician is correct?
    A. A only
    B. B only
    C. Both A and B
    D. Neither A nor B (B10)

248. A Technician is measuring the piston to cylinder clearances and finds that there is not enough clearance during engine reassembly. Technician A says the cylinder liner should be honed out to provide the correct clearance. Technician B says the liner or the piston should be replaced. Which Technician is correct?
    A. A only
    B. B only
    C. Both A and B
    D. Neither A nor B (C12)

249. Technician A says that fuel line icing at the outlet of the high-pressure regulator is not normal and that icing is an indication the temperature of the fuel is dropping too low due to an obstruction in the fuel system or the high-pressure regulator is malfunctioning. Technician B says the cause of icing is normal because as the fuel goes from high pressure to low pressure, the lowering of the pressure has a cooling effect and does not require any attention unless a drivability issue develops. Which Technician is correct?
    A. A only
    B. B only
    C. Both A and B
    D. Neither A nor B (A6)

250. A CNG engine is being tuned up. Technician A says that a different style (hotter) spark plug can be installed in an engine if there is a problem with the spark plugs fouling out due to oil contamination in the combustion chamber from worn valve guide seals. Technician B says that only approved spark plugs can be installed in the engine because the engine's emissions were certified with this particular style plug and a different plug could change the emission levels beyond the legal limits. Which Technician is correct?
    A. A only
    B. B only
    C. Both A and B
    D. Neither A nor B (A17)

251. After the Technician cleans a disassembled engine block, what should be done next?
    A. Paint the exterior.
    B. Pressure test the block to check for cracks.
    C. Remove the cylinder sleeves.
    D. Plug all oil ports. (C2)

252. The LEAST-Likely cause of low compression in a cylinder of a CNG engine is
    A. A leaking intake valve seat.
    B. Worn exhaust valve seals.
    C. A scuffed cylinder wall.
    D. A blown cylinder head gasket. (A20)

253. Technician A says that some coolant filters are only designed to remove sand and grit from the cooling system. Technician B says that some coolant filters are designed to remove sand, grit and also to condition the coolant with SCA. Which Technician is correct?
    A. A only
    B. B only
    C. Both A and B
    D. Neither A nor B (D11)

254. When refueling a Type IV cylinder that has been replaced, all of the following apply, EXCEPT
    A. The cylinder should be filled in increments to allow the liner to expand and conform to the carbon fiber structure.
    B. The cylinder should be leak tested at each increment of pressure increases to ensure there are no leaks.
    C. The cylinder should be pressurized to 500 psi with CNG and then purged out to remove entrapped air from the cylinder.
    D. The cylinder brackets need to be torqued before the cylinder is fueled. (F1.7)

255. Technician A says that the battery term ampere hour refers to stored charge capacity of a battery. Technician B says that a 75-ampere hour charge applied to a 200-ampere hour battery should turn the charge indicator green. Which Technician is correct?
    A. A only
    B. B only
    C. Both A and B
    D. Neither A nor B (G3)

256. Spark plugs are being installed in a CNG engine. Technician A says the spark plug should be torqued to manufacturer's specification, usually around 15 lb/ft, by using an accurate torque wrench. Technician B says the spark plug should be torqued to manufacturer's specification, usually around 28 lb/ft, by using an accurate torque wrench. Which Technician is correct?
    A. A only
    B. B only
    C. Both A and B
    D. Neither A nor B (H4)

257. A CNG engine will crank but will not run. The crankshaft sensor is found to be damaged on the end. Technician A says the crankshaft should be checked for damage as well. Technician B says the end play of the crankshaft should be checked to ensure the end play is not the cause of the sensor striking the target and the damage. Which Technician is correct?
    A. A only
    B. B only
    C. Both A and B
    D. Neither A nor B (H7)

258. All of the following would be used while performing an alternator amperage output test EXCEPT
    A. A carbon pile load tester.
    B. An amp clamp.
    C. A hydrometer.
    D. A voltmeter. (G7)

259. An NGV transit bus is being discussed. The complaint is the engine has difficulty gaining speed and lacks power at higher speeds with no signs of black smoke. Technician A says the air cleaner needs to be checked for excessive restriction with a magnehelic gauge measuring inches of water. Technician B says the spark plugs probably need to be replaced. Which Technician is correct?
    A. A only
    B. B only
    C. Both A and B
    D. Neither A nor B (E1)

260. The maximum OEM recommended length of fuel line from the fuel low-pressure shutoff solenoid and the fuel low-pressure regulator of CNG engines is
    A. 35"
    B. 25"
    C. 30"
    D. 18" (F2.1)

261. Technician A says that when the battery cable terminals are corroded they can be cleaned with baking soda and water solution and reinstalled. Technician B says that the cable may be corroded under the insulation and may not be noticeable and need replacement. Which Technician is correct?
    A. A only
    B. B only
    C. Both A and B
    D. Neither A nor B (G5)

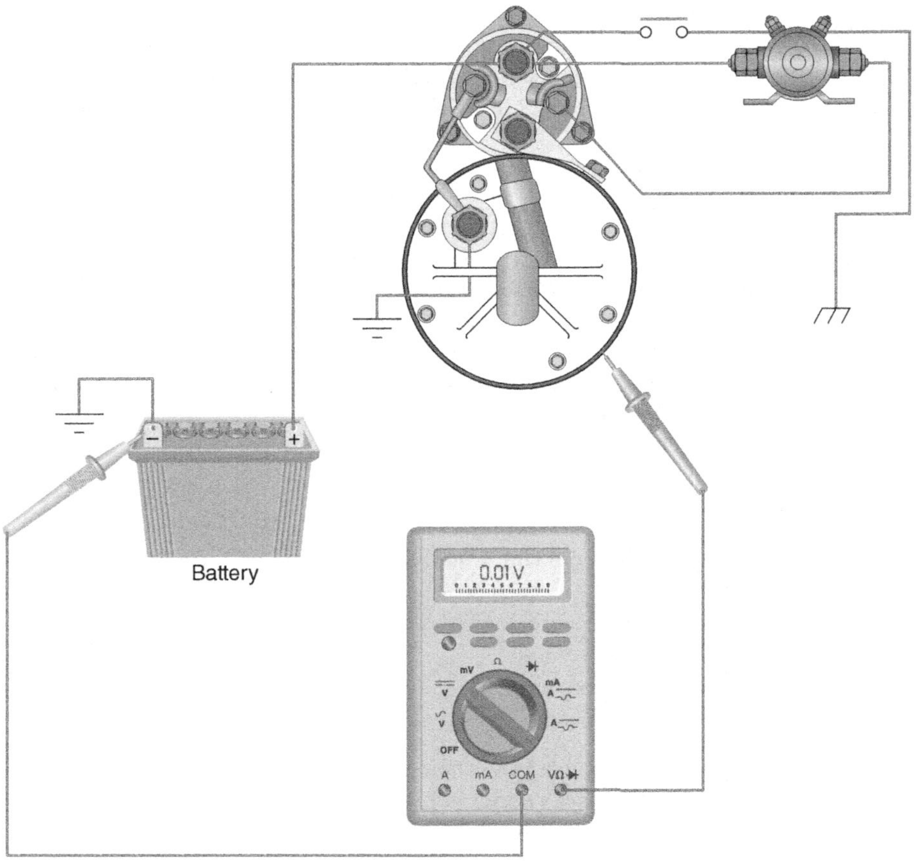

262. What test will be performed in the figure above when the starter is energized?
   A. Starter motor source voltage
   B. Voltage drop of the starter ground circuit
   C. Voltage drop of the starter positive circuit
   D. Resistance in the starter control circuit (G8)

263. A thermostatic fan is being tested for operation. Technician A says that a piece of cardboard can be inserted in front of the radiator to restrict airflow and more rapidly increase the temperature of the coolant to cause the fan to sense the higher temperature and engage more quickly. Technician B says the vehicle should be driven before inserting the cardboard in front of the radiator to preheat the engine and to help raise the temperature of the coolant. Which Technician is correct?
   A. A only
   B. B only
   C. Both A and B
   D. Neither A nor B (D14)

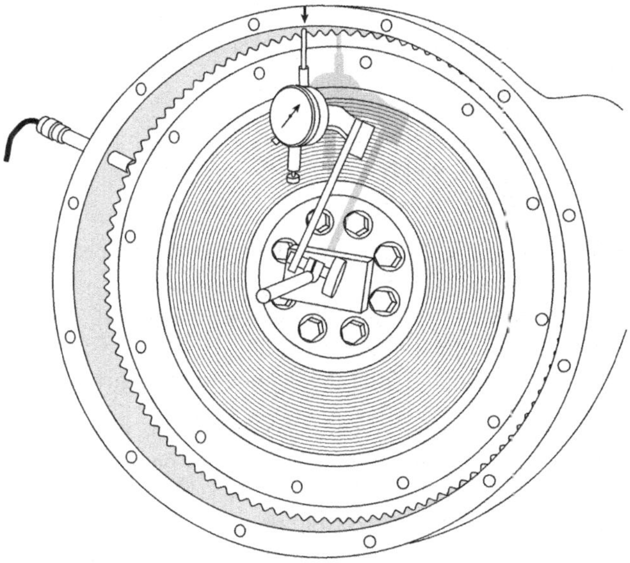

264. The setup shown in the figure above is used to check what on the flywheel housing?
   A. Lateral runout
   B. Face lateral runout
   C. Radial runout
   D. Pilot bore radial runout (C18)

265. Technician A says the spring on the constant torque hose clamp is to allow for expansion and contraction while maintaining constant pressure to prevent leaks or cutting the hoses. Technician B says the spring on the hose clamps is to allow the clamp to be tightened all the way so the hoses will not get cut. Which Technician is correct?
   A. A only
   B. B only
   C. Both A and B
   D. Neither A nor B (D15)

266. The end play of a camshaft is 0.001". The specifications call for a minimum of 0.003". Technician A says this condition is not too critical because the end play is only a little out of specification and the clearance will adjust itself as the engine breaks in. Technician B says the clearance must be corrected but to check the thrust washers to make sure there is no foreign material interfering with the clearances before replacing the thrust washers to adjust the clearance. Which Technician is correct?
   A. A only
   B. B only
   C. Both A and B
   D. Neither A nor B (C7)

267. An NGV engine is being diagnosed with knock codes. Technician A says that engine knock codes can be caused by a defective belt(s). Technician B says that engine knock codes can be caused by a defective tensioner bearing. Which Technician is correct?
   A. A only
   B. B only
   C. Both A and B
   D. Neither A nor B (A19)

268. Technician A says that a micrometer can be used to measure the crankshaft journals. Technician B says that when cleaning crankshaft journals, a wire wheel on a die grinder can be used. Which Technician is correct?
   A. A only
   B. B only
   C. Both A and B
   D. Neither A nor B (C8)

269. All of the following procedures to install an engine oil filter are correct EXCEPT
   A. Apply grease on the gasket or seal.
   B. Hand tighten the filter to the housing.
   C. Pre-fill the filter if possible.
   D. Tighten the filter the amount of turns printed on the filter canister. (D6)

270. To maintain an effective cooling system, all of the following must be done EXCEPT
   A. The radiator and engine must be back-flushed annually.
   B. Coolant should be tested for SCA strength.
   C. Air must be bled from the engine during refilling of the system.
   D. Coolant should be tested for antifreeze protection level. (A14)

271. All of the following are used to determine battery performance and condition EXCEPT
   A. A multimeter.
   B. A carbon pile load tester.
   C. A hydrometer.
   D. A voltage drop from battery post to positive terminal. (G2)

272. Piston rings are being installed onto a piston. The Most-Likely ring gap arrangement is
   A. Staggered 30° apart around the piston crown.
   B. Aligned with the wrist pin.
   C. Aligned in a straight line.
   D. Staggered 90° apart from each other. (C13)

273. Technician A says that by using a pressure cap on the radiator, the boiling point of the coolant is raised. Technician B says that antifreeze raises the boiling point of water. Which Technician is correct?
   A. A only
   B. B only
   C. Both A and B
   D. Neither A nor B (D8)

274. An NGV-powered transit bus comes to the shop with a complaint that the coolant needs to be replenished every day. Upon opening the surge tank, a loud hissing sound is heard and it appears that traces of mercaptan are also coming from the surge tank. Technician A says the pressure cap is defective and should be replaced. Technician B says the cooling system should be tested with an electronic methane leak detector for traces of methane in the cooling system that could be coming from a defective high-pressure regulator. Which Technician is correct?
   A. A only
   B. B only
   C. Both A and B
   D. Neither A nor B (D13)

275. A transit vehicle with a CNG engine is being repaired. The key switch is on, engine off. The ECM has a historical DTC logged in memory. The Technician clears the code and it does not return. Technician A says that because the code was historical, the vehicle is OK to release back in service. Technician B says the engine should be started and run to see if the code returns. Which Technician is correct?
    A. A only
    B. B only
    C. Both A and B
    D. Neither A nor B (F3.2)

276. The Most-Likely cause of scale formation in a cooling system is
    A. Cooling system is running too hot.
    B. Too much SCA in the cooling system.
    C. Too high of a concentration of antifreeze in the cooling system.
    D. Using hard water to replenish the coolant in the radiator. (D11)

277. A CNG engine will crank but will not start. Upon observing the throttle actuator valve, it is found it will not open while the engine cranks. Technician A says to check the fuse for the switched power for the throttle actuator. Technician B says to check the ground connection for the throttle actuator. Which Technician is correct?
    A. A only
    B. B only
    C. Both A and B
    D. Neither A nor B (F3.1)

278. A vehicle is being inspected for preventive maintenance by a Technician. The Technician discovers the engine wiring harness is touching the exhaust manifold and has burnt through several of the wires' insulation. The LEAST-Likely acceptable form of repair is
    A. The defective sections can be cut out and the wires butt-spliced back together as long as the butt-splices have built-in shrink tubing to protect them from water intrusion.
    B. The wires should be separated and taped over to protect the wires from touching together.
    C. Replace the wiring harness.
    D. Cut out the defective sections of the wire(s), solder the two ends together, and apply shrink tubing over the repaired wire(s). (A3)

279. All of the following statements about installing camshaft bushings are true EXCEPT
    A. The bushings need to be aligned with the oil holes in the bores to ensure lubrication will be able to reach the journals.
    B. If the bushings need to be removed from the engine block for any reason, they can be reused if they are not damaged during the removal process.
    C. The correct bushing driver needs to be used while driving the bushings into the bore.
    D. Cam bushings need to be removed from the engine block before hot tanking the block. (C6)

280. A cylinder block main bearing bore is being checked. Technician A says to use a master bar to check the main bore alignment. Technician B says to use an inside micrometer to check the main bore diameter for an out-of-round condition. Which Technician is correct?
    A. A only
    B. B only
    C. Both A and B
    D. Neither A nor B (C8)

281. Technician A says that a diagnostic tool or the appropriate software is the only method to reset DTCs logged in the ECM of CNG engines. Technician B says that turning off the battery disconnect switch will clear the DTCs from the memory of the ECM of CNG engines. Which Technician is correct?
    A. A only
    B. B only
    C. Both A and B
    D. Neither A nor B (F3.2)

282. Which of the following is LEAST-Likely to be used to measure backlash in the timing gear train?
    A. Dial indicator
    B. Feeler gauge
    C. Depth micrometer
    D. Digital caliper (C10)

283. Technician A says that coolant is routed through the fuel high-pressure regulator to raise the temperature of the fuel on cold mornings due to the cooling effect on fuel when the pressure is dropped from high pressure (3600 psi) to low pressure (110 psi). Technician B says that coolant is also routed through the low-pressure regulator to further assist in warming up the fuel on cold mornings.
    A. A only
    B. B only
    C. Both A and B
    D. Neither A nor B (A6)

284. Technician A says that main bearing shells should receive a light coat of oil on both the face and back before final installation. Technician B says that when checking the bearing, install Plastigage in all the bearings, torque to specifications, and rotate the crankshaft. Which Technician is correct?
    A. A only
    B. B only
    C. Both A and B
    D. Neither A nor B (C9)

285. All of the following are correct about vibration dampers EXCEPT
    A. Some dampers are fluid filled.
    B. Some dampers have a rubber ring mounted between the hub and the damper.
    C. A fluid-type damper only needs replacing when the viscous fluid is leaking.
    D. The damper helps to protect the crankshaft from breaking by absorbing damaging vibrations. (C16)

286. The LEAST-Likely method to correct insufficient ring end gap is
    A. File the end gap until the correct gap is achieved.
    B. Replace the rings with new ones with acceptable ring gaps.
    C. Hone the cylinder out until the gaps are within specifications.
    D. Replace the cylinder, piston, and rings as an assembly. (C13)

287. A four-stroke cycle, CNG engine has a vibration occurrence rate that is equal to half the engine rpm. Technician A says that this might be caused by a damaged belt tensioner. Technician B says that this could be caused by a bent connecting rod. Which Technician is correct?
    A. A only
    B. B only
    C. Both A and B
    D. Neither A nor B (A13)

288. Technician A says the amount of metal removed from the engine block deck surface to repair the cylinder counterbores is critical because if too much is removed the pistons could strike the cylinder head. Technician B says the machine shop usually stamps the amount of metal removed from the deck surface on the oil pan rail at the front of the engine block. Which Technician is correct?
    A. A only
    B. B only
    C. Both A and B
    D. Neither A nor B (C3)

289. A vehicle comes to the shop for its preventive maintenance cycle. During the inspection, it is discovered to have a vibration damper with a ding approximately ½" in diameter and ¼" deep. Technician A says this is not a defect that needs to be addressed now unless it leaks fluid. Technician B says this is not a problem unless the engine has a vibration that is more pronounced as the engine rpms are raised. Which Technician is correct?
    A. A only
    B. B only
    C. Both A and B
    D. Neither A nor B (C16)

290. A Technician is checking the oil clearance of a connecting rod bearing with Plastigage and discovers the clearance is too small. Technician A says to clean the rod bore and the back of the shell and recheck the clearance. Technician B says that a piece of crocus cloth can be used to polish and remove metal from the connecting rods bore to allow the clearance to go to specifications. Which Technician is correct?
    A. A only
    B. B only
    C. Both A and B
    D. Neither A nor B (C14)

291. A Technician has found the gear train of a CNG engine has several broken teeth. The Most-Likely cause of the broken teeth is
    A. Lack of lubrication.
    B. Foreign objects falling in the gears.
    C. Normal wear and tear.
    D. Improper assembly. (C10)

292. A CNG engine–powered transit bus has a vibration only when the vehicle is moving and progressively increases as speed increases. Technician A says the flywheel is Most-Likely out of balance and must be checked for damage or cracks. Technician B says the drive shaft and tires need to be checked because it appears to be a speed-induced vibration not related to the engine or transmission operation. Which Technician is correct?
    A. A only
    B. B only
    C. Both A and B
    D. Neither A nor B (A13)

293. Technician A says that if an NGV engine fails to reach operating temperature, excessive condensation can form in the crankcase. Technician B says that if the oil becomes emulsified, this will shorten the life of the oil additives and the by-products of combustion such as corrosion and oxidation can attack and destroy the bearings. Which Technician is correct?
    A. A only
    B. B only
    C. Both A and B
    D. Neither A nor B (D9)

294. Technician A says that most engines require the pistons and connecting rods be aligned in a certain direction and the direction must be observed during the installation. Technician B says the pistons and connecting rod assemblies can be installed in any direction. Which Technician is correct?
    A. A only
    B. B only
    C. Both A and B
    D. Neither A nor B (C15)

7 Appendices

Answers to the Test Questions for the Sample Test Section 5

1. A
2. A
3. C
4. C
5. B
6. C
7. C
8. B
9. A
10. D
11. A
12. B
13. A
14. A
15. B
16. B
17. A
18. B
19. D
20. B
21. C
22. C
23. A
24. B
25. C
26. C
27. A
28. C
29. B
30. C
31. C
32. B
33. C
34. D
35. A
36. B
37. A
38. C
39. C
40. A
41. B
42. A
43. A
44. A
45. B
46. C
47. B
48. A
49. A
50. D
51. C
52. B
53. C
54. A
55. B
56. A
57. A
58. D
59. D
60. B
61. B
62. B
63. B
64. A
65. B
66. A
67. C
68. B
69. D
70. B
71. C
72. B
73. C
74. D
75. D
76. C
77. A
78. B
79. D
80. D
81. B
82. C
83. B
84. A
85. A
86. A
87. B
88. D
89. A
90. D
91. C
92. D
93. A
94. C
95. C
96. B
97. C
98. C
99. C

Explanations to the Answers for the Sample Test Section 5

Question #1
Answer A is correct. Only Technician A is correct. Any exhaust leak in the engine compartment other than from the exhaust outlet pipe above the roof must be corrected.
Answer B is incorrect. Any exhaust leak in the engine compartment must be corrected.
Answer C is incorrect. Only Technician A is correct.
Answer D is incorrect. Technician A is correct.

Question #2
Answer A is correct. Only Technician A is correct in saying that when an oil pump is disassembled all mating surfaces must be checked to specification.
Answer B is incorrect. Before an engine oil pump is replaced during an overhaul, it should be disassembled and thoroughly checked. Gear rotor pumps tend to wear most between the lobes on the impeller and on the apex of the lobes on the rotor ring. These dimensions should be checked to OEM specifications using a micrometer. The gear-to-body clearance should be checked with a thickness gauge sized to the OEM maximum clearance specification.
Answer C is incorrect. Only Technician A is correct.
Answer D is incorrect. Technician A is correct.

Question #3
Answer A is incorrect. Worn valve guides can cause blue smoke.
Answer B is incorrect. Worn valve seals can cause blue smoke.
Answer C is correct. A spark plug broken electrode will not cause blue smoke, and is therefore the exception.
Answer D is incorrect. Worn piston rings can cause blue smoke.

Question #4
Answer A is incorrect. An inside micrometer can be used to detect a cylinder out-of-round condition.
Answer B is incorrect. A telescoping gauge can be used to detect a cylinder out-of-round condition.
Answer C is correct. The LEAST-Likely tool to detect this condition would be a dial indicator.
Answer D is incorrect. A dial bore gauge can be used to detect a cylinder out-of-round condition.

Question #5
Answer A is incorrect. The low-pressure fuel regulator controls the fuel delivery to the engine and is supplied at a pressure of around 100 psi.
Answer B is correct. Only Technician B is correct. The high-pressure fuel regulator regulates the high fuel pressure of as much as 3600 psi to around 100 psi.
Answer C is incorrect. Only Technician B is correct.
Answer D is incorrect. Technician B is correct.

Question #6
Answer A is incorrect. Technician B is also correct.
Answer B is incorrect. Technician A is also correct.
Answer C is correct. Both Technicians are correct. When the intake air temperature is too high, the cylinder temperatures rise. The rise of temperature can become high enough to allow the air fuel mixture to auto ignite. The IAT sensor reading being too low could be a cause of this situation. The diagnostic software will allow a Technician to read the temperature before starting the engine. It should be compared with the ambient air temperature reading or the Engine Coolant Temperature sensor when the engine is at ambient temperature.
Answer D is incorrect. Both Technicians are correct.

Question #7

Answer A is incorrect. Too large a fuse can cause an overload situation and could potentially cause the wire to burn into the harness.

Answer B is incorrect. A loose ground connection causes resistance, which causes heat. The heat can travel though a wire and damage other wires as well as the harness.

Answer C is correct. An insulated clamp provides the best protection from chaffing and rubbing through the insulation.

Answer D is incorrect. A zip tie is a poor device to secure a harness especially in a harsh environment. It does not insulate the harness from chaffing against the component or frame to which it is secured. In an engine compartment, high temperatures can cause the tie to fail, allowing the harness to drop. The harness can fall on a hot exhaust component or into a moving fan or the drive line.

Question #8

Answer A is incorrect. A low fuel pressure condition will not prevent the engine starter from cranking the engine.

Answer B is correct. Only Technician B is correct. Most modern NGV vehicles use some form of programmable controls. Electrical schematics will not describe what safety conditions must be met to allow the starter to engage and crank. Therefore, logic diagrams must be consulted as a diagnostic tool.

Answer C is incorrect. Only Technician B is correct.

Answer D is incorrect. Technician B is correct.

Question #9

Answer A is correct. Only Technician A is correct. Cylinder block warpage is correctly measured with a straightedge and thickness gauges. A typical maximum specification will be in the range of 0.004–0.005 in. (0.10–0.127 mm) overall, and about 0.003 in. (0.076 mm) side to side. Readings beyond this specification indicate the need to resurface the block fire deck. However, you should always refer to the OEM service manual for the correct value. Blocks do not warp nearly as often as cylinder heads. When the block fire deck is warped or not parallel to the main bearing bores, it may be resurfaced with a milling machine or grinder. Next, check the main bearing bore and alignment. It is good shop practice to record the amount of stock removed from the deck by stamping the amount removed on the cylinder block pad.

Answer B is incorrect. A dial indicator is not used to check block deck warpage.

Answer C is incorrect. Only Technician A is correct.

Answer D is incorrect. Technician A is correct.

Question #10

Answer A is incorrect. The guidelines have to be immediately available to the Technician anytime a cylinder is being inspected.

Answer B is incorrect. The minimum recommended tools have to be available to the Technician at all times when a cylinder(s) is being inspected.

Answer C is incorrect. Neither Technician is correct.

Answer D is correct. Neither Technician is correct. The guidelines as well as the minimum recommended tools have to be immediately available to the Technician anytime a cylinder is being inspected.

Question #11

Answer A is correct. Only Technician A is correct. A leak in an air hose upstream from the turbocharger compressor can allow unfiltered (grit-laden) air to be pulled in. Abrasives in the air can wear the compressor wheel vanes to a knife-edge sharpness.

Answer B is incorrect. Technician B is incorrect because the aftercooler is located downstream from the compressor housing.

Answer C is incorrect. Only Technician A is correct.

Answer D is incorrect. Technician A is correct.

Question #12
Answer A is incorrect. Driving conditions such as dusty, sandy, or city driving will affect how long an air cleaner will last but not the use of a cleaner-burning fuel. The air intake is positioned in a manner to prevent the engine from recirculating exhaust. Only by testing the air cleaner element for restriction can a Technician make a decision as to reuse or replace an air cleaner.
Answer B is correct. Only Technician B is correct. The manometer or magnehelic gauge should be used to verify the air cleaner is in need of replacement. Restriction gauges can indicate a potential defect of the air cleaner but are not as accurate and should only be used as a guide.
Answer C is incorrect. Only Technician B is correct.
Answer D is incorrect. Technician B is correct.

Question #13
Answer A is correct. Only Technician A is correct. The engine should be cleaned before an accurate diagnosis can be performed.
Answer B is incorrect. The oil pan gasket is not the only source of leaks in the lower engine area; the front or rear crankshaft seals could also be at fault.
Answer C is incorrect. Only Technician A is correct.
Answer D is incorrect. Technician A is correct.

Question #14
Answer A is incorrect. An air-to-air aftercooler is shown.
Answer B is correct. The diagram shows a typical chassis-mounted charge air cooler that uses ram air as the cooling medium. Ram air is the airflow created as a vehicle moves along the highway. Air-to-air coolers have the appearance of a coolant radiator and are often chassis-mounted in front of the radiator. As the vehicle moves down the highway, ambient air is forced through the fins and element tubing. Ram air is therefore the cooling medium. Cooling efficiencies are highest when the vehicle is traveling at higher speeds. They are least efficient when the airflow through the cooler is minimal. Under optimal conditions, air-to-air heat exchangers have better efficiency ratios than any liquid-cooled heat exchangers. Thermal efficiency is increased by the air-to-air method of charge air cooling. If liquid-cooled heat exchangers are used, actual intake air temperatures are reduced to values only slightly higher than the coolant temperature values.
Answer C is incorrect. This is not a tip turbine type air cooler.
Answer D is incorrect. Exhaust gas routed through the charge air cooler would warm the air, not cool it.

Question #15
Answer A is incorrect. High exhaust back pressure would cause low power and overheating.
Answer B is correct. High manifold boost increases the air density and the potential to produce high power, so this would be the LEAST-Likely cause of a lack of power complaint.
Answer C is incorrect. Clogged cooling fins would cause high air inlet temperatures, which would reduce horsepower.
Answer D is incorrect. A stuck-open waste gate would not allow the turbocharger to spool up and deliver boost to the engine or the intake manifold pressure sensor. The ECM would not allow fuel to be increased sufficiently to allow the engine to develop power.

Question #16
Answer A is incorrect. Removing gasket material is a step.
Answer B is correct. A cylinder head must be removed from the cylinder block to be checked for cracks so checking the torque of the cylinder head bolts would not be part of the inspection procedures.
Answer C is incorrect. Magnetic crack detection is a good detection method.
Answer D is incorrect. Removing carbon buildup is necessary to find cracks.

Question #17
Answer A is correct. Only Technician A is correct. High crankcase pressure should be suspected as the cause of the engine oil dipstick to move out of the tube. A manometer gauge graduated in inches of water should be utilized for accurate measuring. The manufacturer's specifications should be consulted to determine acceptable pressure. Troubleshooting should include: worn piston compression rings, worn valve guides, blown head gasket, clogged breather, etc.

Answer B is incorrect. Wiring the dipstick in place will not repair the cause of high crankcase pressure. If high crankcase pressure is not corrected, gaskets or oil seals could become damaged, and oil could be pushed out of the breather tube as well.
Answer C is incorrect. Only Technician A is correct.
Answer D is incorrect. Technician A is correct.

Question #18
Answer A is incorrect. The block does not need to be replaced when a water pump mounting bolt hole is stripped. However, the manufacturer's service manual should be consulted about cylinder head bolt thread repair.
Answer B is correct. Only Technician B is correct. Stripped threads in noncritical areas can be repaired by several processes, and using a helicoil insert is one of them. So Technician A is incorrect in saying the block has to be replaced. Technician B is correct in identifying a helicoil insert as one possible repair. A common fastening problem is stripped threads. This problem is usually caused by excessive torque or by cross-threading. Rather than replacing the block or cylinder head, the threads can be replaced by the use of threaded inserts. Several types of threaded inserts are available—the helically coiled insert is the most popular.
Answer C is incorrect. Only Technician B is correct.
Answer D is incorrect. Technician B is correct.

Question #19
Answer A is incorrect. The fuel system needs to be depleted of fuel before any attempt can be made to open the fuel filter housings. The high-pressure filter housing is subjected to the fuel system pressure. Any attempt to remove the housing under pressure can lead to serious injuries.
Answer B is incorrect. When the engine is turned off, the fuel filters remain under the full normal fuel system pressures. The ¼ turn valve needs to be turned off and the engine runs until it shuts off. The high-pressure fuel gauge needs to be checked to ensure it reads 0 psi.
Answer C is incorrect. Neither Technician is correct.
Answer D is correct. Neither Technician is correct. The fuel system needs to be depleted of fuel before any attempt can be made to open the fuel filter housings. When the engine is turned off, the fuel filters remain under the full normal fuel system pressures. The ¼ turn valve needs to be turned off and the engine runs until it shuts off. The high-pressure fuel gauge needs to be checked to ensure it reads 0 psi.

Question #20
Answer A is incorrect. The engine as well as the radiator needs to be drained to prevent coolant from going into the combustion chamber.
Answer B is correct. Only Technician B is correct. The cylinder head needs to be drained before attempting to remove the spark plug adaptors. Draining of the engine block will assure the Technician that the cylinder head will also be drained, minimizing the amount of coolant that would otherwise leak into the cylinder and the work involved in removing the coolant out of the cylinder.
Answer C is incorrect. Only Technician B is correct.
Answer D is incorrect. Technician B is correct.

Question #21
Answer A is incorrect. A failed oil cooler would not likely affect oil pressure.
Answer B is incorrect. A failed oil cooler would not likely affect oil pressure.
Answer C is correct. When an engine oil cooler fails, it is more likely that engine lubricating oil gets into the cooling system due to its relatively higher pressure. Usually oil pressure values are not affected, nor would the level in the sump be affected. When coolant mixes with oil, it forms a milky brown sludge that tends to float on top of the coolant. This can be seen in the radiator top tank or surge tank. Coolant in the engine lubricant gives it a milky, cloudy appearance when churned into the oil. After settling, the coolant will usually separate from the oil and collect at the bottom of the oil sump. When the drain plug is removed, the heavier coolant will exit first, as long as sufficient time has passed since running the engine.
Answer D is incorrect. A leaking oil cooler would Most-Likely cause oil to enter the cooling system. This would not raise the oil level.

Question #22
Answer A is incorrect. Technician B is also correct.
Answer B is incorrect. Technician A is also correct.
Answer C is correct. Both Technicians are correct. Technician A is correct. Push tubes should never be straightened because they will fail again, usually sooner than later. Checking ball socket integrity is an essential check to perform before returning push tubes to service, so Technician B is also correct. Both pushrods and tubes seldom fail under normal operation, but they are vulnerable to inaccurate valve lash adjustments and engine overspeeding. Ball and socket wear can be checked visually. Flaking or disintegration of the hard surface indicates that the tube or rod should be replaced.
Answer D is incorrect. Both Technicians are correct.

Question #23
Answer A is correct. Compressed air is an ideal cleaning medium for the gasket groove in the rocker housing cover, so Technician A is correct.
Answer B is incorrect. Oil has the effect of chemically degrading silicone sealants.
Answer C is incorrect. Only Technician A is correct.
Answer D is incorrect. Technician A is correct.

Question #24
Answer A is incorrect. A CNG tank with a damaged label with missing certification information can be returned to service after the tank is repositioned to allow the label to display the necessary information. A new label can also be obtained from the manufacturer by furnishing the serial number of the cylinder. Repositioning of the tank will require that it be defueled prior to handling it. The plumbing and hardware will need to be inspected before releasing the vehicle back to service.
Answer B is correct. Only Technician B is correct. If the tank can be repositioned to allow the certification decal to be read, it can be placed back into service. A new label can also be obtained from the manufacturer by furnishing the serial number of the cylinder. Repositioning of the tank will require that it be defueled prior to handling it. The plumbing and hardware will need to be inspected before releasing the vehicle back to service.
Answer C is incorrect. Only Technician B is correct.
Answer D is incorrect. Technician B is correct.

Question #25
Answer A is incorrect. Technician B is also correct.
Answer B is incorrect. Technician A is also correct.
Answer C is correct. Both Technicians are correct. When the liner checks within serviceability specifications, the Technician deglazes the liner. Deglazing removes the least amount of material using a power-driven flex hone or rigid hone with 200 to 250 grit stones. The best type of glaze buster is the flex hone, typically a conical (Christmas tree) or cylindrically shaped shaft with flexible branches with carbon/abrasive balls. The objective of glaze busting is to machine away the cylinder ridge above the ring belt travel and re-establish the crosshatch. Set the drill at 120 to 180 rpm and use rhythmic, reciprocating thrusts at short sequences; stopping frequently to inspect the finish produces the best results. Observe a 60° crossover angle, 20 to 35 micro-inch crosshatch. Technician A is correct in describing the procedure required to produce a crosshatch in the liner. Technician B is also correct in saying that soap and water is the preferred method of cleaning up a liner after honing.
Answer D is incorrect. Both Technicians are correct.

Question #26
Answer A is incorrect. The oil passage should be open.
Answer B is incorrect. The incorrect oil Most-Likely would not cause the gauge to read zero. It may read low but not zero.
Answer C is correct. The question tells you that if the engine is operating well and the gauge reading is zero, the cause of the problem is likely an electrical open condition in the gauge/sensor circuit. The other answers describe conditions that are unlikely to result in a zero reading although they would

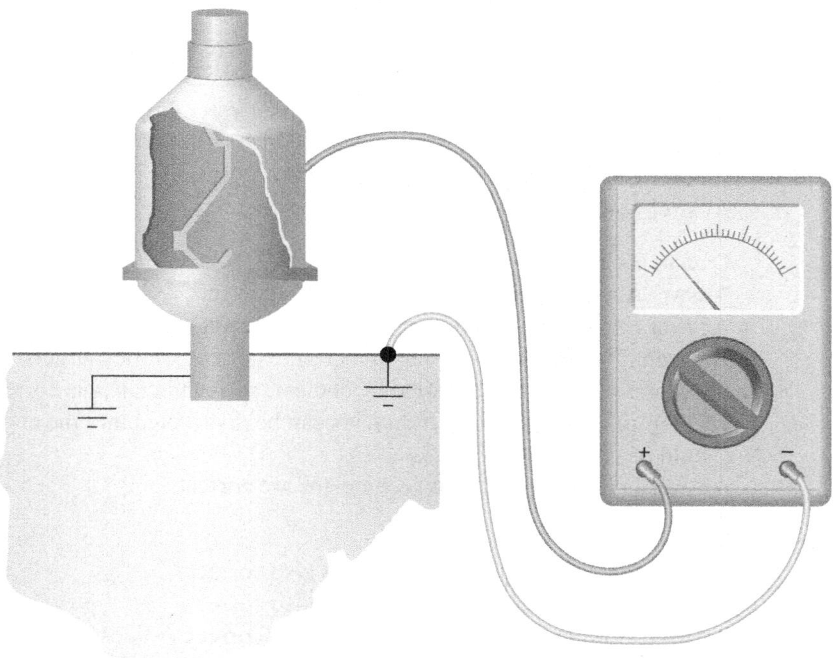

Using an ohmmeter to test a piezoresistive sensor

produce inaccurate/erratic readings. A piezo-resistive sensor is the type of sender often used for oil pressure gauges. An ohmmeter is used to check this type of sender by connecting the leads to the sending unit terminal and ground (see figure D-1). First, check the resistance with the engine off and compare to specifications. Next, start the engine and allow it to idle. Recheck the resistance value and compare it to specifications. Do not condemn this electrical sending unit immediately if the readings are not within specifications. Connect a mechanical pressure gauge to the engine to confirm that it is producing adequate oil pressure.
Answer D is incorrect. The wrong sensor would Most-Likely cause the gauge to read either higher or lower than normal. Most-Likely, it would not read zero.

Question #27
Answer A is correct. All of the conditions listed in the answers have the potential to cause a coolant in oil condition EXCEPT a crack in an exhaust valve seat, which on a CNG engine is usually not integral with the cylinder head. Coolant in the engine lubricant gives it a milky, cloudy appearance when churned into the oil. After settling, the coolant usually collects at the bottom of the oil sump. When the drain plug is removed, the heavier coolant exits first as long as sufficient time has passed since running the engine. When coolant is found in the engine oil, the cylinder head(s) and/or cylinder head gasket(s) are the Most-Likely source.
Answer B is incorrect. A crack in the cylinder liner would cause oil contamination.
Answer C is incorrect. An internal leak in the water pump could cause oil contamination if the pump is gear driven.
Answer D is incorrect. A leaking head gasket can cause oil contamination.

Question #28
Answer A is incorrect. The housing should be visually checked for cracks.
Answer B is incorrect. The gasket surfaces should be checked for nicks.
Answer C is correct. The LEAST-Likely part of the procedure is checking a filter pad housing for cracks with magnetic flux equipment. Visual checking is usually all that is required to properly check a filter pad.
Answer D is incorrect. The passages should be inspected for obstructions.

Question #29
Answer A is incorrect. Improperly installed valve seals can cause excessive oil consumption but are unlikely to be the cause of dropping a valve after an engine overhaul.
Answer B is correct. Only Technician B is correct. Improperly installed keepers could cause a dropped valve.
Answer C is incorrect. Only Technician B is correct.
Answer D is incorrect. Technician B is correct.

Question #30
Answer A is incorrect. Technician B is also correct.
Answer B is incorrect. Technician A is also correct.
Answer C is correct. Both Technicians are correct. Turbochargers are lubricated by engine oil and the center section is also cooled by coolant, so Technician A is correct. Technician B is also correct; when turbine bearings fail, fragments can be distributed into the engine oil sump, so the oil and filters should be changed.
Answer D is incorrect. Both Technicians are correct.

Question #31
Answer A is incorrect. Technician B is also correct.
Answer B is incorrect. Technician A is also correct.
Answer C is correct. Both Technicians are correct. A boost pressure leak can cause low intake manifold pressure. The ECM will interpret the low pressure as a lack of air and remove fuel to prevent the emissions specifications from being exceeded. A partially stuck waste gate or open waste gate will cause lack of intake manifold pressure because the turbocharger will by-pass the exhaust and will not be able to spool up and build up boost. The ECM will remove fuel and as a result, less power will be produced as a symptom.
Answer D is incorrect. Both Technicians are correct.

Question #32
Answer A is incorrect. A punctured air filter would allow some unfiltered air to enter the engine but would not be responsible for high manifold boost.
Answer B is correct. Only Technician B is correct. Excessive manifold boost pressures are the result of high turbine speeds. A stuck turbocharger wastegate could cause this condition by not diverting exhaust gas flow around the turbine.
Answer C is incorrect. Only Technician B is correct.
Answer D is incorrect. Technician B is correct.

Question #33
Answer A is incorrect. Technician B is also correct.
Answer B is incorrect. Technician A is also correct.
Answer C is correct. Both Technicians are correct. Customer concerns must be listened to closely, AND you should always begin with the easiest tests, so BOTH Technicians are correct.
Answer D is incorrect. Both Technicians are correct.

Question #34
Answer A is incorrect. The waste gate stuck closed will cause high boost pressure readings and will log a code as high boost pressure, not high exhaust back pressure.
Answer B is incorrect. The back pressure of the catalyst should be checked with a magnehelic gauge calibrated in inches of mercury (Hg), not inches of water.
Answer C is incorrect. Neither Technician is correct.
Answer D is correct. Neither Technician is correct. The waste gate stuck closed will cause high boost pressure readings and will log a code, as high boost pressure not high exhaust back pressure.

The back pressure of the catalyst should be checked with a magnehelic gauge calibrated in inches of mercury (Hg), not inches of water.

Question #35
Answer A is correct. Only Technician A is correct. When the power connector is disconnected, power flow is interrupted and the engine will turn off as the fuel is used up. If the engine does not turn off the solenoid is leaking internally and it should be replaced.
Answer B is incorrect. The shutoff solenoid cannot be checked by turning off the ignition switch. When the switch is turned off, the ECM loses ignition power and the engine turns off without depleting the fuel and the condition of the valve cannot be determined in this manner.
Answer C is incorrect. Only Technician A is correct.
Answer D is incorrect. Technician A is correct.

Question #36
Answer A is incorrect. The exhaust brake function requires a restriction of the flow of exhaust gases from the engine.
Answer B is correct. Back pressure is not created in the intake system when the exhaust brake is activated and is the EXCEPT statement.
Answer C is incorrect. The exhaust brake system utilizes the compression stroke of the pistons pushing against the pressurized exhaust gases that remain in the engine due to not being able to flow out past the valve that is closed in the exhaust system to help slow the vehicle down.
Answer D is incorrect. The exhaust brake system utilizes the spent exhaust gases that are being restricted from flowing out of the exhaust pipe to resist the upward motion of the piston during the exhaust stroke to slow the vehicle down.

Question #37
Answer A is correct. Only Technician A is correct. The SEL and CEL indicators should be checked to make sure the ECM powers up. Because the vehicle was running when it left the yard, the fuel shutoff valve had to have been on.
Answer B is incorrect. As the ignition switch is being turned on, the Technician will be able to observe the indicators in the dash. The fuel valve should be checked after the indicator lights because the vehicle was running when it left the yard and had to have the valve on.
Answer C is incorrect. Only Technician A is correct.
Answer D is incorrect. Technician A is correct.

Question #38
Answer A is incorrect. Technician B is also correct.
Answer B is incorrect. Technician A is also correct.
Answer C is correct. Both Technicians are correct. The ¼ turn valve handle should be in the vertical, straight up and down, when it is on and horizontal, parallel with the ground, when it is in the off position.
Answer D is incorrect. Both Technicians are correct.

Question #39
Answer A is incorrect. Technician B is also correct.
Answer B is incorrect. Technician A is also correct.
Answer C is correct. Both Technicians are correct. Valve crossheads should be checked for squareness and Magnafluxed to check for cracks during an overhaul. Valve bridges or yokes push down on two valves at a time. Most engines with four valves per cylinder use valve bridges, although some manufacturers are using one rocker arm per valve and no crossheads. They do not usually need to be adjusted during a routine valve adjustment. The bridge must be properly supported to prevent the pedestal shaft from being damaged.
Answer D is incorrect. Both Technicians are correct.

Question #40

Answer A is correct. Only Technician A is correct. Cam profiles are hard surfaced. When the hard surfacing is compromised, the cam is likely to fail. Grooving wear patterns in a camshaft are reason to reject it even if it measures within specification.

Answer B is incorrect. Even if lobe lift measurements are within limits, any grooving or spalling will very quickly turn into a completely ruined camshaft.

Answer C is incorrect. Only Technician A is correct.

Answer D is incorrect. Technician A is correct.

Question #41

Answer A is incorrect. The A/C compressor belt would not be affected during cornering as it is not connected to the power steering.

Answer B is correct. Only Technician B is correct. A belt-driven hydraulic pump for the engine cooling fan is affected more during cornering as well as when the transmission hydraulic brake retarder is engaged.

Answer C is incorrect. Only Technician B is correct.

Answer D is incorrect. Technician B is correct.

Question #42

Answer A is correct. Testing of valve springs is performed as a bench-test procedure and spring height, squareness, and visible damage are tested. Valve harmonic distortion occurs in operation at specific, usually high, rpms, so it cannot be part of an inspection procedure. The question asks for what is LEAST-Likely to be revealed during inspection. Many high-mileage engine overhauls will include new valve springs as a matter of course. If a high-mileage valve spring is reused and then breaks shortly after overhaul, the results can be disastrous.

Answer B is incorrect. Springs are checked for free height.

Answer C is incorrect. Springs are checked for cracks.

Answer D is incorrect. Springs are checked with a square to make sure they are not out of square.

Question #43

Answer A is correct. An out-of-balance turbocharger will usually result in rapid turbocharger failure and, because they are driven by exhaust gas heat (engine-rejected heat), rotational speeds do not correlate with engine rpm. Therefore, an out-of-balance turbocharger would be LEAST-Likely to cause an engine speed-related vibration.

Answer B is incorrect. A bent connecting rod would cause an engine rpm-related vibration.

Answer C is incorrect. An out-of-balance crankshaft would cause an engine rpm-related vibration.

Answer D is incorrect. A dented vibration damper would render it inoperative, which would create an engine rpm-related vibration.

Question #44

Answer A is correct. Shutting off the engine when still hot will not cause the camshaft to have pitting because once the engine stops rotating wear will not continue to occur.

Answer B is incorrect. Not allowing an engine to build up oil pressure before driving off will cause friction to develop on the cam follower roller and cam, causing accelerated wear and pitting to develop.

Answer C is incorrect. Oil with dirt in it will cause friction between the cam follower roller and camshaft causing pitting.

Answer D is incorrect. High mileage can cause pitting of the hard facing on the followers as a result of normal wear and tear.

Question #45

Answer A is incorrect. A loose water pump belt could cause coolant-level surge.

Answer B is correct. A defective spark plug will cause engine performance problems, but it is unlikely this would affect the cooling system, so this would be the exception for the surging condition.

Answer C is incorrect. An internally restricted radiator could cause coolant-level surging.

Answer D is incorrect. A blown head gasket can let compression gases in the cooling system, which would result in surging coolant.

Question #46
Answer A is incorrect. Technician B is also correct.
Answer B is incorrect. Technician A is also correct.
Answer C is correct. Both Technicians are correct. Coolant is used to remove heat from the exhaust before re-introducing it back into the cylinder to cool the combustion temperature and lowering the emission of NOx. Because more heat is introduced into the cooling system the radiator has to be able to remove more heat out of the coolant and usually has more surface area to allow heat to be removed.
Answer D is incorrect. Both Technicians are correct.

Question #47
Answer A is incorrect. V-Ref voltage check is not the first step.
Answer B is correct. When troubleshooting electronic input circuit components, it is essential that the manufacturer's service literature be consulted.
Answer C is incorrect. Simulating V-Ref voltage would be much further down the troubleshooting tree; it would not be the first step.
Answer D is incorrect. Checking resistance of the sensor would be an early troubleshooting step; however, it would not be the first step.

Question #48
Answer A is correct. Only Technician A is correct. The fuel cylinders need to be refueled in increments and leak tested to ensure the system is sound to continue increasing the pressure.
Answer B is incorrect. Frosting at the fuel solenoid(s) is not normal during fueling. Frosting usually occurs during pressure loss not when pressure is increasing. The fuel system has check valves incorporated into the fuel lines to by pass the fuel solenoids. Because the engine must be off during refueling, the solenoids are deactivated and therefore the check valves give an alternate path for the fuel to go into the fuel tanks. Frosting needs to be checked out and its cause repaired.
Answer C is incorrect. Only Technician A is correct.
Answer D is incorrect. Technician A is correct.

Question #49
Answer A is correct. The DMM has been placed in series in the circuit shown; meaning that it has become part of the circuit and full circuit current will pass through the meter. Amperage or current flow is measured. Ammeters are used to measure the amperage or current that is being pushed through the load to do work. When using a direct-reading ammeter, never place the meter leads across a battery or a load. This can, at a minimum, blow the meter's fuse or possibly destroy the meter. It is important to ensure the circuit amperage will not exceed the meter's rating.
Answer B is incorrect. Voltage drop is measured with a voltmeter in parallel to the circuit.
Answer C is incorrect. Circuit resistance is measured with an ohmmeter connected to a circuit that is not energized.
Answer D is incorrect. Wattage is measured by measuring both voltage and amperage.

Question #50
Answer A is incorrect. Handling the spark plugs with bare fingers can cause a spark flashover condition because the oil from the skin can become a conductor to high voltages being created when the engine is under a load.
Answer B is incorrect. Contaminated spark plug tubes can cause a spark flashover condition due to the dirt or oil becoming a conductor when high voltages are being created when the engine is under a load.
Answer C is incorrect. A cracked coil can cause a misfiring cylinder. If the ignition's secondary circuit insulation is not intact, the spark can jump to ground.
Answer D is correct. Cleaning of the spark plug with isopropyl alcohol is the LEAST-Likely condition to cause a misfiring cylinder. Manufacturers highly recommend this practice of cleaning to remove any contamination from the ignition's secondary circuit that could cause the spark to go to ground and cause a misfiring cylinder.

Question #51
Answer A is incorrect. Technician B is also correct.
Answer B is incorrect. Technician A is also correct.
Answer C is correct. Both Technicians are correct. Petroleum jelly will help reduce corrosion on battery cable terminals, so Technician A is correct. Using protective pads and a light coating of grease will also help limit corrosion, so Technician B is also correct.
Answer D is incorrect. Both Technicians are correct.

Question #52
Answer A is incorrect. This is not a flywheel assembly.
Answer B is correct. The figure shows a typical vibration damper assembly that bolts to the front of a crankshaft. A typical vibration damper consists of a damper drive or housing and inertia ring. The housing is coupled to the crankshaft and using springs, rubber, or viscous medium drives the inertia ring. The objective is to drive the inertia ring at average crankshaft speed. The annular housing is hollow and bolted to the crankshaft. Within the hollow housing, the inertia ring is suspended in and driven by silicone gel. The damping action is effected by the shearing of the viscous fluid film between the drive ring and the inertia ring.
Answer C is incorrect. This is not a timing gear assembly.
Answer D is incorrect. This is not a water pump.

Question #53
Answer A is incorrect. Technician B is also correct.
Answer B is incorrect. Technician A is also correct.
Answer C is correct. Both Technicians are correct. Technician A correctly describes the open to fully open specifications on a typical 9-psi-rated radiator cap. Technician B also correctly describes the differential specification that the radiator cap uses to admit air to the cooling system to prevent radiator hose collapse. Great care should be exercised when removing a radiator cap from the radiator. If the system is pressurized, hot coolant may escape from the filler neck with great force. Most filler necks are fitted with double cap lock stops to prevent the radiator cap from being removed in a single counterclockwise motion. If the radiator is still pressurized, the cap will jam on the intermediate stops. Never attempt to remove a radiator cap until the cooling system pressure is equalized. Radiators are usually equipped with a pressure cap designed to maintain a fixed operating pressure while the engine is running. This cap is also equipped with a vacuum valve to admit surge tank coolant (or air) into the cooling circuit (the upper radiator tank) when the engine is shut down to accommodate coolant thermal contraction. Radiator caps permit pressurization of a sealed cooling system. For each 1 psi (7 kPa) above atmospheric pressure, the coolant boiling point is raised by 3° F (1.67° C) at sea level. For every 1000 feet of elevation, the boiling point decreases by 1.25° F (0.5° C). System pressures will seldom be designed to exceed 25 psi (172 kPa); more typically, they will range between 7 psi (50 kPa) and 15 psi (100 kPa).
Answer D is incorrect. Both Technicians are correct.

Question #54
Answer A is correct. Operation of a CNG engine under full load for prolonged periods is exactly what a transit bus engine is designed for. This will not create problems as long as the engine is prevented from overheating. A certain small percentage of crankshaft failures will result from manufacturing and design problems. Design problems usually only occur shortly after an engine series is introduced. No OEM wants to have an engine labeled with a basic engineering fault, so design problems are usually rectified quickly.
Answer B is incorrect. An unbalanced vibration damper can cause crankshaft failure.
Answer C is incorrect. An incorrectly installed flywheel housing can cause crankshaft and bearing damage.
Answer D is incorrect. Loose torque converter bolts can damage the crankshaft.

Question #55
Answer A is incorrect. It would be impossible to measure the ring end gap after the assembly is in the engine.

Answer B is correct. Only Technician B is correct. Piston ring end gap should be measured by installing the ring into the cylinder alone and by measuring the gap with a thickness (feeler) gauge. This is done before fitting the ring to the piston. The specification is usually 0.003 to 0.004 inch per inch of diameter.
Answer C is incorrect. Only Technician B is correct.
Answer D is incorrect. Technician B is correct.

Question #56
Answer A is correct. SAE MID or message identifiers are used to identify each major electronic circuit on a truck chassis. The engine MID is 128 and the transmission MID is 130. SIDs, PIDs, and FMIs all flow from the MID. The MID is used to describe a major vehicle electronic system, usually with independent processing capability.
Answer B is incorrect. The SID is used to identify the major subsystems of an electronic circuit.
Answer C is incorrect. The PID is used to code components within an electronic subsystem.
Answer D is incorrect. Failure Mode Identifiers identify the type of failure (open, short, out of range, etc.).

Question #57
Answer A is correct. Only Technician A is correct. During engine overhaul, all the gears in the engine timing gear train should be thoroughly inspected.
Answer B is incorrect. Rolling or lipping of gear teeth is not acceptable and evidence of either requires the replacement of the gear and its mate(s).
Answer C is incorrect. Only Technician A is correct.
Answer D is incorrect. Technician A is correct.

Question #58
Answer A is incorrect. CNG high-pressure lines are made from stainless steel seamless tubing and are tempered.
Answer B is incorrect. If a line is scratched especially at the fitting area, it should be replaced and not buffed out. Removing more material from the line will leave a wider depression and the ferrule of the fitting will not be able to properly seal, causing a leak.
Answer C is incorrect. Neither Technician is correct.
Answer D is correct. Neither Technician is correct. CNG high-pressure lines are made from stainless steel seamless tubing and are tempered. If a line is scratched especially at the fitting area, it should be replaced and not buffed out.

Question #59
Answer A is incorrect. Back-flushing of NGV transit bus cooling systems is unnecessary if the coolant is maintained and conditioned according to manufacturer instructions.
Answer B is incorrect. It is the correct procedure to refill the cooling system with any bleed valves in the system open.
Answer C is incorrect. Neither Technician is correct
Answer D is correct. Neither Technician is correct. Back-flushing of transit bus cooling systems is unnecessary if the coolant is maintained and conditioned according to manufacturer instructions. EG and PG type systems need the coolant replaced every two years. ELC type coolants need replacing at six-year intervals. It is correct procedure to refill the cooling system with any bleed valves in the system open. This allows air to purge as the coolant enters the system. Most commercially available in-chassis radiator descaling solutions are a poor risk because they are seldom 100 percent effective and may dislodge scale, which will subsequently plug up elsewhere in the cooling circuit. For the same reasons, reverse-flow flushing of the cooling system makes little sense. When OEMs recommend radiator flushing, it is generally performed in the normal direction of flow, often aided by a cleaning solution.

Question #60
Answer A is incorrect. The ITT high-pressure regulator should not be adjusted in the field. The regulator should be replaced as a unit if it goes out of adjustment.
Answer B is correct. Only Technician B is correct. The ITT high-pressure regulator has coolant routed through it to help prevent the fuel from becoming too cold, which would cause icing in the fuel filters and low fuel pressure to the engine.
Answer C is incorrect. Only Technician B is correct.
Answer D is incorrect. Technician B is correct.

Question #61
Answer A is incorrect. Minimum voltage is 9.4 volts at 50° F.
Answer B is correct. The temperature in the question is 50° F. All battery test specifications are temperature dependent, so it is important to be able to interpret this type of chart. This test is referred to as a load test. A load is placed on the battery to see how it functions under a demand condition. Prior to performing this test, the previous hydrometer and/or open circuit voltage test needs to be performed. If the state of charge is found to be low, the battery has to be charged for the load test to be accurate. A battery tester with an adjustable carbon pile is used to perform this test. The tester usually has an ammeter to indicate the load placed on the battery and a voltmeter to indicate the result in volts. The result and specific numbers required for the test can be summed up as: A good 12-volt battery should maintain a voltage level of 9.6 volts or more when a load of 1/2 of its cold cranking amps (CCA) is applied to it for 15 seconds.
Answer C is incorrect. Minimum voltage is 9.4 volts at 50° F.
Answer D is incorrect. Minimum voltage is 9.4 volts at 50° F.

Question #62
Answer A is incorrect. As long as the resistance of a wire is less than 10 Ω, the continuity is considered to be acceptable and no action needs to be taken.
Answer B is correct. Only Technician B is correct. As long as the resistance of a wire is less than 10 Ω, the continuity is considered to be acceptable and no action needs to be taken.
Answer C is incorrect. Only Technician B is correct.
Answer D is incorrect. Technician B is correct.

Question #63
Answer A is incorrect. Belt tensioners can lose their tensioning capabilities over time. They should be closely inspected when a belt is being replaced to prevent the belt from being damaged again from slipping.
Answer B is correct. Only Technician B is correct. Worn pulleys can cause the belt to prematurely wear out. In addition, a slipping belt can start the wearing process and if not corrected when the worn out belt is replaced, the worn pulley(s) will cause the new belt to fail again.
Answer C is incorrect. Only Technician B is correct.
Answer D is incorrect. Technician B is correct.

Question #64
Answer A is correct. Only Technician A is correct. The shoe should fit perfectly into the base of the liner to prevent the block from being damaged.
Answer B is incorrect. If the puller shoe is larger than the liner OD, it will grab on the cylinder bore and not permit the liner to be pulled. Cylinder liners (sleeves) should be removed with a puller and adapter plate or shoe.
Answer C is incorrect. Only Technician A is correct.
Answer D is incorrect. Technician A is correct.

Question #65
Answer A is incorrect. No compound should be applied onto the spark plug boots as they come prelubed from the manufacturer.
Answer B is correct. Only Technician B is correct. The boots come with dielectric compound applied from the manufacturer if it is required and are ready to be installed from the container.

Answer C is incorrect. Only Technician B is correct.
Answer D is incorrect. Technician B is correct.

Question #66
Answer A is correct. Only Technician A is correct. Flywheel ring gears are usually heat shrunk to flywheels. Shrunk-fit to the outer periphery of the flywheel is the ring gear, which provides the means of transmitting cranking torque to the engine by the starter motor during startup. To service the ring gear, first remove the flywheel from the engine. Using an oxyacetylene torch, partially cut through the ring gear, working from the outside, and remove the ring gear from the flywheel. In most cases, this is sufficient to expand the ring gear so that it can be removed using a hammer and chisel. Avoid heating the flywheel any more than absolutely necessary or damaging the flywheel by careless use of the oxyacetylene flame.
Answer B is incorrect. The statement that the entire flywheel should be replaced when the ring gear teeth are damaged is incorrect.
Answer C is incorrect. Only Technician A is correct.
Answer D is incorrect. Technician A is correct.

Question #67
Answer A is incorrect. Technician B is also correct.
Answer B is incorrect. Technician A is also correct.
Answer C is correct. Both Technicians are correct. Bent, cracked, or otherwise damaged fan blades can cause engine imbalance and the cost of repair can be considerably greater than the cost of a fan assembly. Technician A is of the opinion that a damaged fan should be replaced and Technician B says that a bent fan blade can cause engine vibration.
Answer D is incorrect. Both Technicians are correct.

Question #68
Answer A is incorrect. A voltmeter is the best way to check battery potential.
Answer B is correct. Battery capacity means battery potential, which is potential difference or voltage. Voltage is measured with a voltmeter. Some batteries are totally sealed and cannot be checked with a hydrometer. An open circuit voltage test is especially useful in these circumstances.
Answer C is incorrect. An ohmmeter cannot measure battery potential.
Answer D is incorrect. An ammeter cannot measure battery potential.

Question #69
Answer A is incorrect. 1 and 5 are not companion cylinders.
Answer B is incorrect. 4 and 5 are not companion cylinders
Answer C is incorrect. 3 and 6 are not companion cylinders.
Answer D is correct. Cylinders 5 and 2 are companioned. Companion cylinders can be identified by pairs in an inline six-cylinder engine: cylinders 1 and 6, 5 and 2, and 3 and 4 are companioned; that is, they travel through the cycle together. For instance, when piston #1 is at TDC at the completion of its compression stroke, piston #6 is also at TDC, having completed its exhaust stroke. When the crankshaft throws are in line with each other, they are called cylinder throw pairings or companion cylinders. For the firing order 153624, these companion cylinders are 1-6, 5-2, and 3-4. These companion cylinders can be in one of two positions: TDC or TDC overlap. In the case of the 1-6 pair, when #1 piston is at TDC completing its compression stroke, #6 piston, its companion cylinder, is at TDC overlap having just completed its exhaust stroke. Cylinder #1 is at TDC overlap when it has just completed its exhaust stroke and cylinder #6 is at TDC having completed its compression stroke.
If the engine is viewed from overhead with the rocker covers removed, you can identify engine cylinder position by observing the valves over a pair of companion cylinders. For the TDC position, the pistons in cylinders #1 and #6 are approaching TDC when the valves over #6 are both closed and they have lash. At this point, you can rock the valves over #1 cylinder back and forth and you cannot rock the valves on the #6 cylinder. Cylinder #1 TDC is also indicated by the fact that the exhaust has just closed and the intake is about to open. This method of orienting engine location is commonly used for valve adjustment.

Question #70
Answer A is incorrect. Once the belts have become so loose as to be screeching, the belts should be checked to ensure they are not glazed or damaged. In addition, the pulleys should be checked to see if they are worn and to ensure they are properly aligned.
Answer B is correct. Only Technician B is correct. The belts should be checked to ensure they have not become glazed or burned. In addition, the pulleys should be checked to see if they are worn and to ensure they are properly aligned.
Answer C is incorrect. Only Technician B is correct.
Answer D is incorrect. Technician B is correct.

Question #71
Answer A is incorrect. Technician B is also correct.
Answer B is incorrect. Technician A is also correct.
Answer C is correct. Both Technicians are correct. The spark plug gap and condition should be inspected to determine the condition of the plug and to see if anything suspicious is occurring in the cylinder such as oil deposits or metal from the piston on the electrode of the spark plug. It is important to remember that Cummins does not recommend the regapping of spark plugs but rather their replacement. Swapping the problematic cylinder's coil with a coil from an adjacent cylinder that is functioning normally will assist the Technician in determining if the problem is the coil or some other defect.
Answer D is incorrect. Both Technicians are correct.

Question #72
Answer A is incorrect. A series circuit would create a dangerous overvoltage condition.
Answer B is correct. The ground connection should always be last and the boost electrical system should be connected in parallel.
Answer C is incorrect. The negative cable should be last.
Answer D is incorrect. This step is not necessary.

Question #73
Answer A is incorrect. Technician B is also correct.
Answer B is incorrect. Technician A is also correct.
Answer C is correct. Both Technicians are correct. One of the functions of a radiator cap is to relieve pressure by discharging fluid to an expansion chamber when it exceeds a specified value, and most CNG cooling systems function at maximum pressures of 15 psi (100 kPa) or less.
Radiators are usually equipped with a pressure cap whose function is to maintain a fixed operating pressure while the engine is running. This cap is additionally equipped with a vacuum valve to admit surge tank coolant (or air) into the cooling circuit (the upper radiator tank) when the engine is shut down to accommodate coolant thermal contraction. Radiator caps permit pressurization of a sealed cooling system. For each 1 psi (7 KPa) above atmospheric pressure, the coolant boiling point is raised by 3° F (1.67° C) at sea level; for every 1000 feet of elevation, the boiling point decreases by 1.25° F (0.5° C). System pressures are seldom designed to exceed 25 psi (172 KPa) and more typically range between 7 psi (50 KPa) and 15 psi (100 KPa).
Answer D is incorrect. Both Technicians are correct.

Question #74
Answer A is incorrect. The hole and groove are to direct oil evenly around the bearing.
Answer B is incorrect. The tang is not a defect.
Answer C is incorrect. The stamping on the back indicates an oversized or a standard bearing.
Answer D is correct. The notch in the bearing shell is primarily used to position the bearing shell. Because the notch protrudes into a recess in the bearing bore, it also plays a small role in preventing the bearing from turning. Remember, however, that the primary factor in preventing a friction bearing from turning in its bore is bearing crush. Bearings are retained primarily by crush. The outside diameter of a pair of uninstalled bearing shells slightly exceeds the bore to which it is installed. This creates radial pressure that acts against the bearing halves and provides good heat transfer. The bearing halves are also slightly elliptical to allow the bearing to be held in place during installation; this is

known as bearing spread. Tangs in bearings are inserted into notches in the bearing bore to minimize longitudinal movement, prevent bearing rotation, and align oil holes.

Question #75
Answer A is incorrect. Terminal 85 to 30 is incorrect.
Answer B is incorrect. Terminal 86 to 87 is incorrect.
Answer C is incorrect. Terminal 86 to 30 is incorrect.
Answer D is correct. The voltmeter positive lead needs to be connected to terminal 85 and the negative lead connected to a known good chassis ground. The push button would then be depressed and the voltmeter should display the switch voltage. If the voltage is lower than the switch voltage, then a resistance in the switch or wiring should be suspected.

Question #76
Answer A is incorrect. Valve lash will be found on the emission labels because the information is required for certification purposes.
Answer B is incorrect. Timing specifications are required for engine certification purposes.
Answer C is correct. Oil pressure is not provided on the certification labels.
Answer D is incorrect. Emission certification level and certifying agencies as well as the dates are provided on the certification labels.

Question #77
Answer A is correct. Only Technician A is correct. Worn cam bushings increase clearances, which reduce oil pressure.
Answer B is incorrect. When camshaft bushings are removed, they cannot be reinstalled. Once they are driven out, they will not stay snug when reinstalled.
Answer C is incorrect. Only Technician A is correct.
Answer D is incorrect. Technician A is correct.

Question #78
Answer A is incorrect. CNG engine oil will turn dark brown from normal wear.
Answer B is correct. Only Technician B is correct. CNG engine oil rating is 15W-40 GEO with special additives for that particular engine's requirements.
Answer C is incorrect. Only Technician B is correct.
Answer D is incorrect. Technician B is correct.

Question #79
Answer A is incorrect. The ¼ turn fuel shutoff valve has to be turned off when servicing the fuel system to isolate the high-pressure regulator and filters from the fuel tanks and allow the fuel to be depleted.
Answer B is incorrect. The engine has to be run until it shuts off to deplete the fuel from the ¼ turn shut off valve and the engine.
Answer C is incorrect. Both gauges have to be checked to make sure all the fuel is depleted from system and they both read 0 psi. This ensures the system does not have any remaining pressure and is safe. The components between the ¼ turn valve and the engine can be serviced safely.
Answer D is correct. Engine oil is not recommended as an O-ring lubricant for a CNG engine. Engine oil can coat sensitive components of the fuel system such as the GMF sensor, rendering the component ineffective and cause the ECM to log a DTC. This is the EXCEPT action that should not be followed.

Question #80
Answer A is incorrect. Technician A is incorrect. The gap should be set to the required specifications. Improper gapping can decrease the life of a spark plug and increase emissions.
Answer B is incorrect. Technician B is incorrect. The gap should be set to the required specifications. Improper gapping can decrease the life of a spark plug and increase emissions.
Answer C is incorrect. Neither Technician is correct.
Answer D is correct. Neither Technician is correct. The gap should be set to the required specifications. Improper gapping can decrease the life of a spark plug and increase emissions.

Question #81
Answer A is incorrect. Resistance cannot be checked in an energized circuit. This would damage the ohmmeter, if an analog meter was being used.
Answer B is correct. Only Technician B is correct because voltage drop testing identifies high resistances in energized circuits by measuring how much voltage is lost within the cables or connections between the test leads.
Answer C is incorrect. Only Technician B is correct.
Answer D is incorrect. Technician B is correct.

Question #82
Answer A is incorrect. The fuel cylinder that is having its PRD replaced will require defueling.
Answer B is incorrect. The fuel tank needing a manual shutoff valve replacement will require defueling.
Answer C is correct. When inspecting a fuel cylinder, it is LEAST-Likely to require defueling. A CNG cylinder requires defueling at inspection or at re-qualification only when a defect such as deep abrasion marks, delamination, stress corrosion cracking, or apparent burn marks are detected. The cylinder would need defueling because the damage would appear to be Level 3 damage and require further testing to determine its integrity.
Answer D is incorrect. Replacement of a fuel cylinder requires defueling of at least that cylinder.

Question #83
Answer A is incorrect. Although metalizing and machining can be undertaken by machine shops, metalizing repairs on crankshafts are not approved by CNG engine OEM. Crankshaft journals are induction hardened and the hardening would be lost if metalizing and machining was performed.
Answer B is correct. Only Technician B is correct. Grinding the journals to an undersize, providing that undersize bearings are available, is an accepted OEM practice.
Answer C is incorrect. Only Technician B is correct.
Answer D is incorrect. Technician B is correct.

Question #84
Answer A is correct. Only Technician A is correct. A leaking thermostat seal could cause coolant to bypass a closed thermostat. This could cause an engine that would fail to reach operating temperature.
Answer B is incorrect. A loose fan belt would Most-Likely cause the engine to overheat.
Answer C is incorrect. Only Technician A is correct.
Answer D is incorrect. Technician A is correct.

Question #85
Answer A is correct. Service bulletins provide the latest service tips, field repairs, product improvements, and related information of benefit to service personnel. Some service bulletins are updates to information in the service manual. They take precedence over service manual information, until the latter is updated. At that time, the bulletin is usually canceled. The service bulletins manual is usually available only to dealers, who in turn supply the owners of the vehicle. When doing service work on a vehicle system or part, check for a valid service bulletin for the latest information on the subject. Because technical service bulletins are more frequently released, they take precedence over previously published information on the same subject.
Answer B is incorrect. Service manuals are subject to being out of date due to the time required from writing to printing to distribution.
Answer C is incorrect. Parts manuals do not contain service information.
Answer D is incorrect. Operator manuals, typically, only list owner service information.

Question #86
Answer A is correct. Only Technician A is correct. The Cummins spark plug boots come prelubricated from the manufacturer with the correct type and amount of dielectric compound and should be installed as they come from the container.
Answer B is incorrect. The spark plug outer ceramic should be cleaned with alcohol and a clean towel to prevent contamination that could cause the spark to jump to ground on the outside of the ceramic,

causing a misfire in the engine. Cleaning with alcohol will not prevent the boots from sticking to the ceramic; the dielectric compound will do that.
Answer C is incorrect. Only Technician A is correct.
Answer D is incorrect. Technician A is correct.

Question #87
Answer A is incorrect. A partially stuck-open oil pressure regulator valve can lead to low oil pressure at idle only. As the engine rpms rise the engine oil pump volume can increase enough to be able to maintain adequate oil flow.
Answer B is correct. The oil level being too high should not cause low oil pressure at idle and is the LEAST-Likely statement. The oil pump would be supplied with an adequate oil supply to maintain pressure.
Answer C is incorrect. If the engine oil becomes contaminated or diluted the viscosity could be affected and the oil pump may not be able to maintain adequate pressure at idle.
Answer D is incorrect. When oil is not properly cooled because of a clogged oil cooler, the viscosity of the engine oil will decrease and the oil could become thin enough, especially after the engine has been subjected to a heavy load, to cause low oil pressure at idle.

Question #88
Answer A is incorrect. Camshaft lobes should not be ground to repair them.
Answer B is incorrect. Silver solder should not be applied to a camshaft lobe in an attempt to repair it.
Answer C is incorrect. Neither Technician is correct.
Answer D is correct. Neither Technician is correct. Damaged camshaft lobe surfaces should not be turned and CNG engine camshafts are never repaired with silver solder. Replacement of the camshaft is the accepted industry standard. The Technician should be aware that the smallest hard surfacing failure on the cam lobe profile requires little time to advance to a total failure.

Question #89
Answer A is correct. Only Technician A is correct. An improperly installed connecting rod could result in piston skirt cracks.
Answer B is incorrect. Insufficient clearance would cause piston scuffing.
Answer C is incorrect. Only Technician A is correct.
Answer D is incorrect. Technician A is correct.

Question #90
Answer A is incorrect. Technician A is incorrect. Ignition triggering devices are usually located on either the cylinder block, monitoring a pulse wheel on the crankshaft, or on the camshaft.
Answer B is incorrect. Technician B is incorrect. CNG engines do not use distributors.
Answer C is incorrect. Neither Technician is correct.
Answer D is correct. Neither Technician is correct.

Question #91
Answer A is incorrect. Technician B is also correct.
Answer B is incorrect. Technician A is also correct.
Answer C is correct. Both Technicians are correct. Technician A is correct in saying the battery should not be charged if the hydrometer eye is clear or yellow. Technician B is also correct in saying that a battery can be assumed to be close to fully charged when the hydrometer eye is green.
Answer D is incorrect. Both Technicians are correct.

Question #92
Answer A is incorrect. Overfueling of a CNG engine will cause the rings to be overheated due to high cylinder temperatures and pressures and can cause the rings to become stuck in their grooves.
Answer B is incorrect. Overheating of the engine can cause the rings to become stuck in the ring grooves.
Answer C is incorrect. Insufficient lubrication can cause the rings to become stuck in the ring grooves.
Answer D is correct. A cold-running CNG engine is the LEAST-Likely cause for piston rings to become stuck in their grooves.

Question #93
Answer A is correct. Carbon buildup in the combustion would increase compression of the cylinder due to the increase in the compression ratio and would be the LEAST-Likely cause of low compression pressure.
Answer B is incorrect. Worn exhaust valves and/or seats will lower the compression pressure of a cylinder.
Answer C is incorrect. Worn compression rings could allow compression to bypass the rings and cause low compression.
Answer D is incorrect. A cracked piston would allow compression into the crankcase and can cause low compression.

Question #94
Answer A is incorrect. It is recommended the AFR be reset when possible before any diagnostics are performed.
Answer B is incorrect. The ECM should be checked for DTC that could be causing the complaint.
Answer C is correct. Replacing engine components without diagnosing can prove costly and nonproductive and should be the LEAST-Likely action taken. New spark plugs could mask the underlying defect and the real cause could remain undetected.
Answer D is incorrect. Speaking with the operator and checking the vehicle records can alert the Technician to past defects and repairs performed. If the spark plugs have to be replaced prematurely constantly, this could alert the Technician to a more severe existing condition.

Question #95
Answer A is incorrect. An alignment machine is not used to check flywheel housing alignment.
Answer B is incorrect. A micrometer will not check alignment.
Answer C is correct. A dial indicator is used.
Answer D is incorrect. A straightedge and feeler gauge will not check alignment.

Question #96
Answer A is incorrect. The described belt should be changed.
Answer B is correct. A water pump weep hole may produce occasional drops of coolant when the pump is functioning properly. All the other conditions listed can cause cooling system failures because they would cause loss of circulation, loss of coolant, or inhibit the cooling capability of the engine.
Answer C is incorrect. A loose hose clamp should be fixed.
Answer D is incorrect. Rust or scum in the cooling system can cause overheating, and may be an indication of more serious engine problems such as a blown head gasket.

Question #97
Answer A is incorrect. Technician B is also correct.
Answer B is incorrect. Technician A is also correct.
Answer C is correct. Both Technicians are correct. An electronic methane leak detector should be used to verify there are no CNG leaks in the zone of the sensor that has a trace light on the gas detection panel. The ITT high-pressure regulator PRD line should be disconnected to allow testing with a methane detector to ensure the PRD is intact and not starting to leak.
Answer D is incorrect. Both Technicians are correct.

Question #98
Answer A is incorrect. The SCA is already too high and more should not be added.
Answer B is incorrect. If the mixture is extremely overtreated, the engine should be flushed. This might be a suitable answer if the coolant was only slightly overconditioned.
Answer C is correct. Overconditioned coolant can cause a variety of problems, including sludging, accelerated coolant breakdown, and loss of cooling efficiency. When testing reveals overconditioned coolant, the correct remedy is to drain the cooling system and completely replace the coolant mixture. Generally, OEMs recommend that the coolant SCA level be tested at each oil change interval. If there is a substantial loss of coolant and the system has to be replenished, the SCA level must be retested.

Each test provided by an OEM is designed to monitor the SCA package required for its product; it cannot generally be used for another OEM product. Also, the test kits usually consist of test strips that must be stored in airtight containers. They have expiration dates that should be observed. Coolant test kits permit the Technician to test for the appropriate SCA concentration, the pH level, and the TDS. The pH level determines the relative acidity or alkalinity of the coolant. Acids may form in engine coolant exposed to combustion gases or, in some cases, when cooling system metals (ferrous and copper-base) degrade. The pH test is a litmus test in which a test strip is first inserted into a sample of the coolant, and then removed, and the color of the test strip indexed to a color chart provided with the kit. The optimum pH window is defined by each OEM, but normally the window falls between 7.5 and 11.0 on the pH scale. Higher acidity readings (below 7.5 on the pH scale) in tested coolant are indications of corrosion of ferrous and copper metals, coolant exposure to combustion gases, and, in some cases, coolant degradation.
Answer D is incorrect. The engine should not be run without additives.

Question #99
Answer A is incorrect. Technician A is also correct.
Answer B is incorrect. Technician B is also correct.
Answer C is correct. Both Technicians are correct. Piston ring end gaps should be offset on installation to limit blowby, so Technician A is correct. Technician B is also correct because it is important when installing a piston assembly into a cylinder to take measures to protect the studs from damaging the crankshaft journals. Observe OEM instructions. Dividing the number of rings into a 360° circle usually offsets the gaps; so if there were three rings, the stagger would be 120° offset. However, there are other ways of doing this. The ring gaps are usually not placed over the thrust or antithrust side of the piston.
Answer D is incorrect. Both Technicians are correct.

Answers to the Test Questions for the Additional Test Questions Section 6

1. D
2. A
3. C
4. A
5. B
6. C
7. B
8. D
9. B
10. A
11. C
12. B
13. B
14. A
15. C
16. C
17. B
18. D
19. C
20. C
21. C
22. B
23. C
24. B
25. A
26. D
27. D
28. D
29. A
30. D
31. A
32. C
33. C
34. B
35. B
36. D
37. C
38. C
39. C
40. A
41. D
42. D
43. C
44. D
45. B
46. B
47. D
48. B
49. C
50. A
51. C
52. A
53. B
54. D
55. B
56. D
57. A
58. D
59. B
60. D
61. D
62. C
63. A
64. A
65. C
66. B
67. D
68. C
69. D
70. A
71. B
72. B
73. B
74. C
75. B
76. D
77. D
78. C
79. D
80. A
81. B
82. D
83. C
84. A
85. A
86. C
87. C
88. C
89. C
90. A
91. B
92. B
93. C
94. D
95. A
96. B
97. B
98. A
99. B
100. D
101. C
102. A
103. C
104. B
105. A
106. C
107. A
108. C
109. A
110. C
111. C
112. C
113. B
114. C
115. D
116. D
117. B
118. B
119. A
120. A

121. B
122. D
123. C
124. D
125. C
126. D
127. B
128. C
129. C
130. B
131. D
132. D
133. B
134. C
135. D
136. A
137. D
138. D
139. A
140. A
141. D
142. C
143. D
144. C
145. A
146. C
147. D
148. C
149. D
150. A
151. A
152. C
153. B
154. C
155. B
156. D
157. C
158. A
159. C
160. C
161. A
162. C
163. A
164. A
165. A
166. B
167. B
168. A
169. A
170. B
171. B
172. B
173. D
174. B
175. C
176. C
177. C
178. B
179. C
180. D
181. D
182. A
183. D
184. C
185. A
186. D
187. D
188. D
189. D
190. C
191. A
192. C
193. D
194. C
195. C
196. B
197. D
198. C
199. C
200. B
201. C
202. C
203. D
204. D
205. C
206. A
207. B
208. A
209. B
210. B
211. C
212. A
213. D
214. A
215. C
216. D
217. A
218. A
219. D
220. D
221. D
222. C
223. B
224. D
225. B
226. C
227. C
228. C
229. C
230. A
231. D
232. B
233. A
234. A
235. D
236. B
237. A
238. D
239. C
240. B
241. B
242. C
243. B
244. A
245. C
246. A
247. D
248. D
249. A
250. B
251. B
252. B

253. C
254. C
255. C
256. B
257. C
258. C
259. A
260. D
261. C
262. B
263. C
264. C
265. A
266. B
267. C
268. A
269. A
270. A
271. D
272. D
273. C
274. B
275. B
276. D
277. C
278. B
279. B
280. C
281. A
282. D
283. C
284. D
285. C
286. C
287. D
288. C
289. D
290. A
291. B
292. B
293. C
294. A

Explanations to the Answers for the Additional Test Questions Section 6

Question #1
Answer A is incorrect. Lack of air can cause a low boost pressure. The ECM removes fuel to prevent the emissions from being exceeded and a slower acceleration is a result.
Answer B is incorrect. Low power will result from a high inlet restriction. The ECM will remove fuel to prevent the emissions from being exceeded and therefore power will be cut back as well.
Answer C is incorrect. High inlet vacuum can cause rubber boots to be sucked closed.
Answer D is correct. Blue smoke is an indication of oil consumption. Some causes of blue smoke are worn oil control rings, worn or leaking valve seals, and defective turbocharger seals.

Question #2
Answer A is correct. Only Technician A is correct. High crankcase pressures can be caused by a clogged crankcase breather. It is extremely important that the filter element be serviced on a regular schedule to prevent oil from being carried over from the crankcase into the air intake of the engine because eventually the oil will coat the surfaces of critical engine sensors.
Answer B is incorrect. CNG engines that use EGR to control NOx emissions are not affected by the higher combustion chamber pressures. The compression rings can contain the compression in the cylinder. Normally, high crankcase pressures are caused by defective compression rings, scored or cracked pistons and liners, blown head gaskets, and clogged breathers.
Answer C is incorrect. Only Technician A is correct.
Answer D is incorrect. Technician A is correct.

Question #3
Answer A is incorrect. Technician B is also correct.
Answer B is incorrect. Technician A is also correct.
Answer C is correct. Both Technicians are correct. With no visible external signs, the engine may have a leaking head gasket. Draining some engine oil from the oil pan will allow determination of whether the head gasket is leaking the coolant because coolant is heavier than oil and will run out first. A leaking heater core will show up either in the cab or through the A/C drain.
Answer D is incorrect. Both Technicians are correct.

Question #4
Answer A is correct. The display multiplexing controller communication is Most-Likely faulty.
Answer B is incorrect. Since the HVAC and all other accessories are still functioning, the alternator is not the Most-Likely cause of the problem.
Answer C is incorrect. A faulty oil pressure sensor would affect the engine operation; it would Most-Likely cause the engine to shut down. It should also set a DTC in the engine ECM.
Answer D is incorrect. A faulty ECM would probably affect the operation of the engine as well.

Question #5
Answer A is incorrect. This is not an oil filter.
Answer B is correct. Oil coolers are heat exchangers consisting of a housing and a bundle (element/core) through which coolant is pumped and around which oil is circulated. Oil temperatures in CNG engines run higher than coolant temperatures typically around 110° C (230° F). However, the engine coolant reaches its operating temperature more rapidly than the oil and plays a role in heating the oil to operating temperature in cold weather startup/warm-up conditions. The figure shows an oil cooler.
Answer C is incorrect. This is not a filter housing.
Answer D is incorrect. This is not an oil pump.

Question #6
Answer A is incorrect. Technician B is also correct.
Answer B is incorrect. Technician A is also correct.
Answer C is correct. Both Technicians are correct. An indicator lamp can burn out or be unplugged to mask a defect. Because of the unknown buzzer and a drivability issue, a DTC should be suspected as having been logged. Whenever the ignition is energized, a self-check of the bulbs for five seconds is initialized and then lights turn off. If an active DTC exists, then the indicator will turn back on if the bulb is functioning. The DDR and diagnostic software allows a Technician to energize the CEL or SEL indicators to check them.
Answer D is incorrect. Both Technicians are correct.

Question #7
Answer A is incorrect. The CAC should not cause a soot-type material in the intake manifold.
Answer B is correct. The probable source of a soot-type material in the intake manifold of a turbocharged, air-cooled CNG engine is the air intake piping is probably defective and allowing light dirt to pass through into the manifold. Soot-type material in the intake manifold is not normal in most CNG engines.
Answer C is incorrect. The intake manifold should not cause a soot-type material inside itself.
Answer D is incorrect. The head gasket should not cause a soot-type material to build up in the intake manifold.

Question #8
Answer A is incorrect. The IAT sensor senses the temperature of the air as it passes through the intake manifold and transmits the information to the ECM. When the air reaches a predetermined temperature level, the ECM commands the engine fan to turn on. This forces cooler air through the CAC fins to cool the turbocharged air to the engine, therefore the ECM does have control over the CAC air temperature.
Answer B is incorrect. Because the coolant temperature is between 180 and 205° F, coolant is not being used to cool the incoming air. A CNG engine requires cool air (preferably less than 120° F during the intake cycle) to prevent auto ignition of the air/fuel mixture in the combustion chamber as it is being compressed more tightly and to higher temperatures.
Answer C is incorrect. Neither Technician is correct.
Answer D is correct. Neither Technician is correct. The IAT sensor senses the temperature of the air as it passes through the intake manifold and transmits the information to the ECM and the ECM commands the engine fan to turn on forcing cooler air pass the fins of CAC removing heat from the boosted air inside. There is no coolant passing through the CAC.

Question #9
Answer A is incorrect. Gasket failure is not the only cause of oil leaks. Cracked oil pans, loose or defective accessory covers, and loose components can all lead to engine oil leaks. Some engine manufacturers have begun using anaerobic-type sealants with no gaskets being used. Silicone-type seals are also used to prevent oil leaks.
Answer B is correct. Only Technician B is correct. Oil leaks can be found by adding a dye to the engine oil that is visible under ultraviolet light.
Answer C is incorrect. Only Technician B is correct.
Answer D is incorrect. Technician B is correct.

Question #10
Answer A is correct. Using a stethoscope, removing the power at each coil or the individual spark plug wires, and performing an electronic cut-out test using the diagnostic software are all good methods to try and find an engine noise. However, listening to the exhaust to try and determine an engine noise is the LEAST-Likely method to help a Technician discover the origin of the noise as the exhaust is also very noisy and would mask the sound.
Answer B is incorrect. Using a stethoscope is a valid method to identify the origin of a noise.
Answer C is incorrect. Disconnecting each coil power connector or removing individual spark plug wires are valid testing methods.
Answer D is incorrect. Performing a cylinder cut-out with the diagnostic software is a valid testing method.

Question #11
Answer A is incorrect. The indicator gauge mounted on the air cleaner housing or intake piping is an indicator that the air cleaner element could need servicing but is not recommended as a definitive testing device.
Answer B is incorrect. Poor engine performance can be caused by many factors such as poor fuel quality, defective ignition components, and corrupted fuel tables as well as a clogged air cleaner element. Testing of the air cleaner should determine when it needs replacement.
Answer C is correct. A magnehelic gauge and a water manometer are both accurate tools to diagnose if a paper-type air cleaner element needs replacing. Most engine manufacturers recommend the element be replaced when the restriction reaches 20" of water column.
Answer D is incorrect. Oil change intervals are not a good method of determining when a paper-type air cleaner element needs replacing.

Question #12
Answer A is incorrect. A torn air cleaner element can allow foreign objects pass it and cause the debris to strike the compressor wheel blades damaging them.
Answer B is correct. A loose V-band at the compressor turbine housing will cause low power but will not allow debris to come in contact with the compressor wheel and is the LEAST-Likely to cause the condition.
Answer C is incorrect. A disconnected air intake hose can allow debris into the turbocharger compressor wheel damaging the blades when they are struck.
Answer D is incorrect. Mishandling of the turbocharger can cause the compressor wheel blades to be damaged.

Question #13
Answer A is incorrect. The tank should not be condemned unless a new label cannot be obtained. The cylinder should be defueled and closed until a new label can be obtained and permanently secured on the cylinder.
Answer B is correct. Only Technician B is correct. The manufacturer should be contacted with the cylinder serial number so a new label can be obtained and permanently attached to the cylinder. This will allow the cylinder to be placed back into service.
Answer C is incorrect. Only Technician B is correct.
Answer D is incorrect. Technician B is correct.

Question #14
Answer A is correct. Only Technician A is correct. The software will display the intake manifold pressure as absolute, meaning the pressure starts at atmospheric pressure plus or minus the sensed pressure once the engine starts. Because NGV engines have throttle valves (plate), at idle the engine will have a vacuum in the manifold. The ECM must be able to determine this condition to be able to regulate the fuel delivery; therefore, absolute pressure is capable of providing both vacuum at idle and pressure once the engine is placed under a load and the turbocharger is providing boost.
Answer B is incorrect. If the intake manifold pressure sensor read 0 at start-up, the ECM would be unable to differentiate between no pressure and a vacuum. The pressure sensor reads absolute pressure and can subtract from atmospheric pressure for vacuum or add to absolute pressure for a boost condition to enable the ECM to calculate the amount of air being delivered to the cylinders.
Answer C is incorrect. Only Technician A is correct.
Answer D is incorrect. Technician A is correct.

Question #15
Answer A is incorrect. Technician B is also correct.
Answer B is incorrect. Technician A is also correct.
Answer C is correct. Both Technicians are correct. The CAC should be cleaned out with compressed air and the engine and filters should be replaced to ensure that no coolant or metal particles remain in the lubrication system of the engine. Coolant can cause engine damage from the lack of lubrication to the bearings. The filters should also be replaced at the same time because coolant could be trapped inside them, and when the engine is restarted the coolant would get pushed through the engine.
Answer D is incorrect. Both Technicians are correct.

Question #16
Answer A is incorrect. Technician B is also correct.
Answer B is incorrect. Technician A is also correct.
Answer C is correct. Both Technicians are correct. Technician A is correct because worn engine lube pumps can cause low oil pressure. Technician B is also correct because CNG engine oil pumps are positive displacement gear–type pumps.
Answer D is incorrect. Both Technicians are correct.

Question #17
Answer A is incorrect. Fuel filters are made of a coalescing material, not paper.
Answer B is correct. Only Technician B is correct. The fuel filters used in CNG systems are a coalescing type to remove any oil and contaminants that could damage engine sensors or control valves.
Answer C is incorrect. Only Technician B is correct.
Answer D is incorrect. Technician B is correct.

Question #18
Answer A is incorrect. A mercaptan odor signifies a fuel leak. The leak must be detected and repaired. Suspect areas are the Sherex fuel port, fuel lines, and fittings. Pressure relief devices (PRDs) should be tested for possible leakage due to eutectic creep.
Answer B is incorrect. No attempt should be made to introduce any material into the fuel system other than CNG. Dyes could coat sensors and render them ineffective due to contamination. An electronic methane leak detector and/or a soapy water solution or commercially available testing solutions such as snoop are the only approved leak detection devices.
Answer C is incorrect. Neither Technician is correct.
Answer D is correct. Neither Technician is correct. A mercaptan odor signifies a fuel leak. The leak must be detected and repaired. Suspect areas are the Sherex fuel port, fuel lines, and fittings. Pressure relief devices (PRDs) should be tested for possible leakage due to eutectic creep. No attempt should be made to introduce any material into the fuel system other than CNG. Dyes could coat sensors and render them ineffective due to contamination. An electronic methane leak detector and/or a soapy water solution or commercially available testing solutions such as Snoop are the only approved leak detection devices.

Question #19
Answer A is incorrect. Technician B is also correct.
Answer B is incorrect. Technician A is also correct.
Answer C is correct. Both Technicians are correct. Technician A is describing a correct method of aligning cylinder heads on a multicylinder-head engine. Technician B is also correct because bolt threads should be free of dirt (which could bind in the threads) and oil or other fluids, which could cause a hydraulic lock. On inline multicylinder heads, separate heads must be aligned with a straightedge across the intake manifold faces before torquing. Failure to observe torquing increments and sequencing can result in cracked cylinder heads, failed head gaskets, and fire rings that will not seal. Because of the large number of fasteners involved, a calibrated torque wrench should be used. Some OEMs require that template torquing be used. This means setting a torque value first and then turning a set number of degrees beyond that value using a template or protractor. Cylinder head bolts should be installed lightly oiled, although excessive quantities of oil should be avoided because the excess can drain into the bolt hole and cause a hydraulic lock.
Answer D is incorrect. Both Technicians are correct.

Question #20
Answer A is incorrect. Low voltage would cause the gauge to read inaccurate.
Answer B is incorrect. A short in the transducer would cause the gauge to read wrongly.
Answer C is correct. Low voltage, short in the wiring to the transducer, and a defective transducer will all cause inaccurate readings in an electric oil pressure gauge. However, low oil pressure in itself would not cause an inaccurate reading at the gauge. Of all the engine monitoring devices

used on an engine, the oil pressure sensor is one of the most critical. Loss of engine oil pressure in most cases will cause a nearly immediate engine failure. Back in the days when few of an engine's operating conditions were monitored and displayed to the operator, there was always a means of signaling a loss of oil pressure. Several types of sensors are used in today's engines, including variable capacitance (pressure) sensors, piezo-electrics, bourbon gauges, or electrical gauges. Whichever means is used to display oil pressure, none should be used as the only means of diagnosing low oil pressure. Use a good quality master gauge such as a gauge with a fluid-filled display dial. If possible, always connect directly into the main oil galley to diagnose a low oil pressure condition.
Answer D is incorrect. A defective transducer would cause the gauge to read incorrectly.

Question #21
Answer A is incorrect. Technician B is also correct.
Answer B is incorrect. Technician A is also correct.
Answer C is correct. Both Technicians are correct. The CNG engine spark plug external ceramic should be cleaned with alcohol and a clean towel and should not be touched prior to installation. Contamination from skin oil can cause the spark to jump past the boots and down the ceramic to ground and cause a misfire when the engine is under load. The boots come prelubricated with dielectric compound from the manufacturer and no additional application is necessary. If too much is applied, it can lead the compound to migrate up into the coil contacts and eventually to a misfire.
Answer D is incorrect. Both Technicians are correct.

Question #22
Answer A is incorrect. Crankshaft main bearings are unlikely to cause the thumping noise described.
Answer B is correct. The description perfectly describes what is often called "bottom end" knock, which is caused by a failed connecting rod bearing at the big end or at the crankshaft end.
Answer C is incorrect. A worn camshaft bearing would be unlikely to cause the loud thumping noise described, especially under changing engine loads.
Answer D is incorrect. A defective rocker arm is likely to cause more of a clicking noise. Rocker arm clicking is usually not affected by engine loads.

Question #23
Answer A is incorrect. Technician B is also correct.
Answer B is incorrect. Technician A is also correct.
Answer C is correct. Both Technicians are correct. Cylinder hydraulic lock is usually caused by liquid, usually engine coolant, filling the cylinder above the piston, preventing the engine from rotating. A crack in a cylinder head or a cylinder liner is a possible source of coolant that could fill a cylinder.
Answer D is incorrect. Both Technicians are correct.

Question #24
Answer A is incorrect. The oil cooler core does not need to be tested at every PM.
Answer B is correct. Only Technician B is correct. Testing an oil cooler core should be undertaken at a major engine overhaul and would not be a part of routine PM scheduling.
Answer C is incorrect. Only Technician B is correct.
Answer D is incorrect. Technician B is correct.

Question #25
Answer A is correct. The Most-Likely cause is too large of a gap. This will cause very high KV, depending on the amount of excessive gap and the voltage needed to jump it.
Answer B is incorrect. Tracking between the electrodes of the spark plug will cause low KV because of the short to ground of the electrodes and not allowing the coil to build up enough voltage to cause a spark.
Answer C is incorrect. A defective ICM will cause the spark plug to fire weakly or not at all and therefore cause low KV.
Answer D is incorrect. A defective coil will usually have a weak spark or none at all.

Question #26
Answer A is incorrect. The cylinder surface to deck specifications established by the manufacturer need to be strictly followed.
Answer B is incorrect. Cylinder heads can warp side to side, transversely as well as lengthwise.
Answer C is incorrect. Neither Technician is correct.
Answer D is correct. Neither Technician is correct. If the head is allowed to become too thin, the strength of the head will be diminished. Warpage or cracking can develop much sooner as a result. Cylinder heads can warp side to side, transversely as well as lengthwise.

Question #27
Answer A is incorrect. Before condemning the starter, the state of charge of the batteries needs to be verified with a DVOM while attempting to start the bus.
Answer B is incorrect. Because the starter tried to engage, verified by the clicking sound of the starter solenoid, the logic has been completed and therefore the fuel door switch is sending a ground input to the I/O controls.
Answer C is incorrect. Neither Technician is correct.
Answer D is correct. Neither Technician is correct. Before condemning the starter, the state of charge of the batteries needs to be verified with a DVOM while attempting to start the bus. If the batteries cannot maintain a minimum of 9.6 volts while cranking, then they need to be replaced with a fully charged set. The starter should be rechecked for starting. Both the positive and negative battery cables and connections should be checked for corrosion and looseness. Because the starter tried to engage, verified by the clicking sound of the starter solenoid, the logic has been completed and therefore the fuel door switch is sending a ground input to the I/O controls.

Question #28
Answer A is incorrect. Neither Technician is correct. A wet sleeve should never be repaired. It should be replaced.
Answer B is incorrect. Neither Technician is correct. If the cylinder block counterbore is within specifications, it should not require to be recut.
Answer C is incorrect. Neither Technician is correct.
Answer D is correct. Neither Technician is correct. A crack in the counterbore flange of a wet sleeve requires that it be replaced. Because it is possible for a wet sleeve flange to crack and not damage the block counterbore, the block counterbore should be carefully measured and inspected and may not need to be recut.

Question #29
Answer A is correct. Only Technician A is correct. The 12 volts is supplied directly from the batteries to the coils whenever the ignition is on. The ICM will control the ground signal to the coils to enable the coils to discharge the high voltage to the spark plugs.
Answer B is incorrect. The ICM controls the ground signal. The 12 volts is supplied directly from the batteries to each coil whenever the ignition is on.
Answer C is incorrect. Only Technician A is correct.
Answer D is incorrect. Technician A is correct.

Question #30
Answer A is incorrect. External gear–type oil pumps are a positive displacement–type pump and do not need to be submerged in oil to pump. They only require priming.
Answer B is incorrect. Internal gear–type oil pumps are a positive displacement type pump and do not need to be submerged in the oil to pump. They only require being primed to be able to pump.
Answer C is incorrect. Neither Technician is correct.
Answer D is correct. Neither Technician is correct. External or internal gear–type oil pumps are positive displacement types of pumps and do not need to be submerged in oil to be able to pump; they only require priming.

Question #31

Answer A is correct. Only Technician A is correct. Missing connector seals and/or broken locks are the number 1 reason for corrosion occurring on the pins and sockets of sensors. Because the lock is broken or missing, the connector is free to back off, exposing the pins to moisture and dirt that allow corrosion to form.

Answer B is incorrect. A zip tie is not a good locking device. The tie can break or fall off, allowing the condition to repeat itself.

Answer C is incorrect. Only Technician A is correct.

Answer D is incorrect. Technician A is correct.

Question #32

Answer A is incorrect. Technician A is also correct.

Answer B is incorrect. Technician B is also correct.

Answer C is correct. Both Technicians are correct. Reuseable cylinder head bolts have to be inspected to make sure they are not stretched beyond use by using a special bolt gauge. Erosion and pitting are conditions that must be detected and corrected prior to reusing bolts. Stretch-to-yield bolts are usually replaced each time they are removed.

Answer D is incorrect. Both Technicians are correct.

Question #33

Answer A is incorrect. Contact cleaner may not be able to remove all of the contaminants that are on the ceramic of the spark plug.

Answer B is incorrect. Brake cleaner can leave a residue on the ceramic of the spark plug, leading to high-voltage leakage of the secondary circuit.

Answer C is correct. Isopropyl alcohol is the recommended cleaning agent of the secondary ignition circuit.

Answer D is incorrect. Water will not be able to remove the oily deposit that can cause the secondary high voltage to leak to ground, causing a misfiring cylinder.

Question #34

Answer A is incorrect. Compressed air should never be used to vent a cylinder. Compressed air contains oxygen molecules that can react with methane gas and if the correct mixture is achieved combustion can take place should an ignition source be present. Compressed air flowing creates static electricity and could be a source of ignition.

Answer B is correct. Only Technician B is correct. Water can be used to vent or purge a CNG fuel cylinder.

Answer C is incorrect. Only Technician B is correct.

Answer D is incorrect. Technician B is correct.

Question #35

Answer A is incorrect. The coolant probe should only be changed if it is determined to be defective. The coolant level should be checked first. The DTC in the ECM should be checked to determine if it is a coolant level or the ECM has sensed a voltage defect in the circuit.

Answer B is correct. Only Technician B is correct. The coolant level should be checked and corrected if needed before any repairs are initiated. If the coolant is low, then the source of the leak should be found and corrected by pressure testing.

Answer C is incorrect. Only Technician B is correct.

Answer D is incorrect. Technician B is correct.

Question #36

Answer A is incorrect. Blue-colored smoke from the tailpipe usually indicates oil being burned in the engine.

Answer B is incorrect. Excessive rattling of the exhaust system would indicate an exhaust system problem.

Answer C is incorrect. Misfires cannot be determined by tail pipe exhaust temperatures. However, wide temperature differences at each exhaust port immediately after startup or when under a load can indicate a misfiring cylinder.

Answer D is correct. A stumble or miss is indicative of an engine misfire.

Question #37

Answer A is incorrect. Too wide a gap can cause an extremely high KV coming from the coil and lead the spark to follow the path of least resistance, causing flashover on the ceramic of the spark plug as the spark travels externally to ground.

Answer B is incorrect. A dirty spark plug socket usually has some oil residue in it. Oil residue can lead to flashover. As the spark plug and the oil heat up, the oil's insulating capabilities can be compromised and it could conduct electricity, causing flashover on the ceramic of the spark plug. Clean, good quality, and magnetic spark plug sockets are recommended during the installation of spark plugs to prevent the rubber foam that prevents the spark plug ceramic from being cracked or broken during installation from contaminating the new plugs.

Answer C is correct. Isopropyl alcohol will remove all residues from the spark plugs as well as the tools and all associated secondary circuitry components.

Answer D is incorrect. Handling spark plugs with bare hands can leave an oil residue on the spark plug that could cause conductivity and eventually flashover.

Question #38

Answer A is incorrect. Technician B is also correct.

Answer B is incorrect. Technician A is also correct.

Answer C is correct. Both Technicians are correct. Small popping sounds can result from the liner moving back into position after moving out of position after the cylinder was defueled especially if for an extended period of time. Liners used in Type IV tanks are permeable and can allow trace amounts of methane that an electronic leak detector can detect or a Technician can sometimes smell. Leak detection solutions rather than electronic detectors should be used on these types of tanks to accurately determine if there is a methane leak at the end bosses.

Answer D is incorrect. Both Technicians are correct.

Question #39

Answer A is incorrect. Coolant leaking into the combustion chamber from a spark plug adapter, cracked head, or another source will damage the catalyst as well as the O_2 sensor. Coolant leaks must be repaired quickly.

Answer B is incorrect. Worn oil control rings will allow engine oil into the combustion chamber and eventually can coat the catalyst media with deposits and shorten the catalyst's life.

Answer C is correct. A stuck waste gate will deliver low power due to the ECM receiving a low turbo boost signal. In turn, fuel will be reduced to the combustion chambers. This will not damage the catalyst.

Answer D is incorrect. Worn exhaust valve guide seals will allow oil into the combustion chamber and eventually can coat the catalyst media with deposits and shorten the catalyst's life.

Question #40

Answer A is correct. Only Technician A is correct. When a CPS is installed correctly, the gap will be maintained to the manufacturer's specifications.

Answer B is incorrect. Technician B is incorrect. It is not necessary to measure the gap of a sensor unless the ECM logs DTCs relating to the functionality of the CPS.

Answer C is incorrect. Only Technician A is correct.

Answer D is incorrect. Technician A is correct.

Question #41

Answer A is incorrect. The ¼ fuel shutoff valve is not connected to the cooling system.

Answer B is incorrect. The secondary fuel filter is not connected to the cooling system.

Answer C is incorrect. Fuel is not routed through the cylinder head as with a diesel engine. All of the fuel is directed through the external stainless steel lines. The fuel then passes through the intake manifold before entering the combustion chamber.

Answer D is correct. Because the ITT fuel high-pressure fuel regulator has a heat exchanger built into it, it is possible for the diaphragm to become defective and allow fuel to migrate into the coolant surge tank. If this condition is detected, the regulator should be replaced and no attempt should be made to repair it.

Question #42
Answer A is incorrect. A blown power fuse for the coils will prevent the coils from receiving voltage and cause a no-start condition.
Answer B is incorrect. If the ¼ turn valve is turned off and the fuel has been depleted for maintenance purposes, the engine will not receive fuel and will not run.
Answer C is incorrect. If the engine ECM does not receive a crank sensor signal the engine will not run. A DTC should be logged in the ECM along with a Stop Engine Light (SEL).
Answer D is correct. A stuck-open waste gate is the EXCEPT statement because it will not prevent the engine from starting and running. The driver will complain of low power.

Question #43
Answer A is incorrect. Technician B is also correct.
Answer B is incorrect. Technician A is also correct.
Answer C is correct. Both Technicians are correct. Technician A is correct because if a coil becomes cracked, the insulating material of the coil will no longer be intact, allowing the spark to jump to ground, especially when the spark intensity is very high such as when the engine is under a load. Technician B is also correct. When contamination such as engine oil, which also attracts dirt, is allowed to remain in the spark plug well, the oil can become a conductor when it is heated to a high temperature, causing the engine to misfire.
Answer D is incorrect. Both Technicians are correct.

Question #44
Answer A is incorrect. This measurement does not prevent head warpage; however, if a head is too thin it will be prone to warpage.
Answer B is incorrect. Valve guide height is measured on the valve guide.
Answer C is incorrect. Cylinder head warpage is determined with a straightedge and feeler gauge.
Answer D is correct. Minimum head deck-to-deck thickness is critical. If a cylinder head is machined below this minimum specification, it will be much more likely to crack due to its now thinner casting. Additionally, when the head becomes thinner, critical valve train geometry angles are changed. The figure indicates the minimum head deck-to-deck specification.

Question #45
Answer A is incorrect. An Allison World Transmission can be stalled tested.
Answer B is correct. Only Technician B is correct. Due to the increase of torque multiplication in gears 1 through 3, 4th gear, which is a 1:1 ratio, will place the drive train under lower torque stresses. Stall testing allows the Technician to place the engine under full power and simulate high power demands without leaving the shop. Only with a DDR or the Allison Doc software can the clutch-enabled test be performed. Stall testing longer than 30 seconds or a maximum converter out fluid temperature of 300° F without a cool-down period of a 2- or 3-minute interval can result in transmission damage due to elevated transmission fluid temperatures.
Answer C is incorrect. Only Technician B is correct.
Answer D is incorrect. Technician B is correct.

Question #46
Answer A is incorrect. Technician A is incorrect. Voltage for the primary side of the coil is not provided through an ignition switch controlled signal. The ICM sends up to 300 volts to the coil's primary circuit to ensure the coil is capable of have multiple spark discharges (MSD) during the power stroke of the engine for a lean and clean burning cycle.
Answer B is correct. Only Technician B is correct. The ICM sends up to 300 volts to the coil's primary circuit to ensure the coil is capable of have multiple spark discharges (MSD) during the power stroke of the engine for a lean and clean burning cycle.
Answer C is incorrect. Only Technician B is correct.
Answer D is incorrect. Technician B is correct.

Question #47

Answer A is incorrect. The adaptor tubes have to be free of oil and dirt contamination. Any contamination can fall into the combustion chamber when the spark plug is removed. Additionally, contamination can cause the spark to jump the outside of the spark, plug leading to a misfire.

Answer B is incorrect. The adaptor tubes have coolant around them to prevent them from overheating.

Answer C is incorrect. Adaptor tubes have O-ring seals to prevent coolant from leaking out onto the cylinder head and into the engine oil.

Answer D is correct. The threads of the spark plug adaptor have to be in good condition. If the spark plug does not screw down into the tube easily, the tube should be replaced. Because the spark plug cannot be installed easily, the next time the plug needs to be changed it could be very difficult to remove and cause the spark plug adaptor to unscrew out of the cylinder head, resulting in extensive labor to repair the problem.

Question #48

Answer A is incorrect. You can use a snap gauge to determine the ID.

Answer B is correct. The LEAST-Likely tool to be used to check the inside diameter of a valve guide would be a feeler gauge.

Answer C is incorrect. You can check the valve guide bore with a ball gauge.

Answer D is incorrect. You can also use a dial ball gauge to measure valve guide inside diameter. If in need of replacement, the guide must be pressed or driven out with the correct driver or mandrel. Damage to the guide bore may require reaming to fit an oversized guide. Machining for guide sleeves or knurling can repair integral guides. Freezing (dry ice) can facilitate installation of new guides or the use of press-fit lubricant.

Question #49

Answer A is incorrect. Technician B is also correct.

Answer B is incorrect. Technician A is also correct.

Answer C is correct. Both Technicians are correct. An infrared pyrometer is a very good tool to identify a misfiring cylinder because exhaust port temperatures are directly connected to the power output of a cylinder. The diagnostic software will enable the Technician to turn off the cylinders one at a time. A distinct change in sound will be heard from a cylinder firing normally, whereas a weak or dead cylinder will produce no or very little change in sound.

Answer D is incorrect. Both Technicians are correct.

Question #50

Answer A is correct. Only Technician A is correct. A good spark plug that is properly gapped will generally require between 5 and 20 KV to jump the gap, depending on the cylinder pressures and the amount of fuel in the chamber. When the voltage starts to exceed 25 KV, the spark plugs tend to start to misfire when the engine is under maximum load.

Answer B is incorrect. A good spark plug that is properly gapped will generally require between 5 and 20 KV to jump the gap, depending on the cylinder pressures and the amount of fuel in the chamber. When a spark plug requires 40 KV to jump the gap, the cylinder will Most-Likely misfire.

Answer C is incorrect. Only Technician A is correct.

Answer D is incorrect. Technician A is correct.

Question #51

Answer A is incorrect. Technician B is also correct.

Answer B is incorrect. Technician A is also correct.

Answer C is correct. Both Technicians are correct. Technician A is correct. A cylinder block suspected to be cracked can be immersed in hot water for up to two hours when pressure testing. This time period should be sufficient to swell any cracks and to be able to observe a leak. Technician B is also correct. CNG engine manufacturers recommend that a cracked cylinder block be replaced rather than repaired.

Answer D is incorrect. Both Technicians are correct.

Question #52

Answer A is correct. Only Technician A is correct in his description of how manufacturers want cylinder head bolts torqued.

Answer B is incorrect. It is essential that bent push tubes be replaced rather than straightened. When straightened rods are returned to service, they fail quickly and can cause severe and expensive engine damage.

Answer C is incorrect. Only Technician A is correct.

Answer D is incorrect. Technician A is correct.

Question #53

Answer A is incorrect. Radial runout is checked by installing the dial indicator through the oil feed or oil drain hole and moving the shaft up and down.

Answer B is correct. In the figure, the dial indicator is located to check turbine shaft end play.

Answer C is incorrect. Turbine fin wear is checked by visual inspection.

Answer D is incorrect. Compressor fin wear is checked by visual inspection.

Question #54

Answer A is incorrect. Overheating would discolor the electrodes and insulating ceramic.

Answer B is incorrect. An improper gap would not cause the electrodes to become eroded.

Answer C is incorrect. Deposits would contaminate the electrodes and leave a darker residue on the electrode insulator.

Answer D is correct. The electrodes are eroded. The Most-Likely condition is wear through normal usage over a period of the life of the spark plug.

Question #55

Answer A is incorrect. Blue smoke is usually caused by the combustion of engine oil.

Answer B is correct. Only Technician B is correct. Because the carbon chain is smaller (less carbon content in the fuel), the CNG engine will not produce black smoke if there is a lack of air as compared with a Diesel or Gasoline engine.

Answer C is incorrect. Only Technician B is correct.

Answer D is incorrect. Technician B is correct.

Question #56

Answer A is incorrect. Lack of coolant flow through the heat exchanger of the ITT high-pressure regulator can cause the fuel temperature to become extremely low. This can cause icing of the lines and filters. When a gas goes from a high pressure to a low pressure, a cooling effect takes place.

Answer B is incorrect. Fuel pressure sensors do drift out of calibration and can cause a low fuel pressure DTC to be logged.

Answer C is incorrect. A restricted fuel filter due to contamination can cause a low fuel pressure DTC to be logged.

Answer D is correct. The ITT fuel pressure regulator is the LEAST-Likely item to log a DTC for low fuel pressure. All of the above items should be checked first to ensure they are functioning correctly before faulting the high-pressure regulator.

Question #57

Answer A is correct. Misadjusted valves can open at the incorrect timing or not at all, causing the cylinder to misfire.

Answer B is incorrect. A clogged fuel filter will not cause only one cylinder to misfire because all cylinders would be affected.

Answer C is incorrect. Low fuel delivery pressure to the engine will not cause only one cylinder to misfire because all cylinders would be affected.

Answer D is incorrect. A boost pressure leak in the air intake piping could affect engine performance, depending on how large it is but will not cause only one cylinder to misfire because all cylinders would be affected.

Question #58
Answer A is incorrect. This is not a ridge remover.
Answer B is incorrect. This is not a liner remover.
Answer C is incorrect. This tool does not machine the crosshatch.
Answer D is correct. The tool shown is designed to cut a cylinder liner counterbore. Check the cylinder sleeve counterbore for the correct depth and circumference. Counterbore depth should typically not vary by more than 0.001". Counterbore depth can be subtracted from the sleeve flange dimension to calculate sleeve protrusion. The counterbore should be recut using the OEM-recommended tools and specifications.

Question #59
Answer A is incorrect. High-pressure CNG fuel lines are usually manufactured from stainless steel material.
Answer B is correct. Welded tubing is not utilized for high-pressure CNG fuel lines because of strength requirements and the weld bead can lead to leaks.
Answer C is incorrect. CNG high-pressure fuel lines are usually tempered.
Answer D is incorrect. Minimal wall thickness requirements for different tubing diameters need to be followed.

Question #60
Answer A is incorrect. The engine run switch is usually attached to the cover. If a metal cover is fabricated and installed it could cause the switch to ground out and cause the engine to stall. That could cause an unsafe condition. The manufacturer's design should not be altered.
Answer B is incorrect. Duct tape is useful but can fall off or become dislodged, allowing dirt and moisture to enter the rear run box. The correct cover should be installed and secured after cleaning and drying out the box.
Answer C is incorrect. Neither Technician is correct.
Answer D is correct. Neither Technician is correct. The engine run switch is usually attached to the cover. If a metal cover is fabricated and installed it could cause the switch to ground out and cause the engine to stall. That could cause an unsafe condition. The manufacturer's design should not be altered. Duct tape is useful but can fall off or become dislodged, allowing dirt and moisture to enter the rear run box. The correct cover should be installed and secured after cleaning and drying out the box.

Question #61
Answer A is incorrect. Mesh-type elements are not used on CNG engines.
Answer B is incorrect. The crankcase breather pipe is routed through a coalescing filter to remove moisture and oil vapors. The remaining crankcase fumes are then routed into the air intake system prior to the turbocharger. The coalescing filter element is a preventive maintenance item.
Answer C is incorrect. Neither Technician is correct.
Answer D is correct. Neither Technician is correct. Mesh-type elements are not used on CNG engines. CNG engines utilize a replaceable crankcase breather that is coalescing to remove oil and moisture from the crankcase fumes before being introduced into the air intake pipe before the turbocharger to recycle the fumes and improve air quality by not venting to atmosphere. The crankcase breather pipe is routed through a coalescing filter to remove moisture and oil vapors.

Question #62
Answer A is incorrect. Technician B is also correct.
Answer B is incorrect. Technician A is also correct.
Answer C is correct. Both Technicians are correct. Technician A is correct. The turbocharger oil drain line needs to be inspected to make sure no pieces are lodged in the line, restricting the drainage of lube oil. The blockage could eventually build up to the point where the drainage of the turbocharger's lube oil would be insufficient and the new turbocharger's bearings could flood and

cause the oil to be forced past the seals and be ingested by the vacuum generated in the intake side of the turbo. Technician B is also correct in stating the oil pan should be removed, before placing the engine back in service, and checked for any pieces that could have reached the pan. These pieces could be sucked up by the oil pump, even though it is equipped with a pickup-tube screen, locking up or damaging the oil pump and resulting in serious engine damage.
Answer D is incorrect. Both Technicians are correct.

Question #63
Answer A is correct. Only Technician A is correct. When servicing the fuel low-pressure filter of a CNG system, the ¼ turn valve must be turned off and the engine runs until the engine shuts off to deplete the fuel from the system and the low-pressure gauge must record and remain at 0 psi before attempting to remove the plug or filter from the system.
Answer B is incorrect. The fuel system of a CNG-powered vehicle must never be opened when it is under pressure even if it is considered to be low pressure. Any fuel leakage poses a fire hazard.
Answer C is incorrect. Only Technician A is correct.
Answer D is incorrect. Technician A is correct.

Question #64
Answer A is correct. Before installing valves into a cylinder head, seat contact should be tested with machinist's bluing or by the pencil stroke method. The remainder of the procedures would be performed after the cylinder valves had been installed in the head.
Answer B is incorrect. Rocker arms are adjusted after the head is installed.
Answer C is incorrect. Measuring camshaft lift is checked while inspecting the camshaft.
Answer D is incorrect. Head gaskets are checked during disassembly for evidence of leaks. Head gaskets are not reused; they are replaced.

Question #65
Answer A is incorrect. Technician B is also correct.
Answer B is incorrect. Technician A is also correct.
Answer C is correct. Both Technicians are correct. A slight restriction when the engine is under load is an indication the air intake system is probably in good condition. If no restriction is observed, the technician should check the piping and the air cleaner to make sure they are intact. When an air cleaner restriction reaches 20" of water restriction, the air cleaner needs to be replaced or the engine performance will suffer from lack of air.
Answer D is incorrect. Both Technicians are correct.

Question #66
Answer A is incorrect. Too much fuel being fed into the engine will cause high exhaust temperatures as well as a rich running condition that will not be compensated by O_2 sensor.
Answer B is correct. When a CNG engine idles too high, the Most-Likely cause is a bent throttle plate valve that could have been caused by the engine backfiring. Other causes such as intake leaks after the throttle plate should not be overlooked.
Answer C is incorrect. When valves are too tight, the engine will not idle too high.
Answer D is incorrect. Valves set too loose will not cause the engine to idle too high.

Question #67
Answer A is incorrect. Rocker arms are checked prior to installing.
Answer B is incorrect. The rocker shaft is inspected prior to installation.
Answer C is incorrect. The pallet is inspected prior to reassembly.
Answer D is correct. When installing rocker arms, the bushings, shaft, and pallet (actuation end) must be inspected. However, the adjustment screws are not adjusted until the valve lash is set during the tune-up, which happens after they are assembled and torqued to the cylinder head.

Question #68

Answer A is incorrect. Tight exhaust valves can lead to high exhaust temperatures because the valves could start to remain partially open and allow fuel to escape past them and burn in the exhaust system during the compression cycle.

Answer B is incorrect. A defective fuel control valve such as partially sticking open can cause high exhaust temperatures due to the inability of the valve to respond to commands from the ECM and control the flow of the incoming fuel.

Answer C is correct. High ambient temperatures are the LEAST-Likely to affect exhaust temperatures. Normal CNG engine exhaust temperatures are anywhere from 1000° F to 1200° F, much higher than ambient temperatures.

Answer D is incorrect. Lack of air due to a clogged air cleaner can be a cause of high exhaust temperatures because the engine will not be able to breathe properly and therefore not be able to bring in cool air to help in cooling down the exhaust temperatures and eventually exceed the operating range.

Question #69

Answer A is incorrect. The cylinder can be purged with water.

Answer B is incorrect. The cylinder can be drilled with 2, 1/2" diameter holes to render it unusable.

Answer C is incorrect. The local, state, and federal regulations have to be used when disposing of CNG cylinders.

Answer D is correct. CNG cylinders can never be re-used for anything but CNG. Furthermore, by placing air with oxygen in a cylinder that can still have fuel residues in it can be an explosive situation and should never be done. When air is compressed into the cylinder, a by-product is static electricity which can ignite the air fuel mixture.

Question #70

Answer A is correct. Only Technician A is correct. The fuel high-pressure regulator should be checked for coolant leaks during the cooling system pressure test. Coolant is routed through the high-pressure regulator to prevent icing and low temperature of the fuel to occur especially during cold ambient temperatures.

Answer B is incorrect. Coolant is routed through the high-pressure regulator and the hoses and connections can be a source of coolant leaks.

Answer C is incorrect. Only Technician A is correct.

Answer D is incorrect. Technician A is correct.

Question #71

Answer A is incorrect. Because the fault codes are written to the electronic erasable, programmable, read only memory (EEPROM) in the ECM at engine shutdown, turning off the batteries will not erase the codes as in the past with early electronic controlled engines.

Answer B is correct. Only Technician B is correct. Fault codes in the engine's ECM are logged and written in the EEPROM, so the act of disconnecting the vehicle batteries does not erase them. The DDR or the diagnostic software will need to be used to clear codes.

Answer C is incorrect. Only Technician B is correct.

Answer D is incorrect. Technician B is correct.

Question #72

Answer A is incorrect. A chain can damage the epoxy outer layer of cylinder deep enough to cause the tank to be condemned at visual inspection.

Answer B is correct. Only Technician B is correct. A nylon sling strong enough and long enough to support the cylinder at both ends should be used to prevent the cylinder from being damaged by an object such as the link of a chain at it rubs cross the epoxy outer layer.

Answer C is incorrect. Only Technician B is correct.

Answer D is incorrect. Technician B is correct.

Question #73
Answer A is incorrect. The fueling receptacle should prevent any fuel from leaking during or after the fueling process by the check valve. A strong mercaptan odor is not indicative of residual fuel.
Answer B is correct. The receptacle's check valve could be leaking and is the Most-Likely cause of a strong mercaptan odor in this scenario. The check valve can be tested with a soapy water solution or commercially available testing solutions such as Snoop. Any sign of bubbles is an indication of leakage. The seat of the check valve is defective and the receptacle needs to be replaced. Dirt or mishandling during refueling could be the cause of damage. Any time mercaptan is detected, leak testing should be performed to find the source of the leak.
Answer C is incorrect. The CNG fuel cylinders should not vent after fueling. Venting is an indication the Pressure Relief Device (PRD) has failed due to high temperature, above 189° F. If there are any signs of venting, the cause must be determined and corrected before the vehicle can be placed back into service.
Answer D is incorrect. Although the receptacle's rubber cap must be installed and fit snuggly, its purpose is to prevent contaminants from settling on the check valve and being pushed into the fuel cylinders when fueling. It is not designed to prevent fuel from leaking out past it.

Question #74
Answer A is incorrect. Boring the counterbores to manufacturer's specifications is an acceptable method of adjusting cylinder counterbores, especially after resurfacing of the deck surface.
Answer B is incorrect. Resurfacing of the block deck surface to bring the counterbores to specifications is an acceptable method of bringing the counterbores to specifications as long as only the required material is removed and the amount removed does not exceed the amount specified by the engine manufacturer to maintain the center line of the main bearings to the deck surface minimum height.
Answer C is correct. This is the EXCEPT statement. Grinding of the counterbores with a die grinder is not acceptable. Uniformity around the perimeter of the counterbore would not be maintained and the liner flange would not be clamped down uniformly, causing the liner to rock and premature failure.
Answer D is incorrect. Some manufacturers approve the use of shims.

Question #75
Answer A is incorrect. A punctured CAC would allow turbo boost pressure to escape and cause low boost pressure.
Answer B is correct. If the engine is overfueling due to a malfunctioning fuel control valve or some other reason, the engine would develop more power and would not result in lack of power.
Answer C is incorrect. A clogged air cleaner element would prevent the turbocharger from developing boost pressure and the ECM would cut back the amount of fuel delivered to the engine and cause low boost pressure to be observed.
Answer D is incorrect. When the engine logs a diagnostic code related to engine operation that could damage the engine or components, the ECM can de-rate the fueling rate to reduce engine power and result in lower turbo boost.

Question #76
Answer A is incorrect. A restricted inlet would cause insufficient oil to the filter.
Answer B is incorrect. If the oil pump was sucking air, pressure would be low.
Answer C is incorrect. A massive oil leak would lower pressure, not raise it.
Answer D is correct. When the bypass valve trips on an oil filter pad assembly, the oil is routed around the filter assembly—that is, it bypasses it. This will only occur when the filter element has become plugged. Filter-mounting pad bypass valves operate in much the same way as the oil pressure regulating valves, except that their objective is to route the lubricant around a restricted full flow filter to prevent engine damage by oil starvation. When a filter bypass valve is actuated and the check valve unseated, instead of spilling the oil to the crankcase it reroutes the oil directly to the lubrication circuit, effectively shortcutting the filter assembly. So, when a bypass valve trips, unfiltered oil is circulated through the engine. This is not a good condition but it is better than no lubrication.

Question #77

Answer A is incorrect. The engine should not be disassembled or disabled until all other options have been exhausted. The easiest tasks should be performed first.

Answer B is incorrect. An infrared pyrometer will probably not uncover any reason why the engine is consuming oil because the engine will probably still perform well.

Answer C is incorrect. Neither Technician is correct.

Answer D is correct. Neither Technician is correct. The engine should not be disassembled or disabled until all other options have been exhausted. The easiest tasks should be performed first. The engine oil should be checked for contamination or dilution and to confirm it is the correct type. The records should be checked to see when the last oil change was performed to make sure the oil level was adjusted correctly. The history of the vehicle should be checked for previous complaints for the oil level at the same time. If no history was found the level should be corrected and the vehicle should be road tested and observed before any other corrections are undertaken. An infrared pyrometer will probably not uncover any reason why the engine is consuming oil because the engine will probably still perform well. If the pyrometer was able to detect a temperature difference between two or more cylinders, the Technician would probably be able to detect a noticeable change in engine performance as well.

Question #78

Answer A is incorrect. Technician B is also correct.

Answer B is incorrect. Technician A is also correct.

Answer C is correct. Both Technicians are correct. The turbocharger drain line needs to be checked in case that pieces of the failed bearings have fallen into the drain pipe causing an obstruction to the proper drainage of the new turbocharger leading to possible oil consumption. Technician B is also correct because when shutting down an engine without allowing a turbocharger time to slow down will eliminate lubrication to the bearings even though it could be spinning at a very high rate.

Answer D is incorrect. Both Technicians are correct.

Question #79

Answer A is incorrect. An NGV engine uses ram air pass its fins to remove the heat from the boosted air. Coolant is not utilized because the coolant temperature would be around 180° F and that would heat the boosted air, not cool it.

Answer B is incorrect. The CAC can be pressurized to check for leaks but should not be pressurized beyond 40 psi or damage to the CAC can occur; 75 psi is too high of a test pressure.

Answer C is incorrect. Neither Technician is correct.

Answer D is correct. Neither Technician is correct. NGV engines use ram air or air from a fan past the fins of the CAC to cool the boosted air, not coolant. The maximum pressure to be applied to a CAC is 40 psi.

Question #80

Answer A is correct. Only Technician A is correct. The sensing line can become contaminated and should be checked before proceeding to other diagnostics.

Answer B is incorrect. Diagnostics should always be performed before replacing any parts.

Answer C is incorrect. Only Technician A is correct.

Answer D is incorrect. Technician A is correct.

Question #81

Answer A is incorrect. O_2 sensors are sensitive to being contaminated by other substances other than lead.

Answer B is correct. Only Technician B is correct. Coolant, engine oil, and non O_2 sensor safe sealants are capable of destroying O_2 sensors.

Answer C is incorrect. Only Technician B is correct.
Answer D is incorrect. Technician B is correct.

Question #82
Answer A is incorrect. Only fuel cylinder manufacturers can make copies of certification labels and allow these labels to be affixed back onto a cylinder.
Answer B is incorrect. Labels cannot be transferred. Each cylinder has a unique serial number and the serial number has to be printed on the label by the manufacturer.
Answer C is incorrect. Neither Technician is correct.
Answer D is correct. Neither Technician is correct. Only fuel cylinder manufacturers can make copies of certification labels and allow these labels to be affixed back onto a cylinder. Labels cannot be transferred. Each cylinder has a unique serial number and the serial number has to be printed on the label by the manufacturer.

Question #83
Answer A is incorrect. Technician A is also correct.
Answer B is incorrect. Technician B is also correct.
Answer C is correct. Both Technicians A and B are correct. Technician A is correct because the relative movement between the intake system components on the chassis and those on the engine must be accommodated by having flexible rubber couplings. Technician B is also correct. The smallest leak in the intake system components can allow dirt ingestion by the turbocharger and engine so the turbocharger should first be inspected, and if found to be damaged, the engine should also be inspected.
Answer D is incorrect. Both Technicians are correct.

Question #84
Answer A is correct. A fuel fitting should never be loosened until all the pressure has been relieved from the line. Pressure could cause the line to blow off and injure the Technician.
Answer B is incorrect. Approved solutions can and should be used for leak detection.
Answer C is incorrect. The ¼ turn valve should be turned off and the engine runs until it shuts off to deplete the fuel from the system before removing the fuel filter housings.
Answer D is incorrect. Venting of fuel should only be performed in an approved area, well ventilated and in accordance with local, state, and federal regulations.

Question #85
Answer A is correct. Only Technician A is correct. CNG fast fill stations are temperature compensated to allow approximately 4000 psi into the fuel system. As the fuel cools, the charge should settle down to approximately 3600 psi. Although fuel cylinders have a working pressure of 3600 psi they are tested to 1½ times greater than their working pressure.
Answer B is incorrect. Fuel cylinders are tested to 1½ times their working pressure. As the fuel cools, the pressure will return to the cylinder's rated pressure of 3600 psi.
Answer C is incorrect. Only Technician A is correct.
Answer D is incorrect. Technician A is correct.

Question #86
Answer A is incorrect. Technician B is also correct.
Answer B is incorrect. Technician A is also correct.
Answer C is correct. Both Technicians are correct. Technician A is correct in saying that a CNG engine manifold will be under a partial vacuum. The throttle valve controls the air and fuel flow to the engine. At idle the throttle valve is only partially open and will cause the manifold to be at negative pressure (vacuum). Technician B is also correct. The air inlet pressure sensor reads the intake manifold pressure as absolute. Atmospheric pressure at sea level is 14.7 psi. A reading less than a reading of 14.7 psi would be interpreted as less than atmospheric pressure and as a vacuum.
Answer D is incorrect. Both Technicians are correct.

Question #87
Answer A is incorrect. An improperly gapped spark plug can cause intermittent misfiring. If the gap is too long, the plug can misfire when the cylinder pressures are high and requiring a higher than normal voltage to jump the gap. If the gap is too short, the spark intensity may be too low to promote complete combustion. Both of these conditions can cause intermittent misfiring.
Answer B is incorrect. A loose connection can cause intermittent misfiring of the cylinder.
Answer C is correct. When the electrodes are shorted together, a spark cannot be produced. Therefore the cylinder would not fire at all. This condition would not cause a misfire condition and is the EXCEPT condition.
Answer D is incorrect. A fouled spark plug with oil residue can cause an intermittent misfire condition.

Question #88
Answer A is incorrect. Technician B is also correct.
Answer B is incorrect. Technician A is also correct.
Answer C is correct. Both Technicians A and B are correct. If the fuel low-pressure regulator or the fuel control valves malfunction and deliver excessive fuel to the engine, high exhaust temperatures can result. Exhaust valves adusted too tight can also cause high exhaust temperatures as well and corrective action should be taken to prevent damage to the valves, seats, and pistons from occurring.
Answer D is incorrect. Both Technicians are correct.

Question #89
Answer A is incorrect. Technician B is also correct.
Answer B is incorrect. Technician A is also correct.
Answer C is correct. Both Technicians are correct. The markings in the figure show the strength of the bolt and that it is a torque-turn bolt. The dot is marked on the cylinder head after torquing the bolt to specifications and then the bolt is advanced 90°. The marking from the dot previously made on the cylinder head should now be aligned with the window of the < sign on the head of the bolt when advanced correctly.
Answer D is incorrect. Both Technicians are correct.

Question #90
Answer A is correct. Only Technician A is correct. Engine oil can be used to lubricate the spark plug adaptor O-rings before being installed into the cylinder head. Dow Corning 55 or its equivalent can also be used to lubricate the O-rings.
Answer B is incorrect. Coolant would not provide enough lubrication and could cause damage to the O-rings as the tubes are being installed into the cylinder head. Coolant could leak past the O-rings and leak coolant into the engine oil, requiring the O-rings to be replaced again.
Answer C is incorrect. Only Technician A is correct.
Answer D is incorrect. Technician A is correct.

Question #91
Answer A is incorrect. A valve seat can be struck lightly with the ball peen end of a ball-peen hammer. If the seat "rings," it is loose.
Answer B is correct. Using a chisel to remove valve seats is not good practice because the cylinder head can be damaged; the rest of the answers reference appropriate steps when performing cylinder valve service. Most current NGV engines use valve seat inserts rather than integral valve seats machined into the cylinder head. Valve seat inserts are press-fit to a machined recess in the cylinder head. Since most of the heat of the valve must be transferred from the valve to the seat, it is essential that the contact area of the seat and the cylinder head be maximized. A sectored collet that expands into the valve seat is used to remove the valve seat. The seat can either be levered out or be driven out with a slide hammer. When installing new valve seats, the seat counterbores must be first cleaned out using a low-abrasive emery cloth.
Answer C is incorrect. Valve seats are checked for concentricity with the guide.
Answer D is incorrect. Valve grinding stones should be dressed prior to being used.

Question #92
Answer A is incorrect. Emission labels have to be on the engine anytime the vehicle is in service.
Answer B is correct. Only Technician B is correct. Emission labels have to be on the engine anytime the vehicle is in service. The label must also contain the vehicle's serial and model numbers.
Answer C is incorrect. Only Technician B is correct.
Answer D is incorrect. Technician B is correct.

Question #93
Answer A is incorrect. Technician B is also correct.
Answer B is incorrect. Technician A is also correct.
Answer C is correct. Both Technicians are correct. Crosshead guide pins must be perfectly set at right angles to prevent side loading of the valves by the bridge, so Technician A is correct. Technician B is also correct in saying that crosshead guide pins should be measured to specification with a micrometer.
Answer D is incorrect. Both Technicians are correct.

Question #94
Answer A is incorrect. When installing silicone gaskets into a rocker housing cover, the manufacturer service instructions should be observed. This will seldom involve applying adhesive to the silicone gasket.
Answer B is incorrect. Many rocker housing gaskets on modern engines are designed to be reused and their price reflects this.
Answer C is incorrect. Neither Technician is correct.
Answer D is correct. Neither Technician is correct. When installing silicone gaskets into a rocker housing cover, the manufacturer's service instructions should be observed. This will seldom involve applying adhesive to the silicone gasket. Many rocker housing gaskets on modern engines are designed to be reused, and their price reflects this.

Question #95
Answer A is correct. The oil cooler core is being pressure tested. Once the core is pressurized to the manufacturer's specifications, the core should be placed in a bath of hot water to simulate the operating temperature of the engine.
Answer B is incorrect. The picture depicts an oil cooler core.
Answer C is incorrect. The core is being pressure tested.
Answer D is incorrect. The picture depicts an oil cooler core being pressure tested.

Question #96
Answer A is incorrect. Parts manuals usually provide some good information relating to the components. The pictures of the components can be arranged in chronological order for the assembly of parts. This can assist a technician in assembling unfamiliar components.
Answer B is correct. Only Technician B is correct. Newer or the latest versions of the parts manuals usually contain information about updated changes to engine components as well as previous ones.
Answer C is incorrect. Only Technician B is correct.
Answer D is incorrect. Technician B is correct.

Question #97
Answer A is incorrect. Clogged fuel filters can cause low fuel pressure.
Answer B is correct. Because the primary fuel filter is before the high-pressure regulator, no coolant flow through the high-pressure regulator would only affect the fuel downstream from it such as the secondary fuel filter, not the primary one.
Answer C is incorrect. A partially open low-pressure fuel solenoid could cause a low fuel pressure problem.
Answer D is incorrect. A partially open ¼ turn fuel shutoff valve could cause a low fuel pressure problem.

Question #98
Answer A is correct. Only Technician A is correct. The pencil marks on the valve test the valves' seating integrity, and in the test being performed the valve is clearly not seating because some of the pencil marks have not been disturbed. Technician A is correct in saying the valve seat needs to be reground.
Answer B is incorrect. Valve springs are not installed for this test. Technician B is incorrect because valves will not seal even when under spring pressure.
Answer C is incorrect. Only Technician A is correct.
Answer D is incorrect. Technician A is correct.

Question #99
Answer A is incorrect. Gasket scrapers can cause a knick on the fire deck that would resemble a crack that could result in a diagnosis of a cracked head when in fact it is not cracked.
Answer B is correct. Only Technician B is correct. Fine grit sandpaper can be used with a lubricant such as diesel fuel or penetrating oil to assist in removing the deposits from the fire deck area.
Answer C is incorrect. Only Technician B is correct.
Answer D is incorrect. Technician B is correct.

Question #100
Answer A is incorrect. The spark plug should be installed dry.
Answer B is incorrect. The spark plug should be installed dry.
Answer C is incorrect. Neither Technician is correct.
Answer D is correct. Neither Technician is correct. Spark plugs should be installed dry with no lubricant on the threads. Each plug should be installed by hand until it contacts the crush washer and then torqued to the correct setting of 25 lbs/ft. If the plug is too difficult to install the threads should be cleaned or the tube should be replaced. Lubricants such as anti-seize compounds and engine oil can inhibit good conductivity and also can contaminate the electrodes of the spark plugs, leading to misfiring cylinders.

Question #101
Answer A is incorrect. Technician B is also correct.
Answer B is incorrect. Technician A is also correct.
Answer C is correct. Both Technicians are correct. Both Technicians A and B are describing the correct bench-test procedure for testing a cylinder head. Heat a cylinder to swell any cracks by flowing hot water through it for 10 minutes.
Answer D is incorrect. Both Technicians are correct.

Question #102
Answer A is correct. Only Technician A is correct. Cylinder heads should be stored upright on one end to prevent the head from warping.
Answer B is incorrect. If a cylinder head is stored flat on two pieces of wood, it is more prone to warpage and it will require planing.
Answer C is incorrect. Only Technician A is correct.
Answer D is incorrect. Technician A is correct.

Question #103
Answer A is incorrect. Technician B is also correct.
Answer B is incorrect. Technician A is also correct.
Answer C is correct. Both Technicians are correct. Four-valve cylinder heads can use valve crossheads to operate two valves at the same time. Crossheads with an adjustment usually only require adjustment at initial installation when the cylinder head is first installed and only need to be checked at the time of tune-up. Some engines use the floating type crossheads and are self-compensating and do not require adjustment. The floating type crossheads need to be checked at

the regular tune-up intervals to make sure they are not wearing out improperly where the valve stem contacts them.
Answer D is incorrect. Both Technicians are correct.

Question #104
Answer A is incorrect. The waste gate stuck closed would not cause the slow deceleration and stalling condition.
Answer B is correct. Only Technician B is correct. The PSV valve with a piston that is slow in sliding in its bore can cause the slow deceleration and the stalling condition. The valve should be examined for contamination in the piston bore and cleaned if needed.
Answer C is incorrect. Only Technician B is correct.
Answer D is incorrect. Technician B is correct.

Question #105
Answer A is correct. The bolt holes should be checked first; the remaining procedures are performed after the cylinder head has been installed. On inline multicylinder heads, it is usually required that separate heads be aligned with a straightedge across the intake manifold faces before torquing. Failure to observe torquing increments and sequencing can result in cracked cylinder heads, failed head gaskets, and fire rings that will not seal. Some OEMs require that a template torque method be used; this means setting a torque value first and then turning a set number of degrees beyond that value using a template or protractor. Cylinder head bolts should be installed lightly oiled. Excessive quantities of oil should be avoided because the excess can drain into the bolt hole and cause a hydraulic lock.
Answer B is incorrect. Head bolts are torqued after installation.
Answer C is incorrect. Rocker pedestals are checked after head installation.
Answer D is incorrect. Valve bridges are adjusted after installation.

Question #106
Answer A is incorrect. Technician B is also correct.
Answer B is incorrect. Technician A is also correct.
Answer C is correct. Both Technicians are correct. A clogged crankcase breather can cause high crankcase pressure and oil leaks to develop from the oil pan seals or gaskets. To test the crankcase pressure:

- Connect a water manometer (or magnehelic gauge) to the oil dipstick tube.
- Start and run the engine at idle and observe the reading against OEM specifications.
- Run the engine at 2000 rpm and record the reading; compare it to OEM specifications.

Answer D is incorrect. Both Technicians are correct.

Question #107
Answer A is correct. Only Technician A is correct. Oil trapped in a blind head bolt hole can cause hydraulic lock when the bolt is torqued down. This will invalidate the torque setting.
Answer B is incorrect. A cylinder head bolt with any visible damage should be replaced, not repaired.
Answer C is incorrect. Only Technician A is correct.
Answer D is incorrect. Technician A is correct.

Question #108
Answer A is incorrect. Technician B is also correct.
Answer B is incorrect. Technician A is also correct.
Answer C is correct. Both Technicians are correct. Active DTCS should be repaired prior to the inactive codes because the defect is occurring real-time. Prior to clearing any codes they should be recorded for future reference.
Answer D is incorrect. Both Technicians are correct.

Question #109
Answer A is correct. Only Technician A is correct. Loss of engine oil through the crankcase breather is a common indication of high crankcase pressure.
Answer B is incorrect. Current CNG engines run with a specified crankcase pressure that is considered beneficial in purging the engine crankcase of contaminants, providing it is within specification (usually around 2.5" water). Because the piston compression rings create the seal for a piston in a cylinder bore, some blowby is inevitable, due to end-gap requirement. Normally, blowby is a specification determined by the OEM. Piston rings seal more efficiently when cylinder pressures are highest, and engine rpm is high when there is less time for gas blowby.
Some causes of high crankcase pressure are:

- Cracked piston(s)
- Worn, stuck, or broken piston rings
- Glazed liner/sleeve inside wall
- Scored cylinder wall.

Answer C is incorrect. Only Technician A is correct.
Answer D is incorrect. Technician A is correct.

Question #110
Answer A is incorrect. Technician B is also correct.
Answer B is incorrect. Technician A is also correct.
Answer C is correct. Both Technicians are correct. A flattened camshaft lobe or a defective cam follower roller can both cause the rocker arm not to actuate and cause a miss in the engine.
Answer D is incorrect. Both Technicians are correct.

Question #111
Answer A is incorrect. An inside micrometer can be used to detect a cylinder out-of-round condition.
Answer B is incorrect. A telescoping gauge can be used to detect a cylinder out-of-round condition.
Answer C is correct. The LEAST-Likely tool to detect an out-of-round condition would be a feeler gauge.
Answer D is incorrect. A dial bore gauge can be used to detect a cylinder out-of-round condition.

Question #112
Answer A is incorrect. Technician B is also correct.
Answer B is incorrect. Technician A is also correct.
Answer C is correct. Both Technicians are correct. Before installing an intake manifold it should be checked with a straightedge to ensure it is not warped beyond the OEM specifications. An excessively warped manifold may not draw up tightly against the intake port gasket and leak especially when the engine is under a load and the boost pressure is higher than ambient pressure. An intake manifold will be torqued into place by torquing from the center outward to the ends.
Answer D is incorrect. Both Technicians are correct.

Question #113
Answer A is incorrect. Turbochargers are designed so the housings can be removed if necessary to facilitate the mounting of the turbocharger. However, when the housing is removed the turbocharger center section needs to be handled carefully to prevent the blades of the turbine or compressor wheels from being damaged.
Answer B is correct. Only Technician B is correct. The housings are designed to allow the Technician to rotate them to allow alignment of the components during the installation of the turbocharger.
Answer C is incorrect. Only Technician B is correct.
Answer D is incorrect. Technician B is correct.

Question #114
Answer A is incorrect. Technician B is also correct.
Answer B is incorrect. Technician A is also correct.

Answer C is correct. Both Technicians are correct. The exterior of a CNG engine's wet cylinder sleeves can be cleaned by solvent immersion and glass bead blasting.
Answer D is incorrect. Both Technicians are correct.

Question #115
Answer A is incorrect. High air intake temperature will not lead to a catalyst failure.
Answer B is incorrect. Worn spark plugs will not lead to a catalyst failure. Engine performance will be affected by missing or backfiring.
Answer C is incorrect. An air leak in the pressure side of the air intake system after the turbocharger will affect engine performance but not lead to catalyst failure.
Answer D is correct. If the fuel low-pressure regulator is defective and causing the engine to overfuel, the exhaust temperature can increase dramatically. The operator will not complain because high fuel content will increase engine power. As the exhaust temperature remains high, the catalyst will be damaged and fail. Sometimes high exhaust back pressure will be a result of a failed catalyst.

Question #116
Answer A is incorrect. The button should be replaced to prevent the possibility of it falling off again and possibly causing some damage to the cylinder head or its components.
Answer B is incorrect. The button can be serviced and replaced as an item. The complete rocker arm assembly does not require replacement unless further damage is discovered during the button replacement.
Answer C is incorrect. Neither Technician is correct.
Answer D is correct. Neither Technician is correct. The button and adjusting screw need to be replaced as an assembly. The rocker arm does not need to be replaced unless it is defective. The valve adjustment can be completed once the adjuster has been replaced.

Question #117
Answer A is incorrect. An eroded turbocharger compressor wheel is a tell-tale sign that dirt and debris was allowed to pass through the turbocharger and is not a normal high mileage condition.
Answer B is correct. Only Technician B is correct. The erosion of the blade tips is a good indicator that some abrasive material such as dust or sand has entered the air system and the air cleaner and associated air intake piping need to be checked for proper installation.
Answer C is incorrect. Only Technician B is correct.
Answer D is incorrect. Technician B is correct.

Question #118
Answer A is incorrect. Drivability issues can result from a transmission failure. The clutches, the torque converter, or the hydraulic pump could be defective. The fluid level could be low, causing slippage, or high, causing erratic shifting, to name a few defects.
Answer B is correct. Operators can have different driving habits. Many times different routes require different driving styles. The transmission has the capability to perform adaptive learning of shifting patterns. Full throttle applications require stronger clutch engagement while with less throttle the clutches are allowed to trim pressure for smoother shifting. This learning requires approximately 10 complete upshifts and downshifts to be acquired by the ECU. Making the customer aware of the adaptive learn process can many times solve issues.
Answer C is incorrect. Only Technician B is correct.
Answer D is incorrect. Technician B is correct.

Question #119
Answer A is correct. Only Technician A is correct. The higher than 240 psi cylinder pressure can cause pre-ignition of the fuel and cause knock codes. Deposits in the combustion chamber such as from leaking valve seals are usually the cause of the higher pressure.
Answer B is incorrect. The higher pressure is an indication that the cylinder has an issue that needs to be corrected or the knocking condition will continue and possibly damage the engine.
Answer C is incorrect. Only Technician A is correct.
Answer D is incorrect. Technician A is correct.

Question #120

Answer A is correct. Only Technician A is correct. If a rocker arm shaft assembly is installed incorrectly, engine oil will not be fed up to the rocker arm shafts and allow the rocker assembly to be lubricated. If one assembly is installed incorrectly, that assembly will prematurely fail.
Answer B is incorrect. The cause of the failure needs to be corrected. If only one of the rocker arms failed, it could be possible that a defect existed at the time it was placed on the engine. Because the entire rocker arm assembly bushings and shaft failed the Most-Likely cause is lack of lubrication and it needs to be corrected.
Answer C is incorrect. Only Technician A is correct.
Answer D is incorrect. Technician A is correct.

Question #121

Answer A is incorrect. The cam lobe is not being measured.
Answer B is correct. The cam bearing journal is being measured.
Answer C is incorrect. Cam lift is not being measured.
Answer D is incorrect. Cam duration is not being measured.

Question #122

Answer A is incorrect. Loctite is not recommended when installing cam bushings. The interference fit of the bushings is what will prevent the bushing from moving out of position once driven into its bore.
Answer B is incorrect. An indexing mark with a center punch is not necessary when installing new cam bushings. The interference fit of the bushings is what will prevent the bushing from moving out of position once driven into its bore.
Answer C is incorrect. Neither Technician is correct.
Answer D is correct. Neither Technician is correct. Loctite is not recommended when installing cam bushings. The interference fit of the bushings is what will prevent the bushing from moving out of position once driven into its bore. An indexing mark with a center punch is not necessary when installing new cam bushings. The interference fit of the bushings is what will prevent the bushing from moving out of position once driven into its bore.

Question #123

Answer A is incorrect. Technician B is also correct.
Answer B is incorrect. Technician A is also correct.
Answer C is correct. Both Technicians are correct. Main bearings tell a story, so they should be checked on disassembly, as should the crankshaft, for obvious signs of failure. After cleaning, the crankshaft should again be subject to measuring to determine its serviceability. Lubrication-related failures are caused by insufficient or complete absence of oil in one or all of the crank journals. The bearing is subjected to high friction loads and surface-welds itself to the affected crank journal. This may result in a spun bearing in which the bearing rotates with the journal in the bore or continues to scuff the journal until it is destroyed. When a crankshaft fractures as a result of bearing seizure, the surface of the journal is destroyed by excessive heat; it fails because it is unable to sustain the torsional loading.
Answer D is incorrect. Both Technicians are correct.

Question #124

Answer A is incorrect. A feeler gauge should not be used to check main bearing clearances. Damage can occur to the bearing and crankshaft surfaces due to scratches. Because of the curved surfaces of the main bearing and the crankshaft, a feeler gauge would not work. The only approved methods to check the bearing clearances are to use Plastigage or a bore gauge in combination with a micrometer.
Answer B is incorrect. The bearing shell must be installed metal to metal with the bearing saddle. Therefore a feeler gauge should not be inserted between the bearing and the engine block, as this could damage the bearing.
Answer C is incorrect. Neither Technician is correct.
Answer D is correct. Neither Technician is correct. Plastigage or a bore gauge and micrometer are used to determine main bearing clearances. The Plastigage strip will be crushed when the main bearing cap and bearing are torqued into place. After removing the cap and bearing, the crushed remains of the Plastigage can be compared to the gauge scale on the envelope of the Plastigage strip

to determine the clearance. Another method involves using a bore gauge and measuring the bore of the main bearing after being installed and the main cap is torqued to specifications. The bore gauge is used to measure the diameter of the bore and recorded. A micrometer is used to measure the diameter of the crankshaft journal and recorded. The diameter of the journal is subtracted from the bore of main bearing. The difference is the oil clearance. A feeler gauge should not be inserted between the block and the bearing shell to check main bearing clearances as this could damage the bearing. The bearing shell must be installed metal to metal with the bearing saddle.

Question #125
Answer A is incorrect. Technician B is also correct.
Answer B is incorrect. Technician A is also correct.
Answer C is correct. Both Technicians are correct. If the piston cooling nozzle is bent or damaged, oil will not be directed correctly on the underside of the piston, causing the piston to overheat and possibly pre-ignition to occur along with knock codes. In the case of aluminum pistons, the piston could burn a hole and require the piston to be replaced. Some pistons are cooled through oil passing up the connecting rod and through the wrist pin to be sprayed on the underside of the piston crown. Overheated pistons can cause knock codes and piston/liner damage.
Answer D is incorrect. Both Technicians are correct.

Question #126
Answer A is incorrect. Turbochargers can rotate at a high rate of speed especially when the engine is under a load.
Answer B is incorrect. Modern turbochargers are equipped with speed sensors to monitor their rotational speed.
Answer C is incorrect. CNG engine turbochargers are waste gated to prevent the engines from being overboosted. Overboosting cause high cylinder pressures and temperatures that lead to pre-ignition and can damage the engine.
Answer D is correct. Turbochargers require lubrication be added to the lubrication fitting at the top of the housing before starting the engine to prevent the bearings from being damaged at start-up. This is the EXCEPT statement.

Question #127
Answer A is incorrect. This is a sign of a bent connecting rod. A high-mileage engine that does not have a bent connecting rod would show signs or even wear across the bearing shell.
Answer B is correct. Only Technician B is correct. When a connecting rod bearing shell has more wear on one side, this is an indication of a bent connecting rod. A bent connecting rod applies more pressure on one side of the rod bearing than the other and leads to more wear on one edge of the bearing than the other.
Answer C is incorrect. Only Technician B is correct.
Answer D is incorrect. Technician B is correct.

Question #128
Answer A is incorrect. Technician B is also correct.
Answer B is incorrect. Technician A is also correct.
Answer C is correct. Both Technicians are correct. The fuel cylinder's label must contain the expiration date of the cylinder in addition to the date it was placed in service. The serial number of the fuel cylinder must be affixed on the label.
Answer D is incorrect. Both Technicians are correct.

Question #129
Answer A is incorrect. Technician B is also correct.
Answer B is incorrect. Technician A is also correct.
Answer C is correct. Both Technicians are correct. When turbocharger coolant and oil lines and hoses become nicked or frayed, it is good practice to replace them rather than repair them, so Technician A is correct. Technician B is also correct because it is good practice to inspect oil drain tubes for damage and obstructions before returning them to service.
Answer D is incorrect. Both Technicians are correct.

Question #130
Answer A is incorrect. The rubber cap needs to be in place whenever the vehicle is not being refueled to prevent contamination from accumulating on the valve and being pushed into the fuel tanks when refueling is being performed.
Answer B is correct. Only Technician B is correct. The rubber cap needs to be in place whenever the vehicle is not being refueled to prevent contamination from accumulating on the valve and being pushed into the fuel tanks when refueling is being performed.
Answer C is incorrect. Only Technician B is correct.
Answer D is incorrect. Technician B is correct.

Question #131
Answer A is incorrect. Belt alignment is performed with a straightedge or laser sight.
Answer B is incorrect. A belt does not require straightening. If the belt was to become distorted, it requires replacement.
Answer C is incorrect. Belt thickness is usually not checked.
Answer D is correct. The tool pictured in the figure is used to check belt deflection, which determines the tension applied on the belt.

Question #132
Answer A is incorrect. Dirt or contaminants can cause the flywheel to have improper TIR.
Answer B is incorrect. A bent mounting dowel pin(s) can be caused by improper handling and affect the TIR of a flywheel.
Answer C is incorrect. A damaged flywheel housing from improper handling can cause out of specification TIR.
Answer D is correct. Incorrect installation of the crankshaft would not be a cause of incorrect TIR.

Question #133
Answer A is incorrect. Only an approved electronic methane detector or leak-detection solution is to be used for leak detection. Technicians have reported that over time, they can become desensitized to the scent of mercaptan. Smell should not be used as a leak-detection method.
Answer B is correct. Only Technician B is correct. Turning the ¼ turn valve on the fuel system is not always sufficient for a preliminary leak check. The Technician should start the engine to fully pressurize the fuel lines, then turn the engine back off and perform a leak test utilizing either an approved electronic methane leak detector or soapy solution. When assured there is no methane leaks, the vehicle can be released back into service.
Answer C is incorrect. Only Technician B is correct.
Answer D is incorrect. Technician B is correct.

Question #134
Answer A is incorrect. Technician B is also correct.
Answer B is incorrect. Technician A is also correct.
Answer C is correct. Both Technicians are correct. Technician A correctly describes how to remove the surface charge from a battery that has been recently charged. A stored battery does not have a surface charge, so Technician B is also correct.
Answer D is incorrect. Both Technicians are correct.

Question #135
Answer A is incorrect. The sensors depicted in the figure are supplied the +5 volts from the ECM connector 1 pin 36 and 26 not connector 2 pin 36 and 26.
Answer B is incorrect. The sensors depicted in the figure are supplied the +5 volts from the ECM connector 1 pin 36 and 26 not connector 2 pin 48 and 44.
Answer C is incorrect. The sensors depicted in the figure are supplied the +5 volts from the ECM connector 1 pin 36 and 26 not connector 1 pin 37 and 42.
Answer D is correct. The sensors depicted in the figure are supplied the +5 volts from the ECM connector 1 pin 36 and 26.

Question #136

Answer A is correct. Only Technician A is correct. If a thermostat is stuck in the open position, it can cause an engine to fail to reach the normal operating temperature window.

Answer B is incorrect. An overfilled radiator tank is unlikely to prevent an engine from reaching its normal operating temperature. It would most likely result in coolant being sent to the recovery bottle.

Answer C is incorrect. Only Technician A is correct.

Answer D is incorrect. Technician A is correct.

Question #137

Answer A is incorrect. Overly tight valves can cause the valves to remain partially open once the engine reaches operating temperature and could be struck by the pistons, causing damage to the valves.

Answer B is incorrect. Excessively loose valves can cause the valves to open too late and affect engine performance and emissions.

Answer C is incorrect. Neither Technician is correct.

Answer D is correct. Neither Technician is correct. Overly tight valves can cause the valves to remain partially open once the engine reaches operating temperature and could be struck by the pistons, causing damage to the valves. Excessively loose valves can cause the valves to open too late and affect engine performance and emissions. Loose valves tend to chatter and because CNG engines contain a knock sensor(s), the chatter can be picked up by the knock sensor(s) and log a DTC in the ECM

Question #138

Answer A is incorrect. The engine should be run to bring the oil temperature closer to operating temperature. The oil will be thinner, allowing faster draining and help in removing more of the contaminants from the oil pan.

Answer B is incorrect. By draining the engine oil soon after it is turned off, the contaminants remain suspended and drain out with the used engine oil assuring a cleaner sump.

Answer C is incorrect. Neither Technician is correct.

Answer D is correct. Neither Technician is correct. The engine should be run to bring the oil temperature closer to operating temperature. The oil will be thinner, allowing faster draining and help in removing more of the contaminants from the oil pan. By draining the engine oil soon after it is turned off, the contaminants remain suspended and drain out with the used engine oil assuring a cleaner sump.

Question #139

Answer A is correct. Only Technician A is correct. If the inside diameter of a camshaft bushing is not within specifications after checking with a telescoping gauge and micrometer or a bore gauge, the bushing can be reamed to specification using a reamer adjusted to the correct size.

Answer B is incorrect. Emery cloth should not be used on a cam bushing as it will destroy the bushing. The cloth will make large scratches on the bearing surface and could cause the bushing to spin. Also, the diameter of the bushing will not be consistent. This will not provide adequate support for the camshaft, leading to the premature failure of the bearing or the camshaft.

Answer C is incorrect. Only Technician A is correct.

Answer D is incorrect. Technician A is correct.

Question #140

Answer A is correct. While a high crankcase oil level will cause other problems, it is LEAST-Likely to result in cracks and scoring of the piston skirts. The other three problems would be capable of damaging the piston skirts.

Answer B is incorrect. Improper piston clearance can cause damaged piston skirts.

Answer C is incorrect. Overfueling can cause piston damage.

Answer D is incorrect. Engine overheating can cause piston damage.

Question #141
Answer A is incorrect. Engines equipped with catalysts that have pre- and post-O_2 sensors utilize three-way catalysts to limit the emissions of hydrocarbons (HC), carbon monoxide (CO), and oxides of nitrogen (NOX).
Answer B is incorrect. Engines equipped with catalysts that have only pre-catalyst O_2 sensors are considered to have two-way catalysts to limit the emissions of HC and CO only.
Answer C is incorrect. Neither Technician is correct.
Answer D is correct. Neither Technician is correct. Engines equipped with catalysts that have pre- and post-O_2 sensors utilize three-way catalysts to limit the emissions of hydrocarbons (HC), carbon monoxide (CO), and oxides of nitrogen (NOX). Engines equipped with catalysts that have only pre-catalyst O_2 sensors are considered to have two-way catalysts to limit the emissions of HC and CO only.

Question #142
Answer A is incorrect. Methane is produced by decaying organic material.
Answer B is incorrect. Hydrogen is a natural occurring gas.
Answer C is correct. Hydrogen sulfide gas is the by-product of the chemical process that occurs when a lead-acid battery is charging.
Answer D is incorrect. Carbon monoxide is a by-product of incomplete combustion.

Question #143
Answer A is incorrect. Too low of a voltage regulator setting would cause starting and reserve life capacity but would not cause the electrolyte to be lost.
Answer B is incorrect. Driving with the lights on all the time would not cause the symptom described.
Answer C is incorrect. Neither Technician is correct.
Answer D is correct. Neither Technician is correct. When the battery electrolyte is constantly low in all the batteries, an overcharging situation should be suspected and the voltage regulator setting should be checked. If electrolyte is constantly low in only one of the batteries, that battery should be tested for a possible defect.

Question #144
Answer A is incorrect. Technician B is also correct.
Answer B is incorrect. Technician A is also correct.
Answer C is correct. Both Technicians are correct. The stoichiometric ratio of methane or natural gas is 16.5:1 and the electronically controlled CNG engine is capable of operating at a leaner ratio for a cleaner burn and lower emissions.
Answer D is incorrect. Both Technicians are correct.

Question #145
Answer A is correct. A stuck-closed oil pressure relief valve would be unable to release and therefore more likely to cause high oil pressure complaints. It would be LEAST-Likely to cause low oil pressure complaints. To verify low oil pressure:

- Check the oil sump level.
- Install a master gauge (an accurate, fluid-filled gauge).
- Investigate the oil consumption history.
- Determine the cause.

Some possible causes are:

- Contaminated lube oil (fuel)
- Excessive crankshaft bearing clearance
- Excessive camshaft or rocker shaft bearing clearance
- Pump relief valve spring fatigued or stuck open
- Oil pump defect
- Oil suction pipe defect

Answer B is incorrect. A restricted suction screen will reduce flow to the pump, which will cause low pressure.
Answer C is incorrect. Worn main bearings increase clearances, which reduces pressure.
Answer D is incorrect. A clogged oil filter provides restriction after the pump, which increases pressure.

Question #146
Answer A is incorrect. Technician B is also correct.
Answer B is incorrect. Technician A is also correct.
Answer C is correct. Both Technicians are correct. The oil should be completely drained and disposed of according to local, state, and federal regulations. Engine oil is considered hazardous waste. Upon removal of an engine oil pan, the contents can assist the Technician in making a determination of the condition of the engine. Pieces of ring are signs of a broken piston ring(s). Pieces of gear teeth are an indication the gears need a visual inspection for damage as well as clearances.
Answer D is incorrect. Both Technicians are correct.

Question #147
Answer A is incorrect. The idler gear must be timed to the crankshaft to OEM specifications. Timing procedures and timing mark matching must be followed stringently. Any deviation can cause engine performance issues, emission violations, as well as damage to the engine from pistons striking the valves.
Answer B is incorrect. The idler gear must be timed to the camshaft gear to OEM specifications. Timing procedures and timing mark matching must be followed stringently. Any deviation can cause engine performance issues, emission violations, as well as damage to the engine from pistons striking the valves.
Answer C is incorrect. Neither Technician is correct.
Answer D is correct. Neither Technician is correct. Timing procedures and timing mark matching must be followed stringently. Any deviation can cause engine performance issues, emission violations, as well as damage to the engine from pistons striking the valves.

Question #148
Answer A is incorrect. Combustion gases can be identified with test strips.
Answer B is incorrect. Improper antifreeze can be identified with test strips.
Answer C is correct. Coolant SCA test strips indicate the relative acidity or alkalinity of the coolant. Acids may form in the coolant when coolant is exposed to combustion gases, when there is corrosion of ferrous and copper metals, and when there is coolant degradation. The coolant may have excess alkalinity when there is aluminum corrosion, and when low-silicate antifreeze is used where high-silicate antifreeze is required. Coolant temperature is checked with a thermometer.
Answer D is incorrect. Coolant degradation can be determined with coolant test strips.

Question #149
Answer A is incorrect. CNG lines are constructed of stainless steel usually type 306, and meet the minimum burst pressures requirements.
Answer B is incorrect. Welding of a CNG line to correct a pin hole is not acceptable practice because the line's physical and strength characteristics can be changed and the line may not meet the requirements.
Answer C is incorrect. Neither Technician is correct.
Answer D is correct. Neither Technician is correct. Fuel lines need to meet the minimum requirements of the Natural Gas Association and other regulatory agencies. Steel brake lines are not approved for CNG use. CNG lines are constructed of stainless steel usually type 306, and meet the minimum burst pressures requirements. Additionally welding of a CNG line to correct a pin hole is not acceptable practice because the line's physical and strength characteristics can be changed and the line may not meet the requirements.

Question #150

Answer A is correct. Only Technician A is correct. Gently tilting the battery sometimes will allow the hydrometer to free itself and resume working properly.
Answer B is incorrect. If the battery is already fully charged, more charging time will damage the plates and the battery. Gently tilting the battery may free up the hydrometer if it is stuck.
Answer C is incorrect. Only Technician A is correct.
Answer D is incorrect. Technician A is correct.

Question #151

Answer A is correct. Only Technician A is correct. Coolant is routed through the fuel low-pressure regulator to assist in maintaining adequate fuel temperatures to prevent drivability issues that can result from excessively cool fuel and icing.
Answer B is incorrect. Coolant is routed through the fuel low-pressure regulator to assist in maintaining adequate fuel temperatures. The engine compartment heat is not adequate to assist in maintaining the fuel temperature
Answer C is incorrect. Only Technician A is correct.
Answer D is incorrect. Technician A is correct.

Question #152

Answer A is incorrect. Technician B is also correct.
Answer B is incorrect. Technician A is also correct.
Answer C is correct. Both Technicians are correct. A magnetic spark plug socket should be used on CNG engine spark plugs because the rubber insert can cause dirt and oil to be deposited on the sparks plug during installation. Also, the magnet will prevent the plug from dropping during installation in the cylinder head and potential damage to the electrode. The socket should be cleaned with isopropyl alcohol to remove any traces of contamination that could be deposited onto the new plugs.
Answer D is incorrect. Both Technicians are correct.

Question #153

Answer A is incorrect. The engine of the booster vehicle has to be turned off while connecting the jumper cables to prevent potential voltage spikes from damaging electrical/electronic equipment in the vehicle used to jump start with.
Answer B is correct. Only Technician B is correct. The running engine of the booster vehicle can be set to fast idle to allow the batteries of the other vehicle to charge up some before starting is attempted.
Answer C is incorrect. Only Technician B is correct.
Answer D is incorrect. Technician B is correct.

Question #154

Answer A is incorrect. Technician B is also correct.
Answer B is incorrect. Technician A is also correct.
Answer C is correct. Both Technicians are correct. Both situations could result in slow cranking. The ECM has to receive at least 60 rpm signals to energize the fuel control valve or solenoid.
Answer D is incorrect. Both Technicians are correct.

Question #155

Answer A is incorrect. When the intake valve begins to open, especially at the beginning of the intake cycle, the exhaust valve could still be partially open. This method should not be used because it could allow incorrect setting of the valves.
Answer B is correct. Only Technician B is correct. The approved companion valve overlap method allows the valves to be properly positioned when a valve adjustment is being performed. Improperly positioning of the valves can cause serious engine damage when the engine is turned over or started.

Answer C is incorrect. Only Technician B is correct.
Answer D is incorrect. Technician B is correct.

Question #156
Answer A is incorrect. Liner protrusion is being measured, not the concentricity.
Answer B is incorrect. Liner protrusion is being measured, not the counter bore.
Answer C is incorrect. Liner protrusion is being measured, not the cylinder block flatness.
Answer D is correct. Liner protrusion is being measured.

Question #157
Answer A is incorrect. Technician B is also correct.
Answer B is incorrect. Technician A is also correct.
Answer C is correct. Both Technicians are correct. Technician A is correct. The cooling system boiling point is increased by holding it under pressure. If the pressure cannot be maintained, the coolant can boil at a lower temperature. Boiling coolant within a cooling system accelerates an overheat condition because a gas does not conduct heat as well as a liquid. Technician B is also correct because the radiator cap is designed to relieve system pressure when it exceeds its rating.
Answer D is incorrect. Both Technicians are correct.

Question #158
Answer A is correct. Only Technician A is correct. The SCA will form a white powder type residue on the edges of the tubes. Sometimes the radiator can be treated with a mild acid-based radiator to remove the SCA dropout; however, the most reliable method is to remove the radiator and to have the radiator rodded to clean out the tubes completely.
Answer B is incorrect. A black residue can be caused by oil or combustion entering the cooling system and is not usually associated with SCA dropout.
Answer C is incorrect. Only Technician A is correct.
Answer D is incorrect. Technician A is correct.

Question #159
Answer A is incorrect. Technician B is also correct.
Answer B is incorrect. Technician A is also correct.
Answer C is correct. Both Technicians are correct. Depending on the fuel tank manufacturer, the PRD thermal activation point will usually range from 189 to 215° F.
Answer D is incorrect. Both Technicians are correct.

Question #160
Answer A is incorrect. Technician B is also correct.
Answer B is incorrect. Technician A is also correct.
Answer C is correct. Both Technicians are correct. The pressure could be malfunctioning and not relieving the pressure at the predetermined pressure, 9 psi in this case. If the cap tests ok, the vent line should be checked to make sure it is not plugged up or kinked and can vent properly.
Answer D is incorrect. Both Technicians are correct.

Question #161
Answer A is correct. Only Technician A is correct. Slow responding O_2 sensors can cause the AFR of the air/fuel mixture to remain lean for too long a period of time and cause lean misfiring in the cylinders and cause the flame to create multiple fronts. As these fronts collide, oscillations known as knocking will occur and, if severe enough, log a DTC.
Answer B is incorrect. O_2 sensors sometimes become slow in responding or drift out of range. These conditions can still allow the ECM to report the O_2 sensor as functional. O_2 sensors do not always completely fail.
Answer C is incorrect. Only Technician A is correct.
Answer D is incorrect. Technician A is correct.

Question #162
Answer A is incorrect. The temperature sensor only needs replacement when it fails or is damaged.
Answer B is incorrect. The water pump seal only needs replacement when it fails or is damaged.
Answer C is correct. The thermostat seals are the Most-Likely components that will need replacing. They are pressed into the housing and require replacement every time the full-blocking thermostats are replaced or coolant could leak past the seals. If the coolant does not return back to the radiator to be cooled but rather is recirculated back through the engine this can cause an overheating problem.
Answer D is incorrect. The housing plugs only need replacing if they leak.

Question #163
Answer A is correct. To check flywheel runout, mount a dial indicator with a magnetic base on the flywheel housing, and check the flywheel runout at the clutch contact face. The typical allowable runout is 0.001" per inch clutch radius. Measure the radius from the center of the flywheel to the outer edge of the clutch contact face. For example, a 14" clutch would have a 7" radius and allow a 0.007" total indicator reading of the dial indicator. If the measured values fall out of these specifications, repeat the assembly process and remeasure until a cause can be determined. In the figure, the dial indicator is mounted to measure the clutch face surface runout. The clutch bolts to the flywheel.
Answer B is incorrect. The pilot bore is not being measured.
Answer C is incorrect. Crankshaft radial runout is not being checked.
Answer D is incorrect. This is not the flywheel housing concentricity check.

Question #164
Answer A is correct. The throttle plate valve should not be forced open and is the LEAST-Likely course of action to be undertaken when diagnosing an erratic throttle plate DTC. Forcing the throttle plate open could damage it.
Answer B is incorrect. The fuses and wiring connections should be checked to ensure they are intact and not corroded.
Answer C is incorrect. If the throttle plate was not installed following the proper torquing procedures in which bolts are tightened first to the torque required, this could cause the throttle assembly to bind and become erratic.
Answer D is incorrect. If the ignition voltage required for the throttle plate to operate correctly is not supplied, this can cause an erratic DTC to be set.

Question #165
Answer A is correct. The CCA rating of a battery is an indication of the amperage a battery can deliver for 30 seconds at 0° F.
Answer B is incorrect. The CCA rating of a battery is an indication of the amperage a battery can deliver for 30 seconds not 60 seconds at 0° F.
Answer C is incorrect. The CCA rating of a battery is not an indication of the temperature range it can operate within but the amperage a battery can deliver for 30 seconds at 0° F.
Answer D is incorrect. Reserve capacity of battery is an indicator of the length of time a battery will allow the engine to be operated after the alternator fails. Not CCA.

Question #166
Answer A is incorrect. The manufacturer's specifications have to be followed for safety. Over- or undertightening of components can lead to injury and damage to the vehicle.
Answer B is correct. Only Technician B is correct. The manufacturer's specifications are either printed on the filter housing or can be found in the service manual and should be followed. Overtorquing will make it extremely difficult to remove the housing at the next interval. Undertorquing can lead to fuel leaks.
Answer C is incorrect. Only Technician B is correct.
Answer D is incorrect. Technician B is correct.

Question #167
Answer A is incorrect. A zip tie is not a good method to secure a connector with a broken lock.
Answer B is correct. Only Technician B is correct. When a connector has a broken lock, the connector should be replaced to ensure a sound and weathertight connection so moisture will not be able to intrude into the connector.
Answer C is incorrect. Only Technician B is correct.
Answer D is incorrect. Technician B is correct.

Question #168
Answer A is correct. A high battery charging voltage rate, in this case 29.5 volts for a 24-volt system, does not affect the engine operation as related to knock codes and is the LEAST-Likely to cause a knock code. The high charging rate will damage the batteries and will be recognized by the loss of electrolyte in the batteries and can be accompanied by the smell of hydrogen sulfide gas.
Answer B is incorrect. A faulty belt tensioner can cause knock codes at the engine knock sensor and be logged as a DTC in the ECM.
Answer C is incorrect. Poor engine to chassis ground points can cause engine knock codes to be logged in the ECM due to induced erroneous electrical signals into the engine sensors or back-feeding through the wiring caused by higher than normal resistance.
Answer D is incorrect. High intake air temperatures will cause the fuel air mixture to pre-ignite and cause detonation which is logged as a DTC for the engine knocking.

Question #169
Answer A is correct. Only Technician A is correct. The camshaft end play of an overhead cam engine needs to be measured and adjusted before the rocker arms and shafts are installed on the cylinder head. Once these components are installed, the camshaft end play will become very difficult to measure due to the pressure exerted on the camshaft lobes.
Answer B is incorrect. Camshaft end play cannot be properly determined once the rocker arm assemblies are installed and torqued to specifications. The cam must be free to move fore and aft to correctly measure the end play.
Answer C is incorrect. Only Technician A is correct.
Answer D is incorrect. Technician A is correct.

Question #170
Answer A is incorrect. The liner does not need replacement unless the pitting is beyond manufacturer specifications.
Answer B is correct. Only Technician B is correct. The figure directs the Technician to rotate the liners 45° from their original position to move the area of pitting to ensure the coolant will not continue to be directed toward the same area. This will allow a new surface for the coolant to act against as it flows through the engine, lengthening the life of the liners.
Answer C is incorrect. Only Technician B is correct.
Answer D is incorrect. Technician B is correct.

Question #171
Answer A is incorrect. Bluing of a journal is an indication of overheating during the operation of the component. During the manufacturing of a camshaft, the journal will be ground and cleaned, leaving no signs of bluing.
Answer B is correct. Only Technician B is correct. A careful inspection of the camshaft and the bushing associated with the bluing marks is necessary. The camshaft journals need to be checked for roundness, taper, and scoring. The camshaft should also be checked to ensure it is not warped. The bushings need to be visually inspected to make sure the bushing oil hole is aligned with the engine block oil hole. If any discrepancy is found, the camshaft needs to be replaced.
Answer C is incorrect. Only Technician B is correct.
Answer D is incorrect. Technician B is correct.

Question #172
Answer A is incorrect. A total runout reading of 0.005" is within the acceptable specifications.
Answer B is correct. A total runout reading of 0.015" is not within the acceptable maximum of 0.013" and corrective action needs to be taken.
Answer C is incorrect. A total runout reading of 0.009" is within the acceptable specifications.
Answer D is incorrect. A total runout reading of 0.002" is within the acceptable specifications.

Question #173
Answer A is incorrect. Removing and dismantling of the oil pump should only be undertaken after the oil level, viscosity, and its condition are checked. The actual oil pressure needs to be verified with a mechanical gauge directly connected into the main oil gallery.
Answer B is incorrect. Removing the main bearings should only be undertaken after the oil level, viscosity, and its condition are checked. The actual oil pressure needs to be verified with a mechanical gauge directly connected into the main oil gallery.
Answer C is incorrect. Neither Technician is correct.
Answer D is correct. Neither Technician is correct. The engine oil level, condition, and viscosity all should be checked before any dismantling of the engine is undertaken. The oil pressure should also be verified with an accurate gauge directly connected into the main oil gallery. The simplest diagnostic methods should always be employed before the engine is taken apart.

Question #174
Answer A is incorrect. Slipping belts are not always symptomatic with the familiar screeching associated with slipping belts. Sometimes belts such as a power steering and hydraulic fan drive pumps only screech when a large load is applied such as when the steering wheel is turned or the hydraulic fan is actuated.
Answer B is correct. Only Technician B is correct. Slipping belts are not always symptomatic with the familiar screeching associated with slipping belts. Engines equipped with a belt-driven hydraulic fan should always have the belt tension checked as a starting diagnostic point. The pulleys should be checked to make sure they are not worn to the point where the V of the belts bottoms out in the pulley. The alignment of the pulleys should also be checked as this condition will cause premature wear of the belts.
Answer C is incorrect. Only Technician B is correct.
Answer D is incorrect. Technician B is correct.

Question #175
Answer A is incorrect. Technician B is also correct.
Answer B is incorrect. Technician A is also correct.
Answer C is correct. Both Technicians are correct. EG and PG should not be mixed because they are chemically incompatible and once mixing has taken place, it becomes impossible to measure the extent of antifreeze protection provided by the coolant. Propylene and ethylene glycol–based coolants should never be mixed. The mixture in itself will not cause any engine or cooling system problems, but it will be impossible to determine the antifreeze mixture strength with either a refractometer or a hydrometer. If a mixture of EG and PG is known to have taken place and the coolant, for whatever reasons, cannot be immediately replaced, use a refractometer with an EG and a PG scale and average the two readings. If the coolant types are to be switched, the cooling system should be flushed prior to the change.
Answer D is incorrect. Both Technicians are correct.

Question #176
Answer A is incorrect. Technician B is also correct.
Answer B is incorrect. Technician A is also correct.
Answer C is correct. Both Technicians are correct. The radiator should be the first step in determining why an engine overheats especially as the ambient temperatures start to rise. The next step involves checking the speed of the fan with a photo tachometer and commanding the fan on to high speed. If the speed of the fan does not meet manufacturer specifications, the hydraulic pressure to the fan motor needs to be checked and corrected if not within specification. The vehicle

manufacturer's service manual usually gives in-depth details and procedures to follow for the testing.
Answer D is incorrect. Both Technicians are correct.

Question #177
Answer A is incorrect. Technician B is also correct.
Answer B is incorrect. Technician A is also correct.
Answer C is correct. Both Technicians are correct. A reading of 0.8 volts from the O_2 sensor when back probing and monitoring its voltage is an indication of a rich fuel condition. A reading of 0.2 volts is an indication of a lean fuel condition.
Answer D is incorrect. Both Technicians are correct.

Question #178
Answer A is incorrect. An expired fuel cylinder cannot be used as a compressed air tank.
Answer B is correct. Only Technician B is correct. An expired fuel cylinder cannot be used for any other purpose other than to store CNG as depicted on the label. An expired fuel cylinder must be destroyed to render it useless by drilling a 1" diameter hole in the sidewall and disposed of according to local, state, and federal regulations.
Answer C is incorrect. Only Technician B is correct.
Answer D is incorrect. Technician B is correct.

Question #179
Answer A is incorrect. Technician B is also correct.
Answer B is incorrect. Technician A is also correct.
Answer C is correct. Both Technicians are correct. The PRD vent line should be checked for traces of fuel. If fuel is found this is an indication the regulator is starting to fail and should be replaced before releasing the vehicle back into service.
Answer D is incorrect. Both Technicians are correct.

Question #180
Answer A is incorrect. A clear eye does not indicate a fully charged battery.
Answer B is incorrect. A clear eye does not indicate a partially charged battery.
Answer C is incorrect. A clear eye does not indicate a discharged battery.
Answer D is correct. A clear eye indicates the battery electrolyte level is at or below the top of the plates and the battery should be discarded. The battery could still maintain a charge and start an engine. However, if the plates are exposed, the battery is at risk of exploding if a spark occurred due to the buildup of hydrogen sulfide gas.

Question #181
Answer A is incorrect. Using 12 volts to jump start a vehicle with a 24-volt system could be detrimental to the vehicle needing the jump because the starter would not be receiving the required voltage to deliver the amperage for the starter to operate efficiently, causing overheating of the starter. Additionally, if the engine of the dead vehicle did start, as the charging system came on line, the 24 volts of the charging system could damage the electrical/electronics of the vehicle used to jump start with.
Answer B is incorrect. Applying an overvoltage situation to another vehicle, 24 volts to 12 volts, could damage many electronic components such as the ECM, ECU, and any other components not designed to operate at that voltage. This is a very costly and serious damage that should never be allowed to occur.
Answer C is incorrect. Neither Technician is correct.
Answer D is correct. Neither Technician is correct. Using 12 volts to jump start a vehicle with a 24-volt system could be detrimental to the vehicle needing the jump because the starter would not be receiving the required voltage to deliver the amperage for the starter to operate efficiently, causing overheating of the starter. Applying an overvoltage situation to another vehicle, 24 volts to 12 volts, could damage many electronic components such as the ECM, ECU, and any other components not designed to operate at that voltage. Both scenarios could be very costly and should be avoided.

Question #182
Answer A is correct. The LEAST-Likely function of a starter solenoid is to allow the starter drive gear to slip in the event the engine requires too much power to crank over, such as in the case of a tight engine.
Answer B is incorrect. The starter solenoid aids in disconnecting the starter drive gear when the solenoid is de-energized. The secondary voltage produced within the hold-in coil when it is turned off is applied in the opposite direction and pulls the drive gear away from the flywheel ring gear.
Answer C is incorrect. Most starter solenoids allow the transfer of power from the positive battery cable to the starter motor by means of heavy-duty internal contacts.
Answer D is incorrect. The starter drive gear is engaged by the movement of electromagnetic action of the plunger within the solenoid that is connected to a cantilever within the starter motor. The cantilever moves the splined shaft of the starter gear to engage it with the flywheel ring gear.

Question #183
Answer A is incorrect. Before any components are replaced, the proper diagnostic procedures must be followed. The DTC have to be checked.
Answer B is incorrect. Before any components are replaced, the proper diagnostic procedures must be followed. The DTC have to be checked.
Answer C is incorrect. Neither Technician is correct. Before any components are replaced, the proper diagnostic procedures must be followed. The DTC have to be checked.
Answer D is correct. Neither Technician is correct. Before any components are replaced, the proper diagnostic procedures must be followed. The DTC have to be checked. The troubleshooting trees in the service manuals and embedded into the diagnostic software will assist the Technician in following the necessary steps to determine where the defect(s) are located.

Question #184
Answer A is incorrect. Technician B is also correct.
Answer B is incorrect. Technician A is also correct.
Answer C is correct. Both Technicians are correct. If the resistance measurement for a Detroit Diesel Series 50G COP coil between pins 1 and 2 is $\leq 0.55\ \Omega$, the coil needs to be replaced. The resistance reading of a Cummins COP between the spark plug connection and the pins of the connector must be $\leq 20\text{K}\ \Omega$ or the coil must be replaced.
Answer D is incorrect. Both Technicians are correct.

Question #185
Answer A is correct. Only Technician A is correct. A CNG engine should not be cranked for a period exceeding 30 seconds for any reason.
Answer B is incorrect. A starter load test should be performed using the chassis batteries without any kind of external assist.
Answer C is incorrect. Only Technician A is correct.
Answer D is incorrect. Technician A is correct.

Question #186
Answer A is incorrect. The spark plugs should only be changed when they are defective or at the scheduled preventive maintenance interval.
Answer B is incorrect. The valves only require adjustment when engine performance indicators dictate they be adjusted or at the scheduled preventive maintenance interval.
Answer C is incorrect. The fuel filters and air cleaner only require replacement when performance indicators dictate they be replaced or at the scheduled preventive maintenance interval.
Answer D is correct. Resetting the AFR is the Most-Likely action to be taken before continuing to test the engine.

Question #187
Answer A is incorrect. The battery temperature must not exceed 125° F during the charging process. Sealed batteries have built-in check valves that could open due to the increase in pressure and allow electrolyte vapors to escape. Electrolyte cannot be added to the cells of sealed batteries, and

overheating during charging could cause the battery electrolyte level to get low and the battery to require replacement.
Answer B is incorrect. The caps of a conventional battery should be installed tightly during charging to prevent the loss of electrolyte. The caps have a vent to relieve the pressure within the cells and allow the vapors to condensate and return back into the cells.
Answer C is incorrect. The battery should not be charged any longer than necessary to prevent it from being damaged.
Answer D is correct. The open circuit voltage of a fully charged battery is 12.6 volts not 12.4 volts.

Question #188
Answer A is incorrect. It ignites early and combusts very rapidly; starting fluid is unlikely to be the cause of an engine runaway. It will destroy the engine by other means first.
Answer B is incorrect. Starting fluid will cause vapor lock, not hydraulic lock.
Answer C is incorrect. Starting fluid will not damage the fuel system sensors.
Answer D is correct. Excessive use of starting fluid to start an engine causes much higher cylinder pressures, which can cause the piston ring lands to break.

Question #189
Answer A is incorrect. Cold coolant leaks do occur at hose connections when the engine and coolant have cooled down with no apparent leakage when the engine is brought up to temperature. The area where the vehicle is parked would be a good starting point to find the area of the leak and identify the connection that is leaking because the coolant will collect under the vehicle very close to the area that is the problem.
Answer B is incorrect. A small crack in the cylinder head can cause coolant loss and remain hidden until a larger crack develops.
Answer C is incorrect. A loose spark plug adaptor tube in the cylinder head can cause coolant loss and not give any outward appearance of a defect.
Answer D is correct. The LEAST-Likely cause of regular coolant loss is that it evaporates away. Periodically, over periods of days or weeks, some coolant will need to be replenished but not every day.

Question #190
Answer A is incorrect. Technician B is also correct.
Answer B is incorrect. Technician A is also correct.
Answer C is correct. Both Technicians are correct. Fuel cylinders have to be inspected whenever the vehicle has been involved in an accident that could have damaged the cylinder(s) and could have caused them to move in the mounting brackets or affected the plumbing. Whenever a cylinder is disconnected, it must be inspected for leaks and to make sure the plumbing is correctly installed with the correct valves and PRD before it can be placed in service.
Answer D is incorrect. Both Technicians are correct.

Question #191
Answer A is correct. Only Technician A is correct. Catalysts help to reduce the emissions of hydrocarbons, unburned fuel, into the atmosphere. The catalyst will burn off any leftover fuel that did not burn during the combustion process.
Answer B is incorrect. Oxygen is a by-product of complete combustion and therefore the catalyst increases the oxygen content.
Answer C is incorrect. Only Technician A is correct.
Answer D is incorrect. Technician A is correct.

Question #192
Answer A is incorrect. Technician B is also correct.
Answer B is incorrect. Technician A is also correct.
Answer C is correct. Both Technicians are correct. If ELC is purchased as a premix solution it should not be diluted. Silicone hoses should have specially designed spring-loaded constant torque clamps and should be carefully torqued, since silicone hoses may be damaged overtightening.
Answer D is incorrect. Both Technicians are correct.

Question #193
Answer A is incorrect. The contact surfaces should be inspected.
Answer B is incorrect. The O-ring grooves should be inspected.
Answer C is incorrect. The bore should be inspected for scoring.
Answer D is correct. All of the checks are valid EXCEPT for the one referencing the interference fit of the liner into the cylinder bore. Wet sleeves are never interference fitted to the cylinder block bore. They are sealed with O-rings. Dry sleeves are interference fit.

Question #194
Answer A is incorrect. Technician B is also correct.
Answer B is incorrect. Technician A is also correct.
Answer C is correct. Both Technicians are correct. A colder than normal operating engine leads to the formation of higher than normal condensation in the engine. This causes the additives in the engine oil to work harder at preventing the oxidation and corrosion from taking place and wearing out sooner than normal. Cracked cylinder heads can cause small amounts of coolant to leak into the crankcase of the engine. The presence of coolant in the crankcase can lead to the emulsification of the lube oil and damage the internal components of the engine. The engine oil and filters should be replaced prior to placing the vehicle back in service.
Answer D is incorrect. Both Technicians are correct.

Question #195
Answer A is incorrect. Technician B is also correct. CNG engine pistons can be a one-piece design.
Answer B is incorrect. Technician A is also correct. CNG engine pistons can be of a two-piece design.
Answer C is correct. Both Technicians are correct. CNG engine pistons can be a one-piece design.
Answer D is correct. Both Technicians are correct.

Question #196
Answer A is incorrect. By twisting or spiraling the rings onto the ring lands, the rings can become overstressed, distorted, and damaged.
Answer B is correct. Only Technician B is correct. Use of a ring spreader and only spreading the rings as necessary will ensure the rings do not become overstressed. Overstressed rings become deformed and apply excessive pressure against the cylinder walls and can fail by breaking or wearing out faster.
Answer C is incorrect. Only Technician B is correct.
Answer D is incorrect. Technician B is correct.

Question #197
Answer A is incorrect. The Technician failed to make any checks for any defects such as the front or rear mounts, u-joints, or excessive pinion to ring gear free play before condemning the component.
Answer B is incorrect. The Technician failed to make any checks for any defects such as the front or rear mounts, u-joints, or excessive pinion to ring gear free play before condemning the component.
Answer C is incorrect. Neither Technician is correct.
Answer D is correct. Neither Technician is correct. The Technicians failed to check the components for defects, they only condemned them. Other causes of a clunking noise when the vehicle is placed in gear can be worn drive line splines, defective u-joints, and excessive pinion to ring gear backlash, all of which need to be checked to make the correct repair.

Question #198
Answer A is incorrect. Technician B is also correct.
Answer B is incorrect. Technician A is also correct.
Answer C is correct. The pressure cap will maintain the pressure within the cooling system and the water pump is capable of increasing the coolant flow in the engine as engine speed increases to ensure that all the components in the engine are cooled properly.
Answer D is incorrect. Both Technicians are correct.

Question #199
Answer A is incorrect. Technician B is also correct.
Answer B is incorrect. Technician A is also correct.

Answer C is correct. Both Technicians are correct. The communication interface device that is used to communicate between the ECM and the diagnostic software must be powered up. The power LED needs to be illuminated showing that it is receiving power. If the LED fails to light, the software will fail to initialize and will not communicate. Additionally the software has to be configured for the correct driver and port for the PC to interface with the ECM.

Question #200
Answer A is incorrect. Piston cooling nozzles are usually removed prior to removing the pistons and connecting rod assemblies.
Answer B is correct. Only Technician B is correct. If the piston cooling nozzles are not removed prior to removing the connecting rod and piston assembly, the piston cooling nozzle can become distorted if struck and can result in becoming misaligned when reinstalled causing improper spraying of lubricating oil on the underside of the piston crown. The lack of oil cooling oil could cause the piston crown to overheat causing piston damage.
Answer C is incorrect. Only Technician B is correct.
Answer D is incorrect. Technician B is correct.

Question #201
Answer A is incorrect. Technician B is also correct.
Answer B is incorrect. Technician A is also correct.
Answer C is correct. Both Technicians are correct. ELCs are low-maintenance coolants that may require only one SCA treatment in six years. The engine-cooling fan may be engaged for transmission hydraulic braking as part of the braking strategy. ELCs promise a service life of 600,000 miles (960,000 km) or 6 years, with one additive recharge at 300,000 miles (480,000 km) or three years. This compares with a typical service life of two years during which up to 20 recharges of SCA would be required for conventional EG and PGs. ELCs are available as a premixed solution to ensure that the water quality is at the required level. ELCs are also available in full strength, but the make up water should be de-ionized to ensure the cooling system is not contaminated with minerals that can be deposited in the cooling system and clog it up. The pricing of ELCs by quantity is generally comparable with EG and PG, but because of reduced cooling system maintenance and extended service life, they will probably become the coolant of choice of the engine OEMs. No test kits are required to monitor the ELCs' SCA level.
Answer D is incorrect. Both Technicians are correct.

Question #202
Answer A is incorrect. Technician B is also correct.
Answer B is incorrect. Technician A is also correct.
Answer C is correct. Both Technicians are correct. Technician A is correct in describing the alignment specifications for pulley-running centerlines. Technician B is also correct in saying that pulley belts should not contact the bottom of the pulley grooves.

- When two or more identical belts are used on the same pulley, all of the belts must be replaced at the same time.
- Make sure the distance between the pulley centers is as short as possible when you install the belts. Do not roll the belts over the pulley. Do not use a tool to pry the belts onto the pulley.
- Pulleys must not be out of alignment more than 1/16" for each 12" of distance between the pulley centers.
- Belts must not touch the bottom of the pulley grooves, and they must not protrude more than 3/32" above the outside diameter of the pulley.
- When identical belts are installed on a pulley, the protrusion of the belts must not vary more than 1/16".
- Make sure that the belts do not touch or hit against any part of the engine.

Answer D is incorrect. Both Technicians are correct.

Question #203
Answer A is incorrect. All fuel pressure must be removed from the fuel lines before the fittings are manipulated in any manner. The fitting's nut should not be advanced greater than a ¼ of a turn or the ferrules can be destroyed and require to be replaced.
Answer B is incorrect. All fuel pressure must be removed from the fuel lines before the fittings are manipulated in any manner. Once the fuel pressure at the fitting is vented to 0 psi, the fitting's nut can be advanced an additional ¼ turn maximum and repressurized and checked for leakage.
Answer C is incorrect. Neither Technician is correct.
Answer D is correct. Neither Technician is correct. All fuel pressure must be removed from the fuel lines before the fittings are manipulated in any manner. If the pressure is not depleted from the lines and the fittings are loosened, the lines can be blown off the fitting and cause injury to the Technician. The fitting, ferrules and seats need to be checked for scratches, dirt or other defects that can cause fuel to leak and corrected. When installing a pre-swaged fitting, it needs to be installed finger tight and tightened an additional ¼ turn and checked for leaks by pressurizing the fuel lines. If the fitting continues to leak, the fuel pressure needs to be vented before the fitting can be moved. Once the fuel pressure at the fitting is vented to 0 psi, the fitting's nut can be advanced an additional ¼ turn maximum and re-pressurized and checked for leakage. If the fitting still leaks, the fitting and ferules assembly will need to be replaced. Most of the time, the line needs to be replaced as well as the fitting.

Question #204
Answer A is incorrect. The test being performed is a voltage and amperage test.
Answer B is incorrect. The test being performed is a voltage and amperage test.
Answer C is incorrect. The test being performed is not a voltage drop test.
Answer D is correct. The test being performed is a monitoring of the amperage and voltage output of the generator (alternator). The engine should be at fast idle (1000 rpm) with all of the accessories such as the HVAC turned on and the headlamps on high beam to simulate a high demand on the charging system to determine if the generator is capable of meeting the demand placed on it.

Question #205
Answer A is incorrect. Technician B is also correct.
Answer B is incorrect. Technician A is also correct.
Answer C is correct. Both Technicians are correct. The angles of the valve and the seat are usually ground equally to ensure as much metal-to-metal contact between the valve and the seat making the valve and seat more resistant to wear caused by the valve rotator. This also helps to make the contact surfaces larger and ensures that the maximum amount of heat can be transferred from the valve to the seat, into the cylinder head and to the cooling system to cool the valve and prevent it from burning.
Answer D is incorrect. Both Technicians are correct.

Question #206
Answer A is correct. Only Technician A is correct. One lead of the voltmeter is connected to the ground terminal of the battery and the other lead is connected to the base of the starter. Then the voltmeter is read while the engine is cranked. If the voltage reading exceeds 0.2 volts, excessive resistance exists in the ground circuit.
Answer B is incorrect. A defective starter drive is a mechanical problem that would not affect the voltage at the starter.
Answer C is incorrect. Only Technician A is correct.
Answer D is incorrect. Technician A is correct.

Question #207
Answer A is incorrect. A series circuit would produce 24 volts.
Answer B is correct. If both batteries are 12 volts and they supply a 12/12 electrical system, the batteries must be connected in parallel.

Answer C is incorrect. A series parallel circuit would be used with series parallel connected batteries in a 12/24 circuit.
Answer D is incorrect. An isolation circuit typically isolates a ground on a 12/24 circuit.

Question #208
Answer A is correct. Only Technician A is correct. The left figure shows an offset camshaft key: The extent of offset determines how the camshaft gear is phased with the camshaft so it can be called a timing key. Technician A is generally correct in saying that the key must be replaced with the same type, though if a timing change is required, this is the means of accomplishing it. The key must be installed with the arrow pointing in the required direction to achieve correct timing. The letter on the key identifies the amount of offset.
Answer B is incorrect. The key shown on the left is not damaged. This type of key is used to alter the cam gear position on the cam and the timing. Actuation of the valves by the engine timing gear train must be precisely timed in any CNG engine. This process correctly times the intake and exhaust cycles of the engine.
Answer C is incorrect. Only Technician A is correct.
Answer D is incorrect. Technician A is correct.

Question #209
Answer A is incorrect. Worn bearings are a very common cause of low oil pressure.
Answer B is correct. Oil with a high viscosity index changes little over a wide temperature range and, therefore, would be very unlikely to cause low oil pressure. CNG engines require a special 15W-40 GEO oil. The GEO oil type should not be interchanged with another type of engine oil in a CNG engine. The incorrect type of oil in a CNG engine can cause severe, even catastrophic engine damage.
Answer C is incorrect. A restricted oil pump suction tube would cause low oil pressure, especially at higher rpms when oil flow is at the highest.
Answer D is incorrect. A worn oil pump would bypass oil internally and cause low oil pressure.

Question #210
Answer A is incorrect. The tension applied by the tensioner cannot be corrected by repositioning the tensioner differently. Excessive tension can damage the bearings in the components as well as the drive belt.
Answer B is correct. Only Technician B is correct. The correct course of action to be taken is to replace the tensioner.
Answer C is incorrect. Only Technician B is correct.
Answer D is incorrect. Technician B is correct.

Question #211
Answer A is incorrect. Technician B is also correct.
Answer B is incorrect. Technician A is also correct.
Answer C is correct. Both Technicians A and B are correct. Some fasteners, such as head bolts and main bearing cap bolts, will require a specified torque setting, that will be spelled out in the service manual. Some fasteners will only require generic torque settings derived from the grade, size, and thread pitch and will be found in the torque chart.
Answer D is incorrect. Both Technicians are correct.

Question #212
Answer A is correct. The Technician is testing for the opening temperature of the thermostat. The ribbon will be released as the thermostat starts to open and the Technician can read the temperature on the thermometer.
Answer B is incorrect. Although the Technician can check the thermostat if it does stick open after opening during the testing, the Technician is checking the opening temperature.
Answer C is incorrect. Although the Technician can check the thermostat's fully open temperature during the testing, the Technician is checking the opening temperature.
Answer D is incorrect. Although the Technician can check the closing temperature of the thermostat, the Technician is checking the opening temperature.

Question #213
Answer A is incorrect. The fuel gauge has to be a minimum of 5000 psi when the working pressure of the system is 3600 psi.
Answer B is incorrect. The PRD lines need to be secured every 12" when the lines exceed 24" in length.
Answer C is incorrect. An expansion loop is necessary at connection points of a high-pressure CNG fuel system. Expansion loops prevent the lines from being stressed by expansion and contraction of the lines due to cooling or heating of the lines.
Answer D is correct. The fuel tanks should never be unclamped from their securement brackets or moved when under pressure. The cylinder(s) that are going to be manipulated need to be defueled before moving them.

Question #214
Answer A is correct. Only Technician A is correct. When an engine shows wear only in a specific portion and the rest of the engine looks normal, the Technician should concentrate the diagnostic routine on items that affect that part of the engine only. At first glance, both items mentioned would seem to affect the entire engine; therefore, they are not very likely to cause damage to only the rear two cylinders. If a CNG engine was operated with dirty oil, then all the bearing surfaces should be worn. All cylinders would be affected if the water pump produced inadequate flow, but the rear two cylinders would be the most affected due to distance from the pump. Technician A is correct.
Answer B is incorrect. Dirty oil would Most-Likely effect the entire engine.
Answer C is incorrect. Only Technician A is correct.
Answer D is incorrect. Technician A is correct.

Question #215
Answer A is incorrect. Technician B is also correct.
Answer B is incorrect. Technician A is also correct.
Answer C is correct. Both Technicians are correct. OEM mounting requirements should be followed closely. Some ECMs are mounted directly on a grounded surface. Other ECMs require mounting with isolators to prevent the case from grounding out. The isolators need to be inspected to make sure they are in good condition. When the ECM mounting bolts are tightened, caution should be taken to not overtorque them to prevent the isolators from being damaged. The isolators prevent outside interference from being induced into the ECM electronics.
Answer D is incorrect. Both Technicians are correct.

Question #216
Answer A is incorrect. This fan does use a coil spring.
Answer B is incorrect. Air temperature does regulate this fan operation.
Answer C is incorrect. Radiator temperature does affect this fan because radiator temperature affects the temperature of the air of passing through it.
Answer D is correct. A coil spring, radiator temperature, and air temperature can all be used to regulate fan operation in a thermostatically controlled fan. Air pressure is not a factor in controlling fan cycling, so this is the EXCEPT statement. Thermatic viscous drive fan hubs are integral fan drive units with no external controls. They use silicone fluid as a drive medium between the drive hub and the fan drive plate.

Question #217
Answer A is correct. Only Technician A is correct. A PRD vent line must project above the roof line to prevent fuel from accumulating in the roof area.
Answer B is incorrect. A PRD is not a field serviceable item. It must be replaced with another one that is manufacturer approved.
Answer C is incorrect. Only Technician A is correct.
Answer D is incorrect. Technician A is correct.

Question #218
Answer A is correct. Only Technician A is correct. A ball end micrometer is used to check the bearing shell thickness to see if it is within specification.

Answer B is incorrect. A dial caliper is not the tool used to check the bearing shell thickness to see if it is within specification.
Answer C is incorrect. Only Technician A is correct.
Answer D is incorrect. Technician A is correct.

Question #219
Answer A is incorrect. Engine certification labels are required on all CNG engines as well as diesel engines.
Answer B is incorrect. A photocopy of another engine's certification label is not acceptable because the label has to contain the engine's model number and most importantly its serial number. New certification labels can be obtained from a certified engine dealer.
Answer C is incorrect. Neither Technician is correct.
Answer D is correct. Neither Technician is correct. Engine certification labels are required on all CNG engines. A photocopy of another engine's certification label is not acceptable because the label has to contain the engine's model number and most importantly its serial number. New certification labels can be obtained from a certified engine dealer.

Question #220
Answer A is incorrect. The oil pressure needs to be checked with a good known pressure gauge and compared to the oil pressure displayed on the software.
Answer B is incorrect. The main bearings should only be checked after checking the oil level and for contamination or dilution, verifying the accuracy of the oil pressure sensor with a good known oil pressure gauge, checking to make sure the oil pressure regulator valve is not sticking.
Answer C is incorrect. Neither Technician is correct.
Answer D is correct. Neither Technician is correct. Oil pumps, although they do wear out, are rarely a cause of low oil pressure. The oil condition should be checked to make sure it is of the correct type, not diluted or contaminated with other liquids. The oil pressure regulator valve should be checked for sticking due to dirt or contamination. The oil pickup tube needs to be checked to make sure the gaskets or O-rings are not defective and allowing air to be sucked into the pump, causing low oil pressure. A high-mileage engine could have worn main bearings that need replacement. The main bearings should only be checked after checking the oil level and for contamination or dilution, verifying the accuracy of the oil pressure sensor with a good known oil pressure gauge, checking to make sure the oil pressure regulator valve is not sticking.

Question #221
Answer A is incorrect. Reaming of the bearing will increase the inside diameter of the bushing and the correct clearance between the journal and the cam bushing will not be maintained.
Answer B is incorrect. As long as the remainder of the cam bearings are within specification and not damaged, only one of the bearings can be replaced.
Answer C is incorrect. Neither Technician is correct.
Answer D is correct. Neither Technician is correct. The cam bearing should be replaced and checked for the correct diameter to ensure the correct clearance is achieved with the cam. As long as the remainder of the cam bearings are within specification and not damaged, only one of the bearings can be replaced.

Question #222
Answer A is incorrect. Technician B is also correct.
Answer B is incorrect. Technician A is also correct.
Answer C is correct. Both Technicians are correct. Both the key and camshaft keyway conditions have to be checked to ensure they are in good condition to align the two components for timing purposes. If the cam gear should become improperly aligned on the camshaft due to the gear slipping on the camshaft, improper engine timing can occur. If the gear should slip on the camshaft a lot, the valve timing could be severely affected. This could cause the valves to strike the piston with major engine damage and expenses as a result.
Answer D is incorrect. Both Technicians are correct.

Question #223
Answer A is incorrect. Oil should not be placed into a cylinder of a CNG engine. This could lead to hydraulic lock and damage the engine. A leak-down test should be performed instead.
Answer B is correct. Only Technician B is correct. A leak-down test will allow the Technician to narrow the diagnosis of the defect and undertake the necessary repairs. During the leak-down test, if air is heard escaping from the crankcase, the rings or pistons could be the cause. If air leaks from the exhaust ports or intake ports, the exhaust or intake valves respectively are Most-Likely defective.
Answer C is incorrect. Only Technician B is correct.
Answer D is incorrect. Technician B is correct.

Question #224
Answer A is incorrect. The ring gear is not pinned or bolted to the flywheel.
Answer B is incorrect. The ring gear is not welded to the flywheel.
Answer C is incorrect. Neither Technician is correct.
Answer D is correct. Neither Technician is correct. Flywheel ring gears are usually shrink-fitted to the outer periphery of the flywheel is the ring gear, which is the means of transmitting cranking torque to the engine by the starter motor during startup.

Question #225
Answer A is incorrect. The engine should never be run without coolant or the sealing surfaces could be destroyed. The sealing surfaces rely on the coolant for lubrication and to keep them cool.
Answer B is correct. Only Technician B is correct. If scratches are found on the sealing surfaces of the seal, this is usually an indication that sand or debris is in the coolant has been able to travel. The cooling system should be flushed, the coolant filter replaced, and the flow through the filter verified to prevent the problem for reoccurring.
Answer C is incorrect. Only Technician B is correct.
Answer D is incorrect. Technician B is correct.

Question #226
Answer A is incorrect. Technician B is also correct.
Answer B is incorrect. Technician A is also correct.
Answer C is correct. Both Technicians are correct. All of the O-rings should be replaced. Before installing the O-rings they should be coated with a light coat of Dow Corning 55 O-ring lubricant or compatible compound to help protect and seal them. If the O-rings are not replaced there is a possibility of a leak to develop and causing further repairs to be undertaken at a later date. After the repairs are completed, the connections and lines should be tested for leaks with soapy water or other solution such as Snoop.
Answer D is incorrect. Both Technicians are correct.

Question #227
Answer A is incorrect. If there are no other traces of oil leakage from the rocker cover gasket, the gasket is not suspect.
Answer B is incorrect. Traces of oil on the underside of the coil are not an indication of the incorrect type of coil has been installed on the engine.
Answer C is correct. The coil is potentially cracked due to the oil leakage. The coil is filled with a special oil to help keep the coil running cooler.
Answer D is incorrect. Oil traces are not indications of a normal condition.

Question #228
Answer A is incorrect. The bolts should not be removed until the battery ground cable is disconnected.
Answer B is incorrect. The starter solenoid should not be removed until the battery ground cable is disconnected.
Answer C is correct. Knowledge of a procedure is being tested in this question. The disconnecting of the battery ground cable is going to be one of the first steps in any procedure involving the replacement of an electrical component because this renders the component electrically neutral and safe to work on.

Answer D is incorrect. The starter magnetic switch should not be removed until the negative battery cable is removed.

Question #229
Answer A is incorrect. Technician B is correct also.
Answer B is incorrect. Technician A is correct also.
Answer C is correct. Both Technicians are correct. A high intake air temperature and a high boost pressure both can cause pre-ignition of the fuel air mixture and a knocking condition. This would log a DTC and possibly de-rate the engine depending on the severity, affecting engine performance.
Answer D is incorrect. Both Technicians are correct.

Question #230
Answer A is correct. Only Technician A is correct. The R terminal of a 50 DN 24-volt alternator is tapped directly in the center of the rectifier diode bridge assembly and will produce half of the voltage of the output terminal's charging rate. This is utilized to ensure, in the event of an alternator failure, the vehicle can return back to the shop without the batteries being completely depleted by the high power demand of the HVAC, which could be as high as 150 amp hours.
Answer B is incorrect. The charging rate of the alternator should be checked, but because the alternator light was not reported as being activated and because other accessories not associated with the HVAC are not working as well, the Most-Likely cause would be the R terminal output is missing due to a defect in the alternator rectifier bridge assembly.
Answer C is incorrect. Only Technician A is correct.
Answer D is incorrect. Technician A is correct.

Question #231
Answer A is incorrect. Roller cam followers should not be installed dry to prevent damaging them when the engine is started after an overhaul or when they are replaced.
Answer B is incorrect. Grease will not penetrate the rollers and bearings of the followers. Damage to the followers will Most-Likely occur when the engine is started up after replacement or an overhaul.
Answer C is incorrect. Neither Technician is correct.
Answer D is correct. Neither Technician is correct. The cam followers should be soaked in engine oil prior to installing in the engine to allow lubrication to penetrate the bearings. If soaking is not performed, damage can occur at start-up. Grease will not penetrate the rollers and bearings of the followers and damage can also occur.

Question #232
Answer A is incorrect. Installing a shorter belt could cause the tensioner to become too tight and damage the bearings of the components the belt is being driven by or driving. The tensioner is defective and the correct repair procedure is to replace the tensioner.
Answer B is correct. Only Technician B is correct. The tensioner is defective and needs to be replaced. Installing a shorter belt can cause abnormal belt tension on the accessories being driven by the belt and damage their bearings as well as the engine's accessory drive bearings.
Answer C is incorrect. Only Technician B is correct.
Answer D is incorrect. Technician B is correct.

Question #233
Answer A is correct. Most manufacturers recommend that the snap ring gap be in the down position. Most importantly, it never should be located at 90° to the direction of travel. All full-floating piston pins require a means of preventing the pin from exiting the pin boss and contacting the cylinder walls. Snap rings and plugs are used. When installing the internal snap rings used by most engine OEMs, it is important to observe the installation instructions.
Answer B is incorrect. The ring gap should not go up.
Answer C is incorrect. The ring gap should not go horizontal, as this will allow inertia the opportunity to loosen the ring.
Answer D is incorrect. It does matter where the ring gap is installed. Most manufacturers recommend that the snap ring gap be in the down position.

Question #234
Answer A is correct. Only Technician A is correct. The oil filter gasket should be lubricated with clean engine oil prior to mounting it on the filter adaptor.
Answer B is incorrect. It is not necessary to remove and inspect the oil filter bypass valve every time a filter is replaced. When contamination is discovered in the oil sump, the valve should be removed and checked.
Answer C is incorrect. Only Technician A is correct.
Answer D is incorrect. Technician A is correct.

Question #235
Answer A is incorrect. Parts manuals do contain specific information concerning acceptable oil filters, lubricants and, in many cases, preventive maintenance schedules.
Answer B is incorrect. Service manuals do contain specific information concerning acceptable oil filters, lubricants, and preventive maintenance schedules.
Answer C is incorrect. The manufacturer's electronic reference CD is an excellent reference source for specific information concerning acceptable oil filters, lubricants, and preventive maintenance schedules because the CD can contain several reference manuals that will contain the required information.
Answer D is correct. The operator's manual is the LEAST-Likely manual to contain specific information concerning acceptable oil filters, lubricants, and preventive maintenance schedules.

Question #236
Answer A is incorrect. Excessive play would not cause the back of the bearing to be discolored.
Answer B is correct. Friction-bearing shells must be able to dissipate the heat they are subjected to in the bore into which they are inserted. Bearing crush helps ensure good heat transfer by making it possible to transfer heat. When bearing crush is insufficient, there is less force loading the bearing into its bore and it does not transfer heat as efficiently. Dirt or grease on the back of the bearing would have the same effect. Bearings are retained primarily by crush. The outside diameter of a pair of uninstalled bearing shells slightly exceeds the bore to which it is installed. This creates radial pressure that acts against the bearing halves and provides good heat transfer. The bearing halves are also slightly elliptical to allow the bearing to be held in place during installation; this is known as bearing spread. Tangs in bearings are inserted into notches in the bearing bore to minimize longitudinal movement, prevent bearing rotation, and align oil holes.
Answer C is incorrect. A belled crankshaft journal would establish a wear pattern on the bearing front, not back.
Answer D is incorrect. Degraded lube would establish wear on the front of the bearing, not the back.

Question #237
Answer A is correct. The piston illustrated was subject to excessive cylinder temperatures. Because no other piston showed this wear, the problem must be a single cylinder issue. A partially worn exhaust lobe would prevent all of the exhaust gasses from escaping the cylinder on the exhaust stroke, thus creating excessive cylinder temperatures in a single cylinder.
Answer B is correct. A restricted exhaust system could cause excessive cylinder temperatures; however, all cylinders would be affected.
Answer C is incorrect. A restricted air cleaner could also cause excessive cylinder temperatures; however, it too would affect all cylinders.
Answer D is incorrect. An overfueling condition of the engine would not affect only one cylinder, rather it would cause high temperature in all of the cylinders because the fuel control valve controls the fuel to all the cylinders from a single point.

Question #238
Answer A is incorrect. The cap should be replaced.
Answer B is incorrect. The coolant should be tested.
Answer C is incorrect. The radiator hoses should be checked.

Answer D is correct. If a radiator cap is tested to be defective, the cap should be replaced, the coolant should be tested, and the radiator hoses should be checked for leakage and collapsing. Attempting to adjust or replace the spring is not good practice.

Question #239
Answer A is incorrect. Technician B is also correct.
Answer B is incorrect. Technician A is also correct.
Answer C is correct. Both Technicians A and B are correct. Any energy absorbed by a fluid type or viscous damper is converted into heat. This is a good method of determining if it is working correctly. Because the crankshaft is constantly speeding up as the engine is firing and slowing down before the next ignition cycle, the damper should be absorbing energy and creating heat. If one or more bolt holes of the damper become elongated, the damper should be replaced to ensure it can continue to absorb energy without being able to move on it mounting bolts.
Answer D is incorrect. Both Technicians are correct.

Question #240
Answer A is incorrect. If the test was performed correctly, then the battery was already adequately charged. Recharging the battery and placing it in service will place a defective battery in service.
Answer B is correct. The battery should be replaced because it was already fully charged and it failed the load test.
Answer C is incorrect. Although it is always a good idea to check the alternator drive belt, in this situation, the belt is not the cause of a defective battery. Most-Likely the plates are worn out.
Answer D is incorrect. There is no reason to believe the voltage regulator is faulty. Replacing the regulator will not fix a faulty battery. The charging rate should be checked before releasing the vehicle into service to make sure the charging system is functioning properly.

Question #241
Answer A is incorrect. 50 to 150 Ω is incorrect.
Answer B is correct. The resistance reading must be 500 to 1.5K Ω.
Answer C is incorrect. 5K to 15K Ω is incorrect.
Answer D is incorrect. 500 to 1.5K Ω per foot of wire is incorrect.

Question #242
Answer A is incorrect. A defective coil can cause a misfiring cylinder.
Answer B is incorrect. Too large of a spark plug gap can cause a misfiring cylinder.
Answer C is correct. A loose valve adjustment is the LEAST-Likely cause of a misfiring cylinder.
Answer D is incorrect. A cracked piston can cause a misfiring cylinder due to low compression.

Question #243
Answer A is incorrect. The battery to chassis ground connection is the first connection to remove.
Answer B is correct. All of the steps listed are required when installing a new starter relay EXCEPT for disconnecting the starter positive cable at the starter solenoid.
Answer C is incorrect. Once the ground connection at the chassis to battery has been broken the starter relay or mag switch can be replaced.
Answer D is incorrect. The ground cable would need to be installed on the battery-to-chassis last.

Question #244
Answer A is correct. Only Technician A is correct. When jump-starting one vehicle using another, it is good practice to switch off the engine of the boost vehicle while making the battery connections to prevent electrical/electronic damage.
Answer B is incorrect. The negative cable clamps should be connected last and disconnected first.
Answer C is incorrect. Only Technician A is correct.
Answer D is incorrect. Technician A is correct.

Question #245

Answer A is incorrect. Technician B is also correct.

Answer B is incorrect. Technician A is also correct.

Answer C is correct. Both Technicians are correct. The concentricity of the flywheel housing has to be checked no matter if a flywheel or flexplates are installed. The crankshaft wobble in relationship to the housing would affect the transmission and the engine. Flexplates would help to absorb some runout, but the crankshaft seal would be affected and premature failure would Most-Likely occur.

Answer D is incorrect. Both Technicians are correct.

Question #246

Answer A is correct. The engine will not usually shut down after the warning light illuminates and this is the EXCEPT statement. Rather a predetermined period of time usually 30 seconds will be allowed so the driver can safely pull the vehicle onto the shoulder of the road and bring it to a stop.

Answer B is incorrect. The ECM can be programmed for a gradual ramp down of power to allow the driver to safely bring the vehicle to a stop or until the overheating condition no longer exists.

Answer C is incorrect. The ECM can be programmed to not shut down the engine and continue to run normally. The warning light will be illuminated.

Answer D is incorrect. The ECM can be programmed to shut down after a period of 30 seconds the condition starts. The warning light will be illuminated.

Question #247

Answer A is incorrect. The valve adjustment must be performed using the manufacturer's specifications.

Answer B is incorrect. Most manufacturers recommend the engine temperature be less than 140° F before performing a valve adjustment.

Answer C is incorrect. Neither Technician is correct.

Answer D is correct. Neither Technician is correct. The valve adjustment must be performed using the manufacturer's specifications. Failure to follow the correct procedures can result in engine failure and void the engine warranty if the engine is still under warranty. Temperature cannot be compensated for. Most manufacturers recommend the engine temperature be less than 140° F before performing a valve adjustment.

Question #248

Answer A is incorrect. Honing of a cylinder liner to provide more liner to piston clearance is not an acceptable method. The liner and piston should be replaced as a unit to bring the cylinder back to within the manufacturer's specifications.

Answer B is incorrect. The liner and piston should be replaced as a unit to bring the cylinder back within the manufacturer's specifications.

Answer C is incorrect. Neither Technician is correct.

Answer D is correct. Neither Technician is correct. Honing of a cylinder liner to provide more liner to piston clearance is not an acceptable method. The liner and piston should be replaced as a unit to bring the cylinder back to within the manufacturer's specifications.

Question #249

Answer A is correct. Only Technician A is correct. Icing is a direct result of the pressure being too low. As a result of a loss of pressure, a cooling effect occurs. For ice to from, the temperature has to be lower than 32° F. CNG engines will develop performance issues when the fuel is too cold. Normally, the manufacturers will specify a temperature of 55° F as the minimum operating fuel temperature. The reason for the large temperature drop needs to be detected and repaired. A partially clogged high-pressure fuel filter, a defective high-pressure regulator, or coolant not circulating through the high-pressure regulator could all be causes of icing.

Answer B is incorrect. The high-pressure regulator is designed to maintain a minimum pressure above the point where icing should develop. Coolant is circulated through the high-pressure regulator to assist in keeping the fuel temperature above freezing. If ice appears, the fuel is very cold, below 32° F, and it is only a matter of time before the engine malfunctions.
Answer C is incorrect. Only Technician A is correct.
Answer D is incorrect. Technician A is correct.

Question #250
Answer A is incorrect. If a spark plug is not approved by the engine manufacturer, it should not be installed in the engine. The emission levels could be altered and engine damage could occur.
Answer B is correct. Only Technician B is correct. Only spark plugs that are approved for the engine model should be installed in the engine. Inappropriate spark plugs could alter the emission beyond legal limits as well as potentially damage the engine due to providing too hot of a spark and burning a hole in a piston. Too long of a tip length can also cause the piston to strike the electrode and damage the engine.
Answer C is incorrect. Only Technician B is correct.
Answer D is incorrect. Technician B is correct.

Question #251
Answer A is incorrect. The painting would take place after it has passed pressure testing for cracks.
Answer B is correct. When a major CNG engine overhaul is performed, it is good practice to thoroughly inspect the cylinder block using pressure testing.
Answer C is incorrect. The cylinder sleeves would be removed before cleaning.
Answer D is incorrect. Plug installation should take place only after pressure testing has been done.

Question #252
Answer A is incorrect. A leaking intake valve seat can cause low compression in that particular cylinder.
Answer B is correct. Worn exhaust valve seals will usually leak oil into the combustion chamber but will not cause low compression and are the LEAST-Likely cause of low compression.
Answer C is incorrect. A scuffed cylinder wall can cause compression to leak past the compression rings and low cylinder compression pressure will result.
Answer D is incorrect. A blown cylinder head gasket will allow compression to escape past it and low compression pressure in the cylinder will result.

Question #253
Answer A is incorrect. Technician B is also correct.
Answer B is incorrect. Technician A is also correct.
Answer C is correct. Both Technicians are correct. Some coolant filters are only designed to remove sand and grit from the cooling system while some coolant filters are designed to remove sand, grit and also to condition the coolant with SCA. It is important for the Technician to check the condition of the coolant before servicing a filter that has SCA. Too much SCA can drop out of the coolant and clog the radiator.
Answer D is incorrect. Both Technicians are correct.

Question #254
Answer A is incorrect. The cylinder should be filled in increments after it has been replaced or empty for an extended period of time to allow the liner to conform to the carbon structure. This will ensure that no air bubbles are entrapped between the liner and the carbon fiber structure.
Answer B is incorrect. The cylinder should be leaked tested at each increment of the pressurizing process in case a leak does develop. It is easier to remove a small amount of fuel than a large quantity.
Answer C is correct. The cylinder does not require being purged of air when being filled after replacement. The air volume is negligible and will be blended with the methane as the fuel enters. As the fuel is being consumed, the air will be removed along with it.
Answer D is incorrect. The cylinder hold-down brackets should be torqued while the cylinder is empty.

Question #255
Answer A is incorrect. Technician B is also correct.
Answer B is incorrect. Technician A is also correct.
Answer C is correct. Both Technicians are correct. Technician A correctly defines the term ampere hour and Technician B is correct in saying that a 75-ampere hour charge to a 200-ampere hour battery should be sufficient to turn the charge indicator green, providing the battery is in good condition. Batteries with charge indicators are sufficiently charged when the colored dot in the hydrometer is visible. Gently shake or tilt the battery at hourly intervals during charging to mix the electrolyte and to see if the colored dot appears. Do not tilt the battery beyond a 45° angle. If the colored dot does not appear after a 75-ampere hour charge, continue charging for another 50 to 75 ampere hours. If the colored dot still does not appear, replace the battery.
Answer D is incorrect. Both Technicians are correct.

Question #256
Answer A is incorrect. The correct torque is 28 lbs/ft
Answer B is correct. Only Technician B is correct. An accurate torque wrench is the accepted method to ensure a spark plug is tightened enough but not too much. A loose spark plug can cause leakage of the air/fuel mixture past the threads during the compression cycle, leading to improper combustion due to low compression pressure and overheating of the coil and extenders/boots, resulting in damaged components. An overtorqued spark plug can be difficult to remove and lead to a damaged adaptor tube, requiring the replacement of the adaptor at a considerable amount of labor. The appropriate torque for a spark plug is 28 lbs/ft.
Answer C is incorrect. Only Technician B is correct.
Answer D is incorrect. Technician B is correct.

Question #257
Answer A is incorrect. Technician B is also correct.
Answer B is incorrect. Technician A is also correct.
Answer C is correct. Both Technicians are correct. If a sensor shows signs of damage it can be caused by excessive end play of the crankshaft allowing the target to strike the sensor. The target wheel should be checked to make sure it did not get damaged and repaired if necessary as well as the sensor getting replaced.
Answer D is incorrect. Both Technicians are correct.

Question #258
Answer A is incorrect. A carbon pile tester is used to create a load on the alternator.
Answer B is incorrect. An amp clamp can be used to measure alternator output.
Answer C is correct. To perform an alternator amperage output test, an amp clamp is placed around the wire attached to the alternator's output terminal, engine rpm is increased to high idle, and a load is gradually placed across the battery terminals with a carbon pile load tester until maximum amperage output is reached without reducing battery voltage below 12 volts. A hydrometer is used to test specific gravity of the electrolyte.
Answer D is incorrect. A voltmeter is used to measure alternator voltage.

Question #259
Answer A is correct. Only Technician A is correct. A lack of power with no complaints about the engine misfiring is an indication that the basics need to be addressed first. The air cleaner should be one of the first things checked even though there are no outward signs of black smoke. Methane has a lower carbon content and the engine usually will not smoke unless some other material is mixed in such as engine oil. A magnehelic gauge or a manometer graduated in inches of water is used to check air inlet restriction.
Answer B is incorrect. Replacing the spark plugs without any reason should be the last course of action undertaken.
Answer C is incorrect. Only Technician A is correct.
Answer D is incorrect. Technician A is correct.

Question #260
Answer A is incorrect. The maximum recommended length of line is 18" in length.
Answer B is incorrect. The maximum recommended length of line is 18" in length.
Answer C is incorrect. The maximum recommended length of line is 18" in length.
Answer D is correct. The maximum recommended length of line from the fuel low-pressure shutoff solenoid and the fuel low-pressure regulator is 18" in length. If greater that 18", excess fuel will remain in the line and low fuel pressure regulator and the engine will experience backfiring when the ignition switch is turned off.

Question #261
Answer A is incorrect. Technician B is also correct.
Answer B is incorrect. Technician A is also correct.
Answer C is correct. Both Technicians are correct. Corrosion can be removed with a baking soda and water solution. However, sometimes the corrosion can creep up the cables and be unnoticeable. The cables should be checked out carefully and replaced if their condition is questionable.
Answer D is incorrect. Both Technicians are correct.

Question #262
Answer A is incorrect. This is a voltage drop test on the ground side of the starter.
Answer B is correct. A voltage drop test is performed by placing the voltmeter leads across the portion of the circuit being tested and observing voltage with the circuit operating. In the figure, one lead is on the battery negative post, and the other one is attached to the starter case. The groundside of the circuit is between the voltmeter leads. The starter motor must be cranking when performing this test.
Answer C is incorrect. This is a voltage drop test on the ground side of the starter.
Answer D is incorrect. This is a voltage drop test on the ground side of the starter.

Question #263
Answer A is incorrect. Technician B is also correct.
Answer B is incorrect. Technician A is also correct.
Answer C is correct. Both Technicians are correct. Technician B is correct in that if the vehicle is driven, the higher load will cause the engine to warm up more quickly and when the cardboard is inserted in front of the radiator, the fan will engage more rapidly. Technician A is also correct in recommending placement of cardboard in front of the radiator because the amount of time required for the fan to engage can assist in determining how quickly the engine warms up. If it takes a long time, the thermostats could be partially stuck open. If it takes a very short period of time, the radiator should be checked to make sure it is not clogged internally reducing the reserve capacity of the coolant to assist in cooling the engine.
Answer D is incorrect. Both Technicians are correct.

Question #264
Answer A is incorrect. This is not the lateral runabout test.
Answer B is incorrect. The dial indicator is not on the housing face.
Answer C is correct. In the figure, the flywheel housing is being indicated for radial runout—a critical measurement that should be performed whenever flywheel housing is removed and reinstalled from the engine block. If this measurement is out of specification, the result can be crankshaft damage, engine mount damage, flywheel housing damage, and transmission failures.
Answer D is incorrect. The dial indicator is not on the pilot bore.

Question #265
Answer A is correct. Only Technician A is correct. The spring allows the hose clamp to expand and contract with the change in temperature while maintaining constant pressure on the hose and prevent leaks from occurring.
Answer B is incorrect. The constant torque clamps require to be torqued correctly to prevent the clamp from being damaged. Overtorquing of the spring will not allow the spring to compensate for expansion and the hose could get cut or the clamp could fail.
Answer C is incorrect. Only Technician A is correct.
Answer D is incorrect. Technician A is correct.

Question #266
Answer A is incorrect. If the clearance is not corrected and the engine is run with too small end play clearance, the thrust washer and thrust surfaces will become overheated. This will damage the surfaces and eventually premature failure would Most-Likely result leading to expensive repairs.
Answer B is correct. Only Technician B is correct. The thrust washer and thrust surfaces should be checked for foreign material before continuing on to correcting the thrust clearances. Sometimes the thrust washers can move and cause lack of clearance. The clearances must be corrected before continuing with the reassembly of the engine. If left uncorrected, expensive repairs could result.
Answer C is incorrect. Only Technician B is correct.
Answer D is incorrect. Technician B is correct.

Question #267
Answer A is incorrect. Technician B is also correct.
Answer B is incorrect. Technician A is also correct.
Answer C is correct. Both Technicians A and B are correct. A defective belt or a defective tensioner can cause engine knock codes, and they should not be overlooked as a source of defect.
Answer D is incorrect. Both Technicians are correct.

Question #268
Answer A is correct. Only Technician A is correct. A micrometer is commonly used to measure crankshaft journals. Specifications listed here are typical maximums. The OEM workshop manual should always be consulted when determining the serviceability of a specific crankshaft. Journals should be measured across axes 90° apart in three places, usually at either end and in the center of the journal, and the results matched to OEM specification. Using a measuring chart is the preferred method of organizing the data and retaining it.

- Journal out of roundness: 0.001"–0.002".
- Journal taper: 0.0015".
- Bent crankshaft: Check in V-blocks with a dial indicator while smoothly rotating by hand—out-of-service specifications differ widely due to differing crankshaft materials and lengths.

Answer B is incorrect. Using a wire wheel to clean up crankshaft journals would be disapproved of by most CNG engine manufacturers.
Answer C is incorrect. Only Technician A is correct.
Answer D is incorrect. Technician A is correct.

Question #269
Answer A is correct. No grease should be applied to the gasket or seal. Only clean engine oil.
Answer B is incorrect. The filter should be hand tight to begin the remaining torquing requirements.
Answer C is incorrect. The filter should be pre-filled before installing on the engine whenever possible to prevent setting a DTC in the ECM.
Answer D is incorrect. The filter will be tightened the required amounts of turns printed on the filter canister. Failure to follow the instructions can lead either to oil leaks or to difficulty in removing the filter at the next oil change.

Question #270
Answer A is correct. All the steps listed are required to maintain effective cooling in a CNG engine EXCEPT flushing and refilling of the cooling system annually. Some extended-life cooling systems are designed to last for up to six years and back-flushing is seldom recommended by engine manufacturers unless contamination from oil has occurred.
Answer B is incorrect. Coolant should be tested regularly to ensure adequate SCA strength.
Answer C is incorrect. Air must be bled from the cooling system during refilling to prevent engine-damaging air pockets.
Answer D is incorrect. Coolant must be tested for antifreeze protection level. This is usually done by test strips or a refractometer.

Question #271
Answer A is incorrect. A multimeter can be used to determine state of charge.
Answer B is incorrect. A carbon pile load tester can be used to test the ability of a battery to deliver sufficient amperage.
Answer C is incorrect. A hydrometer can be used to test a battery.
Answer D is correct. All of the above EXCEPT the voltage drop test described are used to check the condition of the connection between the battery post and positive terminal, so this is not a battery test.

Question #272
Answer A is incorrect. The rings gaps require to be staggered a minimum of 90° apart from each other.
Answer B is incorrect. The rings gaps require to be staggered a minimum of 90° apart from each other.
Answer C is incorrect. The rings gaps require to be staggered a minimum of 90° apart from each other.
Answer D is correct. The rings gaps require to be staggered a minimum of 90° apart from each other and the Most-Likely answer. If the ring gaps are staggered too close to each other, combustion gases could travel through the gaps into the crankcase and causing high crankcase pressure.

Question #273
Answer A is incorrect. Technician B is also correct.
Answer B is incorrect. Technician A is also correct.
Answer C is correct. Both Technicians are correct. When a liquid is held under pressure, its boiling point is raised, so Technician A is correct.
Antifreeze, when mixed with water, lowers the freezing point and raises the boiling point, so Technician B is also correct.
Answer D is incorrect. Both Technicians are correct.

Question #274
Answer A is incorrect. A hissing sound coming from the cooling system is normal as the pressure is being released. The hissing sound signifies the cap is able to maintain pressure. Before being replaced, the cap's pressure setting should be tested to determine that it is correct for the cooling system requirements.
Answer B is correct. Only Technician B is correct. Whenever a scent of mercaptan is found coming from the surge tank, the cooling should be tested with a methane detector to see if there is a presence of methane in the cooling system. If methane is detected, the ¼ turn fuel shutoff valve should be turned off to allow the fuel to vent. The system should be checked for a defective high-pressure regulator that could be passing the methane into the cooling system through the heat exchanger lines routed through it. If the component is defective, the component should be replaced before placing the vehicle back into service.
Answer C is incorrect. Only Technician B is correct.
Answer D is incorrect. Technician B is correct.

Question #275
Answer A is incorrect. Turning the key on with the engine off may not meet the criteria to set the DTC and make the code historical. A new DTC will not log until the engine is running, making the code active again.
Answer B is correct. Only Technician B is correct. Some codes only become active and set a DTC when the engine is running. Turning the key on with the engine off may not meet the criteria to set the DTC and make the code historical. A new DTC will not log until the engine is running, making the code active again.
Answer C is incorrect. Only Technician B is correct.
Answer D is incorrect. Technician B is correct.

Question #276
Answer A is incorrect. If the coolant was running too hot due to a malfunctioning thermostat or defective fan drive or some other reason the coolant would not deposit scale in the cooling system.
Answer B is incorrect. Too much SCA would cause a white powder residue in the radiator.
Answer C is incorrect. Too high of a concentration of antifreeze could cause the cooling system to create a slimeike gel in the cooling system but not scale.
Answer D is correct. Using hard water to make up for coolant loss is the Most-Likely cause that can lead minerals to separate from the coolant and leave a scale-type deposit in the cooling system and the engine. This scale inhibits heat transfer. Whenever possible, de-ionized or distilled water should be used to replenish the coolant.

Question #277
Answer A is incorrect. Technician B is also correct.
Answer B is incorrect. Technician A is also correct.
Answer C is correct. Both Technicians are correct. If the power fuse or the ground circuit for the throttle actuator is defective, the throttle actuator motor will not be able to operate even though it will receive a pulse width modulated command from the ECM.
Answer D is incorrect. Both Technicians are correct.

Question #278
Answer A is incorrect. Using butt splices with built-in shrink tubing is a method of wiring repair approved by most engine manufacturers.
Answer B is correct. The LEAST-Likely approved repair method of simply separating and taping over the exposed conductor of the wiring can leave defective wires with higher than acceptable resistance. Taping the wiring harness will hide the repaired area and make further diagnosis of electrical defects much more difficult at a later date.
Answer C is incorrect. The wiring harness can be replaced and is an approved repair method by the manufacturer.
Answer D is incorrect. The bad section of wiring can be removed with a new section of wiring soldered in its place and shrink tubing applied over each repaired section. The harness should then be taped up and secured with appropriate clamps to secure it and prevent it from falling onto the hot exhaust again.

Question #279
Answer A is incorrect. This statement is true, so it is not the exception. The bushings have to be aligned with the engine block lubrication hole to ensure that lubrication will reach the bushings and the camshaft journals.
Answer B is correct. This statement is not true, therefore it is the EXCEPT condition. A camshaft bushing should never be reused once the bushing has been installed into the engine block. The bushing gets compressed and material also is transferred to the engine block during the installation and removal process. If a bushing is reused and a new one is not installed, the reused bushing could spin in its bore. The oil hole would not be aligned, preventing lubrication from reaching the camshaft bearing journal and causing premature failure.
Answer C is incorrect. The statement is true so this is not an except statement. The correct bushing driver needs to be used when removing and installing the camshaft bushings to prevent from damaging the new bushings and the engine block.
Answer D is incorrect. The statement is true so this is not an except statement. The cam bushings need to be removed from the engine block when hot tanking it. The solution will attack the bushing material and can destroy the bushings.

Question #280
Answer A is incorrect. Technician B is also correct.
Answer B is incorrect. Technician A is also correct.
Answer C is correct. Both Technicians are correct. Whenever a crankshaft is damaged, the cylinder block should be checked out. The master bar is designed to check the main align bore. It is installed

without bearing shells with the main caps snug. If it does not rotate freely, the block align bore requires machining. The main bores should also be checked for out of round and this can be done with an inside micrometer.
Answer D is incorrect. Both Technicians are correct.

Question #281
Answer A is correct. Only Technician A is correct. Because the DTCs are written in the long-term memory of the ECM, the EEPROM, only a DDR or the appropriate software can be used to clear the DTCs logged in the ECM of a CNG engine.
Answer B is incorrect. Turning off the battery disconnect switch will not clear DTCs.
Answer C is incorrect. Only Technician A is correct.
Answer D is incorrect. Technician A is correct.

Question #282
Answer A is incorrect. Dial indicators can be used to measure backlash between mating gears.
Answer B is incorrect. Feeler gauges can be used to measure backlash between mating gears.
Answer C is incorrect. A depth micrometer can be used to measure backlash between mating gears.
Answer D is correct. A dial indicator, feeler gauges, and a depth micrometer can all be used to measure backlash between mating gears, but doing this with a digital caliper would be difficult so this would be the LEAST-Likely. Gear backlash should always be checked using a dial indicator or thickness gauge when the engine gear train has been assembled. A typical gear train backlash specification would be in the region of 0.008", but the OEM data should be consulted. An out-of-specification backlash specification that is higher than the OEM-permitted parameters is an indication of gear contact face wear and usually requires that the gear or gears be replaced. A tighter-than-specified backlash factor often indicates an assembly or timing problem.

Question #283
Answer A is incorrect. Technician B is also correct.
Answer B is incorrect. Technician A is also correct.
Answer C is correct. Both Technicians are correct. Coolant is routed through the high-pressure regulator and the low-pressure regulator to help raise the temperature of the temperature of the fuel due the effect of cooling whenever fuel is dropped from a high pressure to a low pressure.
Answer D is incorrect. Both Technicians are correct.

Question #284
Answer A is incorrect. Bearing shells should be installed to the bore clean and dry to optimize heat transfer between the bearing material and the bore. A thin coat of oil may be applied to the bearing face only.
Answer B is incorrect. When Plastigage is used to check bearing clearance, it is important that the bearings be installed dry and the crankshaft NOT be rotated.
Answer C is incorrect. Neither Technician is correct.
Answer D is correct. Neither Technician is correct. Bearing shells should be installed to the bore clean and dry to optimize heat transfer between the bearing material and the bore. A thin coat of oil may be applied to the bearing face only. When Plastigage is used to check bearing clearance, it is important that the bearings be installed dry and the crankshaft NOT be rotated. Plastigage is a soft plastic thread that easily deforms to conform to whatever clearance space is available when compressed between a bearing and journal. The crushed width can then be measured against a scale on the Plastigage packaging. The less clearance available will result in the Plastigage being flattened to a wider dimension. A short strip of Plastigage should be cut and placed across the center of the bearing in line with the crankshaft. The bearing cap with the bearing shell in place should then be installed and torqued to specification in the incremental steps outlined in the workshop manual. Do not rotate the engine with the Plastigage in place. Next, the bearing cap and shell should be removed and the width of the flattened Plastigage checked against the dimensional gauge on the packaging. If clearance is within specifications, carefully remove the Plastigage from the journal before reinstalling the cap and shell assembly. Oil should be added to the bearing surface prior to final assembly.

Question #285
Answer A is incorrect. Some dampers are fluid filled.
Answer B is incorrect. Some dampers have a rubber ring mounted between the damper ring and the hub.
Answer C is correct. Fluid-type dampers need replacing when they are dented or show signs of overheating as well as leaking the viscous fluid.
Answer D is incorrect. The damper protects the crankshaft by absorbing damaging vibrations.

Question #286
Answer A is incorrect. Filing the end gap of rings sometimes is acceptable to correct insufficient gaps. If the gaps are not corrected, the ends of the rings could touch together once the engine reaches operating temperature and cause them to break. The Technician should suspect that the incorrect rings are the possible reason for incorrect ring gap before any action is taken to correct the ring gaps.
Answer B is incorrect. Replacing the rings with rings that are within end gap specifications is usually the corrective action taken.
Answer C is correct. Honing out a cylinder liner is not recommended because the inside diameter could become out of specifications. The liner could also become out of round, and tapered as well. Some manufacturers require the cylinder liners to not be honed or de-glazed because the cross-hatch pattern cannot be accurately replicated in the field and therefore the liner should be replaced
Answer D is incorrect. Replacing the complete cylinder kit would probably the best method of correcting incorrect ring end gap.

Question #287
Answer A is incorrect. A damaged belt tensioner can be the source of vibrations but not usually at half of the engine rpm. Because of the ratios involved due to pulley sizes the vibrations would tend to be more random.
Answer B is incorrect. A bent connecting rod would produce a vibration incidence that would be equal with the engine speed.
Answer C is incorrect. Neither Technician is correct.
Answer D is correct. Neither Technician is correct. A damaged belt tensioner can be the source of vibrations but not usually at half of the engine rpm. Because of the ratios involved due to pulley sizes the vibrations would tend to be more random. A bent connecting rod would produce a vibration incidence that would be equal with the engine speed.

Question #288
Answer A is incorrect. Technician B is also correct.
Answer B is incorrect. Technician A is also correct.
Answer C is correct. Both Technicians are correct. Removing excessive metal from the deck surface area causes the cylinder head to move closer to the centerline of the crankshaft. This causes the pistons to move closer to the cylinder head and possibly striking it. For this reason, the amount of metal removed must be stringently monitored. The machinist will usually imprint the amount of material removed from the deck surface on the engine block's oil pan rail at the front of the engine for later referencing, making it easier to determine if the block is reusable or not at a later rebuild.
Answer D is incorrect. Both Technicians are correct.

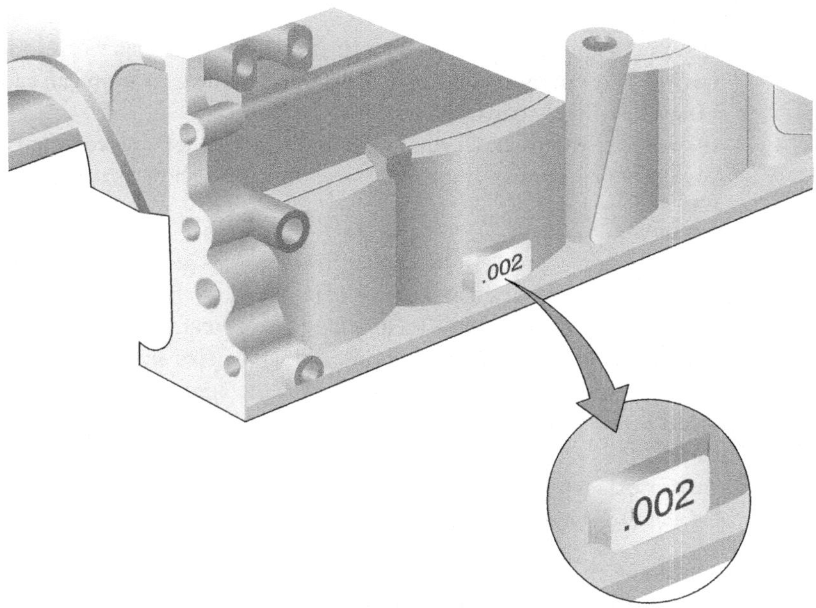

Question #289

Answer A is incorrect. When a dent is discovered in a vibration damper, the only corrective action to be taken is to replace it as soon as possible to prevent damage from occurring to the engine crankshaft.

Answer B is incorrect. If the free-floating disc is squeezed from a dent on the damper, the damper may not function correctly and cause a slight vibration that could not be discernable.

Answer C is incorrect. Neither Technician is correct.

Answer D is correct. Neither Technician is correct. When a dent is discovered in a vibration damper, the only corrective action to be taken is to replace it as soon as possible to prevent damage from occurring to the engine crankshaft. The free floating disc could be squeezed to the point that it will not be able to move and absorb vibrations even though the fluid is not leaking. If the free floating disc is squeezed from a dent on the damper, the damper may not function correctly and cause a slight vibration that could not be discernable. Although a small vibration does not appear as problematic, the crankshaft would be affected and eventually fail.

Question #290

Answer A is correct. Only Technician A is correct. The bearing shell and connecting rod bore should be cleaned and rechecked to ensure the clearance is within specifications. If the specified clearance cannot be attained, the connecting rod bore will either need to be honed or the connecting rod replaced.

Answer B is incorrect. The connecting rod bore should never be polished to remove metal in an attempt to increase the bore.

Answer C is incorrect. Only Technician A is correct.

Answer D is incorrect. Technician A is correct.

Question #291
Answer A is incorrect. Lack of lubrication would show as worn and damaged gears, not just teeth.
Answer B is correct. Foreign debris falling in the gear train can cause one or more broken teeth and is the Most-Likely cause.
Answer C is incorrect. Normal wear and tear would be represented as worn gear teeth.
Answer D is incorrect. Improper assembly would show as defective and worn gear teeth.

Question #292
Answer A is incorrect. If the engine flywheel was at fault, the vibration would Most-Likely change as the engine rpm is varied and the vehicle would not have to be moving.
Answer B is correct. Only Technician B is correct. Because the vehicle has to be moving for the vibration to occur and change in intensity, the drive shaft condition and/or phasing of the u-joints should be suspected as well as the tires. If the engine or the flywheel were at fault, the vibration would Most -Likely occur and change when the engine speed is increased with the transmission in neutral.
Answer C is incorrect. Only Technician B is correct.
Answer D is incorrect. Technician B is correct.

Question #293
Answer A is incorrect. Technician B is also correct.
Answer B is incorrect. Technician A is also correct.
Answer C is correct. Both Technicians are correct. If the engine cannot reach operating temperature due to a partially or fully stuck-open thermostat, moisture vapors will not be expelled from the crankcase and will collect. Eventually the engine oil will emulsify and the additives will be depleted fighting the damaging effects of corrosion on the bearings and they will be attacked removing copper and other metals.
Answer D is incorrect. Both Technicians are correct.

Question #294
Answer A is correct. Only Technician A is correct. Most CNG engines require the pistons and connecting rod assemblies be assembled in a specific direction, facing the front of the engine, or the numbers stamped on the connecting rod face the oil cooler side of the engine.
Answer B is incorrect. Most CNG engines require the pistons and connecting rod assemblies be assembled in a specific direction, facing the front of the engine, or the numbers stamped on the connecting rod face the oil cooler side of the engine.
Answer C is incorrect. Only Technician A is correct.
Answer D is incorrect. Technician A is correct.

Glossary

Absolute Pressure The zero point from which pressure is measured.

Actuator A device that delivers motion in response to an electrical or hydraulic signal.

A/D Converter Acronym for Analog-to-Digital Converter.

Additive An ingredient intended to improve a certain characteristic of a material or fluid.

AFR Air/Fuel Ratio. The relationship of the amount of oxygen to the fuel that is being consumed by the engine.

After-Cooler Any of a number of types of charge air cooler.

Air Compressor An engine-driven mechanism for supplying compressed air to the truck brake system.

Air-Conditioning The control of air movement, humidity, and temperature by mechanical means.

Air Dryer A unit that removes moisture.

Air Filter A device that minimizes the possibility of impurities entering the induction system.

Altitude Compensation System An altitude barometric switch and solenoid used to provide better drivability at altitudes exceeding 1000 feet.

Ambient Temperature Temperature of the surrounding or prevailing air. Normally, it is considered to be the temperature in the service area where testing is taking place.

Amp Abbreviation for ampere.

Ampere The unit for measuring electrical current or the number of electrons that are flowing at a point in a circuit.

Analog Signal A voltage signal that varies within a given range (from high to low, including all points in between).

Analog-to-Digital Converter (A/D converter) A device that converts analog voltage signals to a digital format; this is located in a section of the processor called the input signal conditioner.

Analog Volt/Ohmmeter (AVOM) A test meter used for checking voltage and resistance. Analog meters should not be used on solid state circuits.

Antifreeze A compound, such as alcohol or glycerin, added to water to lower its freezing point.

Antirust Agent Additive used with lubricating oils to prevent rusting of metal components when the engine is not in use.

Armature The rotating component of a (1) starter or other motor, (2) generator, or (3) compressor clutch.

ASE Automotive Service Excellence. A trademark of National Institute for Automotive Service Excellence.

ATDC After Top Dead Center.

Atmospheric Pressure The weight of the air at sea level; 14.696 pounds per square inch (psi) or 101.33 kilopascals (kPa).

Auxiliary Filter Any type of supplementary filter.

Axis of Rotation The center line around which a gear or part revolves.

Back Pressure The amount of restriction present in the exhaust system. Some backpressure is normal, but excess backpressure reduces the efficiency of the engine.

Battery Terminal A tapered post or threaded studs on top of the battery case, or internally threaded provisions on the side of the battery for connecting the cables.

BDC Bottom Dead Center. The lowest point that a piston, connecting rod, and crankshaft throw will travel.

Bellows A movable cover or seal that is pleated or folded like an accordion to allow for expansion and contraction.

Bimetallic Two dissimilar metals joined together that have different bending characteristics when subjected to changes of temperature.

Blade Fuse A type of fuse having two flat male lugs sticking out for insertion in the female box connectors.

Block Diagnosis Chart A troubleshooting chart that lists symptoms, possible causes, and probable remedies in columns.

Blowby The combustion gases that leak past the piston rings and enter the crankcase. Most blowby is contained by the PCV system, so excessive blowby is an indication of worn rings or cylinders.

Blower Fan A fan that pushes or blows air through a ventilation, heater, or air-conditioning system.

Boss A race of a bearing or shaft bore.

Bottoming A condition that occurs when the teeth of one gear touch the lowest point between teeth of a mating gear.

Bracket An attachment used to secure components to the body or frame.

British Thermal Unit (Btu) A measure of heat quantity equal to the amount of heat required to raise one pound of water one degree F.

BTDC Before top dead center.

Btu British Thermal Unit.

CAC Charge Air Cooler. Radiator type device that cools the turbocharged air before passing through the intake manifold and the engine's cylinders.

CAN Controller Area Network. Part of the diagnostic link of a computer-controlled vehicle. Also allows multiple computers to communicate with each other to share information.

Cartridge Fuse A type of fuse having a strip of low melting point metal enclosed in a glass tube. If an excessive current flows through the circuit, the fuse element melts at the narrow portion, opening the circuit and preventing damage.

Cavitation A condition caused by vapor bubble collapse associated with the cooling system.

CCA Cold Cranking Amps. Measurement of the amperage a battery can deliver for 30 seconds at 0° F and maintain 7.2 volts.

CEL Check Engine Light. Indicator in the driver's panel that illuminates to notify of a minor engine or sensor(s) malfunction has occurred that will require attention. This indicator alone will not usually cause the engine to turn off.

Ceramic Fuse A fuse found in some import vehicles that has a ceramic insulator with a conductive metal strip along one side.

Charging Circuit The alternator (or generator) and associated circuit used to keep the battery charged and power the vehicle electrical systems when the engine is running.

Charging System A system consisting of the battery, alternator, voltage regulator, associated wiring, and the electrical loads of a vehicle. The purpose of the system is to recharge the battery whenever necessary and to provide the current required to power the electrical components.

Check-Valve A valve that allows airflow in one direction only.

Circuit The complete path of an electrical current, including the generating device. When the path is unbroken, the circuit is closed and current flows. When the circuit continuity is broken, the circuit is open and current flow stops.

CNG Compressed Natural Gas, pressurized as high as 3600 psi.

CNG Vehicle A vehicle that is powered by compressed natural gas.

Coefficient of Friction A measurement of friction developed between two objects in contact when one of the objects is drawn across the other.

Combination A truck coupled to one or more trailers.

Compression Applying pressure to a spring or any springy substance, thus causing it to reduce its length in the direction of the compressing force.

Compressor (1) A mechanical device that increases pressure within a container by pumping air into it. (2) That component of an air-conditioning system that compresses low temperature/pressure refrigerant vapor.

Condensation The process by which gas (or vapor) changes to a liquid.

Condenser A component in an air-conditioning system used to cool a refrigerant below its boiling point, causing it to change from a vapor to a liquid.

Conductor Any material that permits electrical current flow.

Coolant Liquid that circulates in an engine cooling system.

Coolant Heater A component used to aid engine starting and reduce wear caused by cold starting.

Coolant Hydrometer A tester designed to measure coolant specific gravity and determine the amount of antifreeze protection.

Cooling System A complete system for circulating coolant.

COP Coil Over Plug. An ignition system in which the ignition coil is mounted directly on top of the spark plug.

Crankcase A housing within which the crankshaft and other components of the engine operate.

Crankcase Pressure The result of combustion pressure that escapes past the rings. Crankcase pressure should be removed by a properly functioning PCV system and there should be a slight vacuum.

Cranking Circuit The starter and its associated circuit, including battery, relay (solenoid), ignition switch, neutral start switch (on vehicles with automatic transmission), cables, and wires.

Cycling (1) Repeated on-off action of an air or NC compressor. (2) Repeated full electrical discharge and recharge of a battery that can cause the positive plate material to break away from its grids and fall into the sediment chambers at the base of the battery case.

Dampen To slow or reduce oscillations or movement.

Dash Control Valves A variety of hand-operated valves located on the dash. They include parking brake valves, tractor protection valves, and differential lock.

Data Links Circuits through which computers communicate with other electronic devices such as control panels, modules, some sensors, or other computers in the form of digital signals.

DDL Diagnostic Data Link. The wires that allow the diagnostic tools to communicate with the engine, transmission, and other computers installed in the vehicle. The data links are usually referred to as the J-1708, J-1939, or the CAN network.

DDR Diagnostic Data Reader. A tool that can interface to the ECM/ECU or other controllers to allow the Technician to perform diagnostics.

Deadline To take a vehicle out of service.

Deburring To remove sharp edges from a cut.

Deflection Bending or moving to a new position as the result of an external force.

Department of Transportation (DOT) The U.S. government agency that establishes vehicle standards.

Derating The ability of the engine's ECM to remove fuel to prevent damage.

Detergent Additive An oil additive that helps keep metal surfaces clean and prevents deposits. Such additives suspend particles of carbon and oxidized oil in the oil.

Detonation A destructive condition in the cylinder that occurs when multiple flame fronts collide, producing a metallic knock during combustion. The initial flame front is produced by the firing of the spark plug and secondary flame fronts are the result of either spontaneous or compression ignition. Detonation causes high cylinder pressure spikes that can cause serious damage to the engine.

Diagnostic Flow Chart A chart that provides a systematic approach to electrical system and component troubleshooting and repair. Found in service manuals and vehicle make and model specific.

Dial Caliper A measuring instrument capable of taking inside, outside, depth, and step measurements.

Digital Binary Signal A signal that has only two values: on and off.

Digital Volt/Ohmmeter (DVOM) A type of test meter recommended by most manufacturers for use on solid state circuits. Can also be referred to as a DMM, Digital Multi-Meter.

Diode The simplest semiconductor device formed by joining P-type semiconductor material with N-type semiconductor material. A diode allows current to flow in one direction, but not in the opposite direction.

Dispatch Sheet A form used to keep track of dates when work is to be completed. Some dispatch sheets follow the job through each step of the servicing process.

DOT *See* Department of Transportation.

Driven Gear A gear that is driven or forced to turn by a drive gear, a shaft, or some other device.

Drive or Driving Gear A gear that drives another gear or causes another gear to turn.

Drive Line The propeller or drive shaft and universal joints that link the transmission to the axle pinion gear shaft.

DTC Diagnostic Trouble Code

DVOM Digital Volt/Ohmmeter

ECM Electronic Control Module. A sensor usually associated with the engine's electronic controller.

ECU Electronic Control Unit. A sensor usually associated with the transmission's electronic controller.

EEPROM Electronically Erasable Programmable Read Only Memory

EG Exhaust Gases

EGR Exhaust Gas Recirculation. The process by which exhaust gases are cycled back into the engine's air intake system. The introduction of this inert gas allows cooler combustion temperatures and lowers the production of NOx.

ELC Extended Life Coolant

Electricity Movement of electrons from one place to another.

Electric Retarder Electromagnets mounted in a steel frame. Energizing the retarder causes the electromagnets to exert a dragging force on the rotors in the frame and this drag force is transmitted directly to the drive shaft.

Electromotive Force (EMF) The force that moves electrons between atoms. This force is the pressure that exists between the positive and negative points (the electrical imbalance). This force is measured in units called volts.

Electronic Control Unit (ECU) An electronic system controller usually associated with electronic transmissions.

Electronics The technology of moving electrons through hard wire, semiconductor, gas, and vacuum circuits.

Electrons Negatively charged particles orbiting around every atomic nucleus.

EMF Electromotive force. Usually measured in volts.

Engine Brake A hydraulically operated device that controls the operation of either the intake or exhaust valves to convert the vehicle's engine into a power-absorbing retarding mechanism. Commonly referred to as Jake-Brakes.

Engine Stall Point The point, in rpms, under load as compared to the engine manufacturer's specified rpm for the stall test.

Environmental Protection Agency (EPA) The U.S. government agency charged with the responsibilities of protecting the environment and enforcing the Clean Air Act (CAA) of 1990.

EPA *See* Environmental Protection Agency.

EPROM Electronically Programmable Memory.

Exhaust Brake A valve in the exhaust pipe between the manifold and the muffler. A slide mechanism that restricts the exhaust flow, causing exhaust backpressure to build up in the engine's cylinders. The exhaust brake actually transforms the engine into a low-pressure air compressor driven by the wheels that aids the vehicle to slow down minimizing the use of the service brakes.

False Brinelling Polishing of a surface that is not damaged.

Fatigue Failures Progressive destruction of a shaft or gear teeth material, usually caused by overloading.

Fault Code A code recorded to computer memory. A fault code can be read by a digital diagnostic reader. Also referred to as a DTC.

FCV Fuel Control Valve.

Fixed Value Resistor An electrical device designed to have only one resistance rating, which should not change, for controlling voltage.

Flammable Any material that will easily catch fire or explode.

Flare To spread gradually outward in a bell shape.

Flex Disc A term often used for flexplate.

Flexplate A component used to mount the torque converter to the crankshaft in order to transfer crankshaft rotation to the torque converter assembly.

Float A cruising drive mode in which throttle setting matches engine speed to road speed, neither accelerating nor decelerating.

FMI Failure Mode Identifier. A description of the type of failure the component has experienced, such as below normal: FMI 4; or above normal: FMI 3. There are 15 different FMIs.

Foot-Pound An English unit of measurement for torque. One foot-pound is the torque obtained by a force of one pound applied to a foot-long wrench handle.

Fretting A result of vibration causing the bearing outer race to pick up the machining pattern.

Fuse Link A short length of smaller gauge wire installed in a conductor, usually close to the power source.

Fusible Link A term often used for fuse link.

GCW Gross combination weight.

Gear A disc-like wheel with external or internal teeth that serves to transmit or change motion.

Gear Pitch The number of teeth per given unit of pitch diameter; an important factor in gear design and operation.

Gross Combination Weight (GCW) The total weight of a fully quipped vehicle including payload, fuel, and driver.

Gross Vehicle Weight (GVW) The total weight of a fully equipped vehicle and its payload.

Ground The negatively charged side of a circuit. A ground can be a wire, the negative side of the battery, or the vehicle chassis.

Grounded Circuit A shorted circuit that causes current to return to the battery before it has reached its intended destination.

GVW *See* Gross vehicle weight.

Harness and Harness Connectors The organization of the vehicle's electrical system, providing an orderly and convenient starting point for tracking and testing circuits.

Hazardous Materials Any materials including substances that are flammable, explosive, or known to produce adverse health effects in people or the environment upon exposure during use.

Heater Control Valve A valve that controls the flow of coolant into the heater core from the engine.

Heat Exchanger A device used to transfer heat, such as a radiator or condenser.

High-Resistant Circuits Circuits that have an increase in circuit resistance, with a corresponding decrease in current.

High-Strength Steel A middle-alloy steel that is much stronger than the hot-rolled or cold-rolled sheet steels normally used in the manufacture of car body frames.

Hinged Pawl Switch The simplest type of switch; one that makes or breaks the current of a single conductor.

Hydraulic Retarder A device that absorbs energy from the vehicle as it deaccelerates, dissipating it as heat through the compression and friction of either the engine oil or the transmission fluid, depending on the type of retardation used. Transit buses are increasingly utilizing transmission hydraulic retarders. The internal components of a hydraulic retarder resemble the components of a torque converter except they remove energy rather than multiply energy.

Hydrometer A tester designed to measure the specific gravity of a liquid.

IAT Intake Air Temperature.

ICM Igniter Control Module.

Inboard Toward the centerline of the vehicle.

In-Line Fuse A fuse in series with the circuit in a small plastic fuse holder, not in the fuse box or panel. Used as a protection device for a portion of the circuit even though the entire circuit may be protected by a fuse in the fuse box or panel.

Insite A software that allows the Technician to monitor, adjust, and recalibrate ECMs that are used on the Cummins brand of engines.

Installation Templates Drawings supplied by some vehicle manufacturers to allow the Technician to install an accessory correctly. The templates available can be used to check clearances or to ease installation.

Insulator A material, such as rubber or glass, that offers high resistance to the flow of electrons.

Integrated Circuit A component containing diodes, transistors, resistors, capacitors, and other electronic components mounted on a single piece of material and capable of performing numerous functions.

Jumper Wire A length of wire with an alligator clip at each end, used to temporarily bypass a circuit or components for electrical testing.

Jump-Start A procedure in which a booster battery is used to start a vehicle having a discharged battery.

Kinetic Energy Energy in motion.

Lateral Runout The wobble or side-to-side movement of a rotating wheel or of a rotating wheel and tire assembly.

Linkage A system of rods and levers used to transmit motion or force.

Low-Maintenance Battery A conventionally vented, lead/acid battery, requiring normal periodic maintenance.

Magnetorque An electromagnetic clutch.

Maintenance-Free Battery A battery that is sealed and does not require the addition of water during normal service life.

Maintenance Manual A publication containing routine maintenance procedures and intervals for vehicle components and systems.

Margin The area between the valve face and the head of the valve.

Mercaptan An odorant that is added to natural gas to allow the detection of possible leaks.

MSD Multiple Spark Discharges. Type of coil utilized to deliver multiple sparks per combustion cycle of a cylinder.

NATEF *See* National Automotive Technicians' Education Foundation.

National Automotive Technician's Education Foundation (NATEF) The foundation providing certification for secondary and post secondary automotive and heavy-duty truck training programs.

National Institute for Automotive Service Excellence The nonprofit organization providing an established certification program (under the ASE trademark) for automotive, heavy-duty truck, auto body repair, engine machine shop technicians, and parts specialists.

NGV Natural Gas Vehicle. A vehicle that uses methane as a fuel.

NOx Oxides of Nitrogen. A smog-forming agent.

OEM Original Equipment Manufacturer.

Off-road Any unpaved, rough, or ungraded terrain on which a vehicle will operate. Any terrain not considered part of the public highway system falls into this category.

OHC Overhead Cam.

Ohm A unit of measured electrical resistance.

Ohm's Law The basic law of electricity stating that in any electrical circuit, current, resistance, and pressure work together in a mathematical relationship.

OHV Overhead Valve.

On-road Any paved or smooth-graded surface terrain on which a vehicle will operate, generally considered part of the public highway system.

Open Circuit An electrical circuit whose path has been interrupted or broken either accidentally (a broken wire) or intentionally (a switch turned off).

Oscillation The rotational movement in either fore/aft or side-to-side direction about a pivot point.

Oval A condition that occurs when a tube or bore is not round but somewhat egg-shaped.

Overspeed Governor A governor that shuts off the fuel or stops the engine when excessive speed is reached. This is usually an adjustable parameter for electronically controlled engines utilizing the diagnostic software and appropriate permission (password).

Oxidation Inhibitor (1) An additive used with lubricating oils to keep oil from oxidizing even at very high temperatures. (2) An additive for gasoline to reduce the chemicals in gasoline that react with oxygen.

Parallel Circuit An electrical circuit that provides two or more paths for current flow.

Parts Requisition A form used to order parts, on which the technician writes the names of the part(s) needed along with the vehicle's VIN or company's identification folder.

Payload The weight of the cargo carried by a vehicle, not including the weight of the body.

PCM Powertrain Control Module. A module that typically controls multiple tasks required by a vehicle. Usually all the components are manufactured or designed by the same manufacturer.

PID Parameter Identifier. An indicator in an operating system such as oil pressure or coolant temperature of a DTC.

Pitting A surface irregularity resulting from corrosion.

PMI Preventive Maintenance Inspection.

Polarity The particular state, either positive or negative, with reference to the two poles of a battery or to electrification.

Pole The number of input circuits made by an electrical switch.

Pounds per Square Inch (psi) A unit of English measure for pressure.

Power A measure of work being done factored with time.

PRD Pressure Relief Device. A device in a CNG system that is activated by temperature to allow natural gas to vent safely above the roof line of a vehicle to prevent the fuel cylinder(s) from being damaged from overpressurization that could be caused by exposure of the cylinder to excessive heat usually around 189° F.

Pressure The amount of force applied to a definite area measured in pounds per square inch (psi) English or kilopascals (kPa) metric.

Pressure Differential Difference in pressure between two points of a system or a component.

Printed Circuit Board An electronic circuit board made of nonconductive material onto which conductive metal, such as copper, has been deposited. Parts of the metal are then etched away by an acid, leaving metal lines that form the conductors for the various circuits on the board. A printed circuit board can hold many complex circuits in a very small area.

Programmable Read-Only Memory (PROM) An electronic component that contains program information specific to different vehicle model calibrations.

PROM *See* Programmable Read-Only Memory.

psi Pounds per square inch.

psia Pounds per square inch absolute. Indicates ambient pressure plus the system pressure. Utilized to measure the air intake pressure of CNG engines.

P-type Semiconductors Positively charged materials that enable them to carry current. They are produced by adding an impurity with three electrons in the outer ring (trivalent atoms).

Radial Load A load that is applied at 90 degrees to an axis of rotation.

RAM *See* Random Access Memory.

Ram Air Air that is forced into the engine or passenger compartment by the forward motion of the vehicle.

Random Access Memory (RAM) The memory used during computer operation to store temporary information. The microcomputer can write, read, and erase information from RAM in any order, which is why it is called random.

Rated Capacity The maximum recommended safe load that can be sustained by a component or an assembly without permanent damage.

Reactivity The characteristic of a material that enables it to react with air, heat, water, or other materials.

Read Only Memory (ROM) A type of memory used in microcomputers to store information permanently.

Recall Bulletin A bulletin that pertains to special situations that involve service work or replacement of components in connection with a recall notice.

Reference Voltage The voltage supplied to a sensor by the computer, most of the time at 5 volts, which acts as a baseline voltage modified by the sensor as a varying voltage to act as an input signal.

Relay An electric switch that allows a small current to control a much larger one. It consists of a control circuit and a power circuit.

Reluctor A type of wheel or gear with teeth that rotate with a component such as a crankshaft to provide signals in relationship to the position of the component to determine timing.

Reserve Capacity Rating The ability of a battery to sustain a minimum vehicle electrical load for a determined period of time, usually rated in the number of minutes a battery can deliver 25 amps at 80° F and maintain 10.2 volts, in the event of a charging system failure.

Resistance The opposition to current flow in an electrical circuit.

Retarder A device that dissipates energy from the vehicle when decelerating. A retarder can be either electric or hydraulic and is usually connected to the drive shaft.

Revolutions Per Minute (rpm) The number of complete turns a member makes in 1 minute.

Right to Know Law A law passed by the federal government and administered by the Occupational Safety and Health Administration (OSHA) that requires any company that uses or produces hazardous chemicals or substances to inform its employees, customers, and vendors of any potential hazards that may exist in the workplace as a result of using the products.

ROM *See* Read Only Memory.

Rotation A term used to describe the fact that a gear, shaft, or other device is turning.

Rotor (1) A part of the alternator that provides the magnetic fields necessary to create a current flow. (2) The rotating member of an assembly.

rpm *See* Revolutions per minute.

Runout Deviation or wobble in a shaft or wheel as it rotates. Runout is measured with a dial indicator.

SCR Silicon-Controlled Rectifier. A type of diode that works like a normal diode but that can also flow current in both directions when a specific voltage is applied to its junction.

Screw Pitch Gauge A gauge used to provide a quick and accurate method of checking the threads per inch of a nut or bolt.

SEL Stop Engine Light. Indicator in the driver's panel that illuminates when a serious engine malfunction has occurred that can damage the engine or its components. This indicator usually warns the driver the engine will turn off in 30 seconds to 1 minute.

Semiconductor A solid-state device that can function as either a conductor or an insulator, depending on how its structure is arranged.

Sensing Voltage The voltage that allows the regulator to sense and monitor the battery voltage level.

Sensor An electronic device used to monitor conditions for computer control requirements.

Series Circuit A circuit that consists of two or more resistors or loads connected to a voltage source with only one path for the electrons to follow.

Series/Parallel Circuit A circuit designed so that both series and parallel combinations exist within the same circuit.

Service Bulletin A publication that provides the latest service tips, field repairs, product improvements, and related information of benefit to service personnel.

Service Manual A manual, published by the manufacturer, that contains service and repair information for vehicle systems and components.

SID System Identifier. An indicator of a major component such as a throttle plate assembly or a fuel control valve.

Short Circuit An undesirable connection between two worn or damaged wires.

Solenoid An electromagnet used to perform mechanical work, made with one or two coil windings wound around an iron tube.

Solid Wire A single-strand conductor.

Solvent A substance that dissolves other substances.

Spade Fuse A term used for blade fuse.

Spalling A surface fatigue that occurs when chips, scales, or flakes of metal break off due to fatigue rather than wear.

Specialty Service Shop A shop that specializes in areas such as engine rebuilding, transmission/axle overhauling, brake, air-conditioning/heating repairs, and electrical/electronic work.

Specific Gravity The measurement of a liquid based on the ratio of the liquid's mass to an equal volume of distilled water.

Spontaneous Combustion A process by which a combustible material ignites by itself.

Stall Test A test performed when there is a malfunction in the vehicle power package (engine and transmission) to determine which of the components is at fault.

Starter Circuit The circuit that carries the high current flow within the system and supplies power for the actual engine cranking.

Starter Motor The device that converts electrical energy from the battery into mechanical energy for cranking the engine.

Starter Safety Switch A switch that prevents vehicles with automatic transmissions from being started in gear, also known as a neutral switch.

Static Balance Balance at rest, or still balance.

Stepped Resistor A resistor designed to have two or more fixed values, available by connecting wires to one of several taps.

Stoichiometric The amount of air required to burn a given amount of fuel. A CNG engine requires approximately 16.5 lbs or air to burn 1 lb of fuel.

Storage Battery The source of direct current electricity for both the electrical and electronic systems.

Stranded Wire Wire that is made up of a number of small solid wires, generally twisted together, to form a single conductor that is usually more flexible than a single-strand conductor of larger diameter.

Sulfation A condition that occurs when sulfate is allowed to remain in the battery plates for a long time, causing two problems. (1) It lowers the specific gravity levels, increasing the danger of freezing at low temperatures. (2) In cold weather a sulfated battery may not have the reserve power needed to crank the engine.

Swage To reduce or taper.

Switch A device used to control on/off and direct the flow of current in a circuit. A switch can be under the control of the driver or can be self-operating through a condition of the circuit, the vehicle, or the environment.

Tachometer An instrument that indicates rotating speeds, sometimes used to indicate crankshaft rpm.

Tapped Resistor A resistor designed to have two or more fixed values, available by connecting wires to one of several taps.

TDC Top Dead Center. When the piston connecting rod, and crankshaft throw are at the highest point.

Throw (1) The offset of a crankshaft. (2) The number of output circuits of a switch.

TIK Technical Information Kiosk. A personal computer that has the ability to connect to an engine, transmission, and other control modules to perform diagnostics. Usually the TIK will contain the service manuals relating to the OEM products that require servicing at their locations.

Time Guide A prepared reference material used for computing compensation payable by the OEM manufacturer for repairs or service work to vehicles under warranty, or for other special conditions authorized by the company.

Timing (1) Procedure of marking the appropriate teeth of a gear set prior to installation and placing them in proper mesh while in the transmission. (2) Combustion spark delivery in relation to the piston position.

Torque Twisting force to tighten a fastener to a specific degree of tightness, generally in a given order or pattern.

Toxicity A statement of how poisonous a substance is.

Transistor Any of three terminal semiconductors used for switching and amplifying electronic circuits.

Trouble Tree Diagnosis Chart A chart used to provide a sequence for what should be inspected or tested when troubleshooting a defective system.

TTY Torque to Yield. Generally refers to head bolts, which are torqued to just below their yield point. These types of bolts are not usually reused.

Vacuum Absence of matter but often used to describe any pressure condition below atmospheric pressure.

Validity List A list supplied by the manufacturer of valid bulletins.

Valve Spring Installed Height The distance from the bottom of the valve spring retainer and the spring seat on the cylinder head. It should be checked while reconditioning the cylinder head since a valve spring installed height in excess of specifications lengthens the effective length of the spring, reducing pressure.

VIN Vehicle Identification Number. A set of numbers that is unique to each vehicle.

Volt A unit of electromotive force.

Voltage-Generating Sensors Devices that produce their own voltage signal.

Voltage Limiter A device that provides protection by limiting voltage to the instrument panel gauges to approximately five volts.

Voltage Regulator A device that controls the amount of current produced by the alternator or generator and thus the voltage level in the charging circuit.

Watt A measure of electrical power or the work done.

Watt's Law A basic law of electricity used to find the power of an electrical circuit expressed in watts. It states that power equals the voltage multiplied by the current, in amperes.

Windings (1) The three separate bundles in which wires are grouped in the stator. (2) The coil of wire found in a relay or other similar device. (3) That part of an electrical clutch that provides a magnetic field.

Work (1) Forcing a current through a resistance. (2) The product of a force.

Yield Strength The highest stress a material can stand without permanent deformation or damage, expressed in pounds per square inch (psi).

Zener Diode A variation of the diode, this device functions like a standard diode until a certain voltage is reached. When the voltage level reaches this point, the zener diode will allow current to flow in the reverse direction. Zener diodes are often used in electronic voltage regulators.